SOLUTIONS MANUAL

FOR

AN INTRODUCTION TO GENETIC ANALYSIS

Seventh Edition

William D. Fixsen
Diane K. Lavett

W. H. Freeman and Company
New York

ISBN 0-7167-3525-3

Printed in the United States of America

First printing 2000

Contents

1 GENETICS AND THE ORGANISM

1. Genetics is the study of genes and genomes: their biochemical basis, how they function, how they are controlled, how they are organized, how they replicate, how they change, how they can be manipulated, and how they are transmitted from cell to cell and generation to generation.

   The ancient Egyptian racehorse breeders can be only loosely classified as geneticists because their interests were highly focused on producing fast horses, rather than on attempting to understand the mechanisms of heredity. Their understanding of the processes involved in producing fast horses was very incorrect, and their methods were not analytic in the modern sense of the word. Nevertheless, they did produce very fast horses through a combination of observation, trial and error, and artificial selection.

2. DNA determines all the specific attributes of a species (shape, size, form, behavioral characteristics, biochemical processes, etc.) and sets the limits for possible variation that is environmentally induced.

3. Properties of DNA that are vital to its being the hereditary molecule are its ability to replicate, its informational content, and its relative stability while still retaining the ability to change or mutate. Alien life-forms might utilize RNA, just as some viruses do, as a hereditary molecule. However, of the types of molecules that can exist on earth, only the nucleic acids possess the necessary characteristics.

4. There are four possible nucleotide pairs at each position (A–T, T–A, G–C, or C–G). Therefore, the general formula for calculating the different possibilities

is 4^n (where n = the number of pairs). In this case, there are 4^{10} or 1,048,576 possible DNA molecules!

5. If the DNA is double-stranded, A = T and G = C and A + T + C + G = 100%. If T = 15% then C = [100 - 15(2)]/2 = 35%.

6. If the DNA is double-stranded, G = C = 24% and A = T = 26% .

7. **a. and b.** There are many ways to indicate the polarity of DNA in a simple way. The point of this exercise is to realize that the polarity is based on how the sugar (deoxyribose) is oriented within the backbone of each strand. The "backbone" is the side of the molecule and consists of repeating units of deoxyribose and a single phosphate group. The 5´ carbon is attached to the phosphate group while the 3´ carbon has a hydroxyl group to which a new nucleotide may be added (by its phosphate group). In double-stranded DNA, the two strands are antiparallel, meaning that the sugars of each are oriented in opposite directions and the bases associate by hydrogen bonds to hold the two strands together.

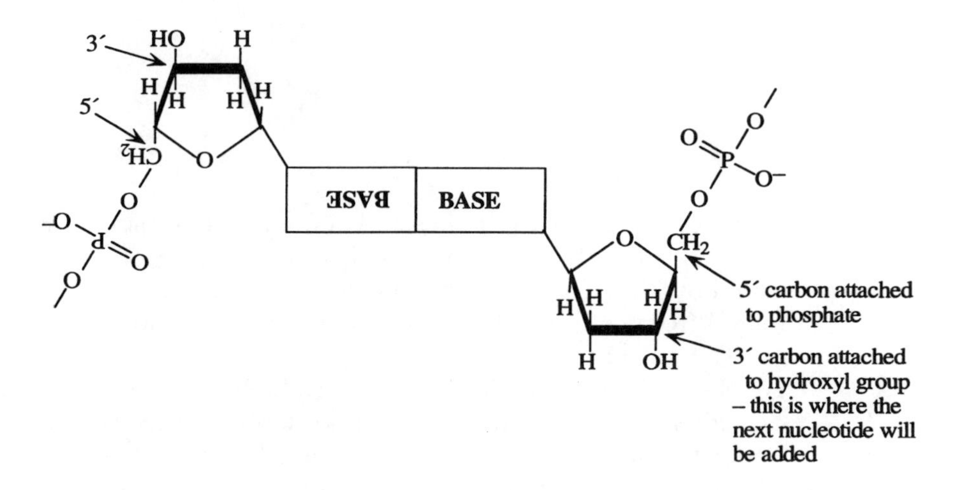

8. Human somatic cells are diploid ($2n$) and the haploid number is 23 (n = 23). Each chromosome is a double-stranded DNA molecule so there are 46 molecules of DNA. It is generally stated that the haploid number represents the number of different types of DNA molecules but that does not take into account the difference between the X and Y chromosome in males. So females have 23 different types (called 1–22 and X) and males have 24 (1–22, X and Y).

9. Keeping this strand antiparallel to the strand given, the sequence would be:

3´ — TAACCACGTAATGAAGTCCGAGA — 5´

10. Yes. There are no sequence restrictions in single-stranded DNA. The percentage of A must equal the percentage of T only in double-stranded DNA.

11. **a.** Yes. Since A = T and G = C, the equation A + C = G + T can be rewritten as T + C = C + T by substituting the equal terms.

b. Yes, the percentage of purines will equal the percentage of pyrimidines in double-stranded DNA.

12. For the following, normal typeface represents previously polymerized nucleotides and italic typeface represents newly polymerized nucleotides.

3′ — TTGGCACGTCGTAAT — 5′
5′ — AACCGTGCAGCATTA — 3′

3′ — TTGGCACGTCGTAAT — 5′
5′ — AACCGTGCAGCATTA — 3′

13. Remember, the transcript will be antiparallel to the DNA template.

3′ — UUGGCACGUCGUAAU — 5′

14. There are several outcomes possible depending on the exact cause of the null allele. For the representation below, it is assumed that the null allele is still transcribed and translated but that the protein product of the allele is altered in a way that affects its mobility within the gel. (For example, it is slightly smaller and migrates to a position that is lower.) For both gels, lane 1 represents two normal alleles, lane 2 represents one normal and one null allele, and lane 3 represents two null alleles. One other possibility is that the mutant allele is neither transcribed nor translated, in which case lane 3 of both blots would be "blank."

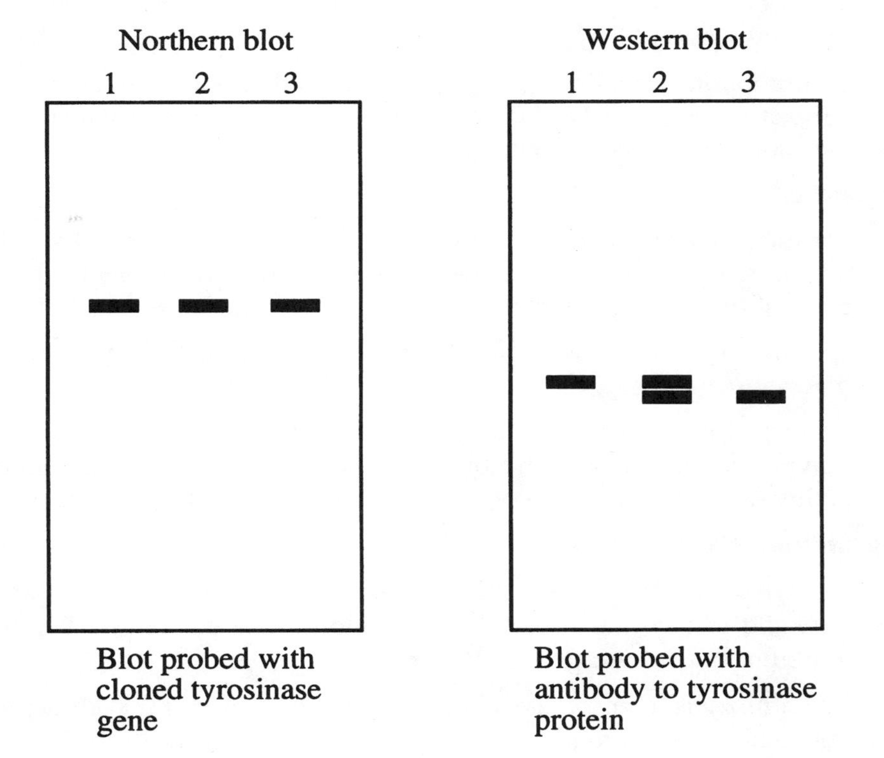

Blot probed with cloned tyrosinase gene

Blot probed with antibody to tyrosinase protein

15. Sulfur. The sugar contains carbon, oxygen, and hydrogen; the phosphate group contains phosphorus and oxygen; and the base contains carbon, oxygen, hydrogen, and nitrogen.

16. The simplest definition is that a *gene* is a chromosomal region capable of making a functional transcript. However, this does not take into account the regulatory regions near the gene necessary for the proper expression of the gene nor the regions that help control transcription that can be quite distant. Also, many eukaryotes have large regions of noncoding sequences (introns) interspersed within the regions that encode the product (exons.)

17. Many eukaryotes have intervening sequences called *introns* which are transcribed but then spliced out (removed) prior to translation. In this case, the 25-kb primary transcript has all but 2.1 kb removed to generate the proper mRNA necessary for translation.

18. PFGE (pulsed-field gel electrophoresis) separates DNA molecules by size. When DNA is carefully isolated from *Neurospora* (which has seven different chromosomes), seven bands should be produced using this technique. Similarly, the pea has seven different chromosomes and will produce seven bands (homologous chromosomes will co-migrate as a single band). The housefly has six different chromosomes and should produce six bands.

19. mRNA size = gene size – (number of introns × average size of introns)

20. The anticodon associates with the codon in an antiparallel orientation. Thus, to pair with the codon 5′ — UUA — 3′, the anticodon must be 3′ — AAU — 5′.

21. **a.** Transcription occurs in the 5′ to 3′ direction that is antiparallel to the DNA template. If the top DNA strand is the template, RNA polymerase would transcribe from right to left.

b. and c. 5′ — UUGGGAAGC — 3′

22. **a.** For the fungus to be orange (as in mutant 1), orange pigment would have to accumulate and then not be converted into red pigment. This would occur if enzyme C was defective.

b. Mutant 2 is yellow. Using similar logic as in part (a), enzyme B must be defective.

c. An organism defective in both enzyme B and C would still have the phenotype associated with the block at the earlier step of this biochemical pathway. In this case, it would be yellow owing to the lack of enzyme B.

23. ***Unpacking the Problem***

1. *Lathyrus odoratus* (sweet peas) are grown for their fragrant, colorful flowers, and in fact, their seeds are poisonous. The edible pea belongs to a different but related genus (*Pisum*).

2. A *pathway* is a set of biochemical reactions that begin with a precursor molecule and, through a series of steps, convert it into some product. In this case, a precursor molecule is converted into an anthocyanin pigment.

3. There are two pathways described.

4. Yes, as described, the two pathways are independent.

5. In biology, a *pigment* is any molecule that reflects or transmits specific wavelengths of light — its color depends on which wavelengths of light are absorbed or reflected.

6. In this case, a plant unable to synthesize pigment would be "colorless," or white. Many common solutes are colorless such as salt (NaCl) and sucrose (table sugar).
7. As stated above, white.
8. Yes. The almost infinite variety of paint colors are obtained by mixing a very few primary pigments into a white base. In sweet peas, purple is achieved through the mixing of red and blue pigments.
9. For now, a *mutation* is any alteration of the DNA that leads to a different and discontinuous variant compared with the more common "normal" phenotype.
10. A *null mutation* is a change in the DNA that prevents the production of a functional product (in this case, enzyme).
11. Null mutations may arise through many types of changes: small deletions or duplications leading to frameshift mutations; nucleotide-pair substitutions that destroy the active site, lead to premature stop codons, or affect proper splicing (in eukaryotes); insertions of DNA; etc.
12. The organism is diploid. It has two copies of each of its chromosomes and therefore two copies of each of its genes.
13. The enzymes encoded by the genes are proteins. Whether or not the proteins are properly encoded (and the enzymes active) is the basis of this analysis.
14. No. The genetic locations of these genes are not relevant to this problem.

    In the following schematics, the "nucleus" separates the DNA and its transcription from translation. Also, the "defect" within the gene that results in the production of nonfunctional enzyme is marked by an "x."

15.

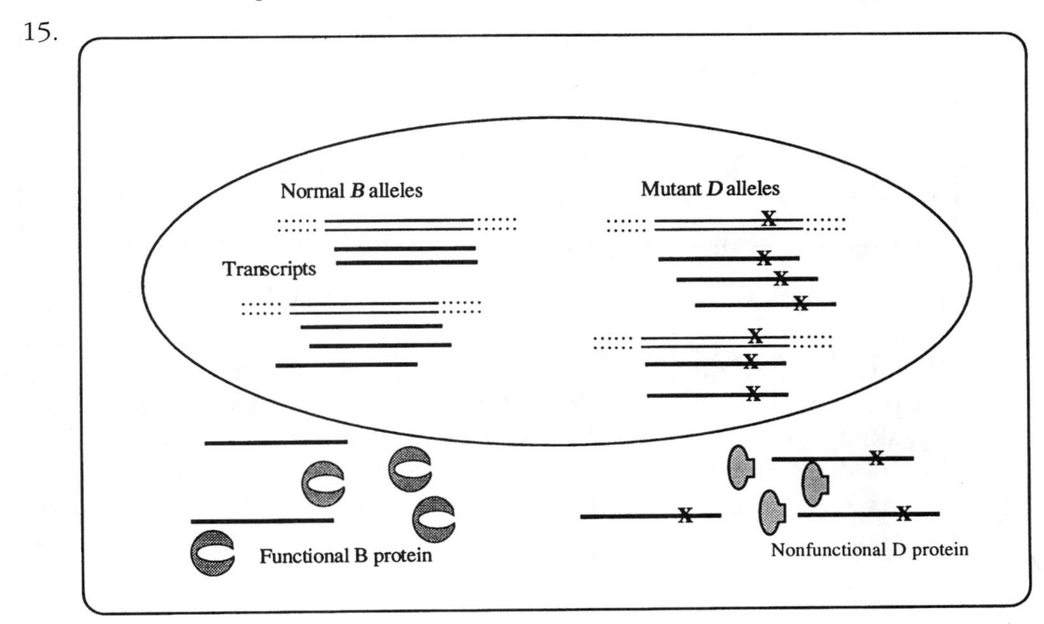

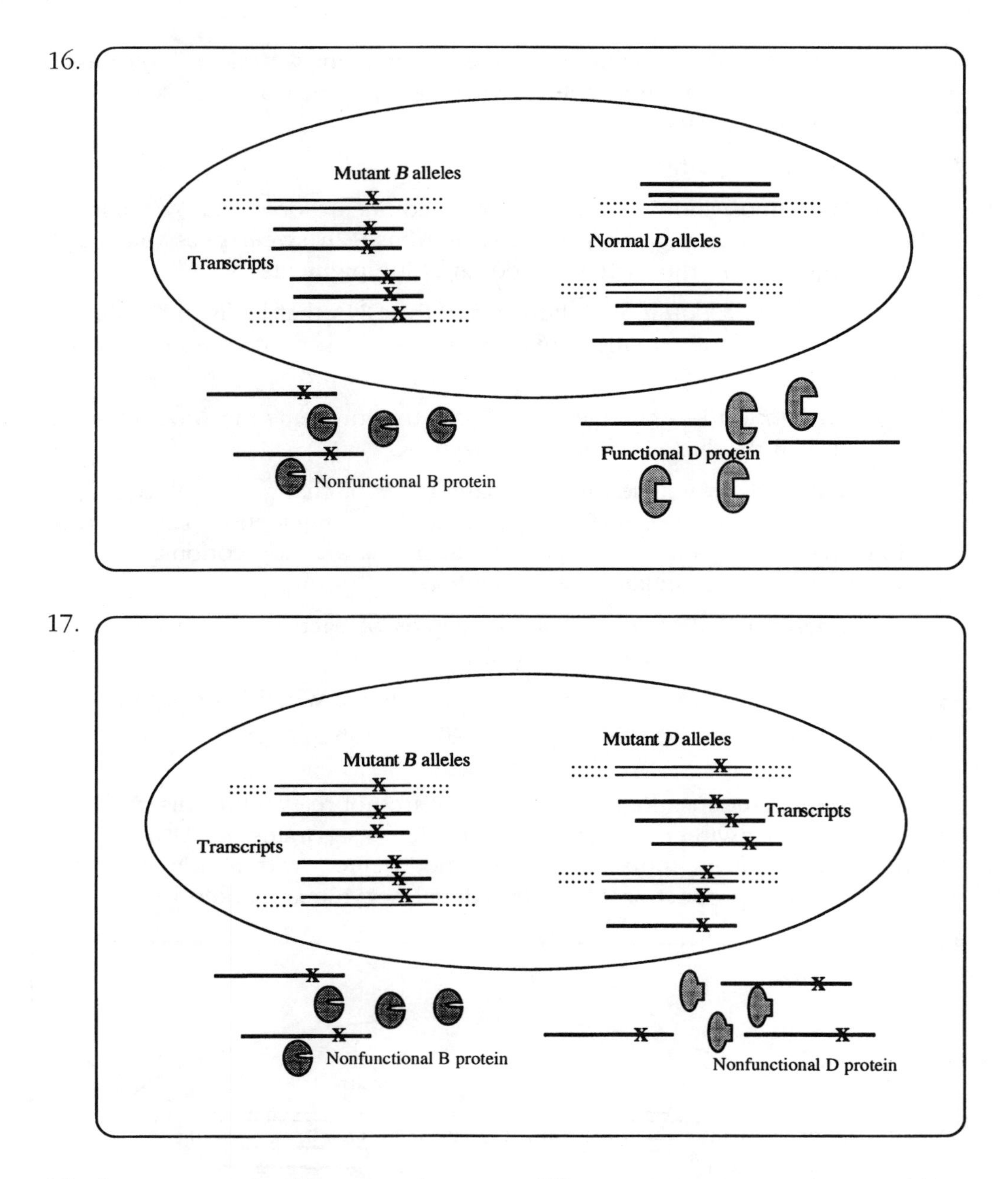

18. Sweet peas are able to produce two different pigments or colors. Just like mixing paints or crayons, these two colors can be blended to produce a third. Only those plants that inherited the ability to make both blue and red pigments have purple flowers. Some plants, however, do not inherit this ability. Some have lost the ability to produce red and therefore have blue flowers, and conversely, some have lost the ability to produce blue and have red flowers. If both of these abilities are lost (or not inherited), the plant has white flowers.

Solving the Problem

a. A plant unable to synthesize red pigment would be blue.

b. A plant unable to synthesize blue pigment would be red.

c. A plant unable to synthesize both blue and red pigments would be white.

24. There are a number of explanations that might explain the positional clustering of null mutants in this *Neurospora* gene. However, based on the material covered in this chapter, it is possible that the gene codes for an enzyme that uses its central amino acids (encoded by the central region of the gene) to form its active site. Mutations that alter the active-site amino acids are most likely to destroy enzymatic function and would be classified as null (loss-of-activity) mutations.

25. Protein function can be destroyed by a mutation that causes the substitution of a single amino acid even though the protein has the same immunological properties. For example, enzymes require very specific amino acids in exact positions within their active site. A substitution of one of these key amino acids might have no effect on overall size and shape of the protein while completely destroying its enzymatic activity.

26. a. The mothers had an excess of phenylalanine in their blood, and that excess was passed through the placenta into the fetal circulatory system, where it caused brain damage prior to birth.

b. The diet had no effect because the neurological damage has already happened *in utero* prior to birth.

c. A fetus with two mutant copies of the allele that causes PKU makes no functional enzyme. However, the mother of such a child is heterozygous and makes enough enzyme to block any brain damage; the excess phenylalanine in the fetal circulatory system enters the maternal circulatory system and is processed by the maternal gene product. After birth, which is when PKU damage occurs in a PKU child, dietary restrictions block a buildup of phenylalanine in the circulatory system until brain development is completed.

The fetus of a mother with PKU is exposed to the very high level of phenylalanine in its circulatory system during the time of major brain development. Therefore, brain damage occurs before birth, and no dietary restrictions after birth can repair that damage.

d. The obvious solution to the brain damage seen in the babies of PKU mothers is to return the mother to a restricted diet during pregnancy in order to block high levels of exposure to her child.

e. PKU is characterized as a rare recessive disorder. A child with PKU has two parents that carry a mutant allele for the metabolism of phenylalanine. When two individuals who are heterozygous for PKU have a child, the risk that the child will have PKU is 25%. A PKU child is unable to make a functional enzyme that converts phenylalanine to tyrosine. As a result, an excess level of phenylalanine is found in the blood, and the excess is detected as an increase in phenylpyruvic acid in both the blood and the urine. The excess phenylpyruvic acid blocks normal development of the brain, resulting in retardation.

27. a. Recessive. The one normal allele provides enough enzyme to be sufficient for normal function (the definition of haplosufficient).

b. There are many ways to mutate a gene to destroy enzyme function. One possible mutation might be a frameshift mutation within an exon of the gene. Assuming that a single base pair was deleted, the mutation would completely alter the translational product 3′ to the mutation.

c. Hormone replacement could be given to the patient.

d. If the hormone is required before birth, it would be supplied by the mother.

28. DNA has often been called the "blueprint of life," but how does it actually compare to a blueprint used for house construction? Both are abstract representations of instructions for building three-dimensional forms, and both require interpretation for their information to be useful. But real blueprints are two-dimensional renderings of the various views of the final structure drawn to scale. There is one-to-one correlation between the lines on the drawing and the real form. The information in the DNA is encoded in a linear array — a one-dimensional set of instructions that becomes three-dimensional only as the encoded linear array of amino acids fold into their many forms. Included in the informational content of the DNA are also all the directions required for its "house" to maintain and repair itself, respond to change, and replicate. Let's see a real blueprint do that!

29. Other than the human ability to synthesize and wear synthetic materials, living systems are "proteins or something that has been made by a protein."

30. The norm of reaction is the phenotypic variation that exists for a species of a fixed genotype within a varying environment. The variability itself can become vital to a species with a change in environment, for it allows for the possibility that some individuals may be able to survive under the new environmental conditions long enough to reproduce.

31. Phenotypic variation within a species can be due to genotype, environmental effects, and pure chance (random noise). Showing that variation of a particular trait has a genetic basis, especially for traits that show continuous variation, therefore requires very carefully controlled analyses. This is discussed in detail in Chapter 25 of the companion text.

32. The formula genotype + environment = phenotype is not quite accurate. While phenotype is a product of the genotype and the environment, there is also an inherent variation due to random noise. Given complete information regarding both genotype and the environment, it is still impossible to specify the phenotype completely, although a close approximation can be achieved.

2 Patterns of Inheritance

1. Mendel's first law states that alleles are in pairs and segregate during meiosis, and his second law states that genes assort independently during meiosis.

2. Do a testcross (cross to *a/a*). If the fly was *A/A*, all the progeny will be phenotypically A; if the fly was *A/a*, half the progeny will be A and half will be a.

3. The progeny ratio is approximately 3:1, indicating classic heterozygous-by-heterozygous mating. Since Black (*B*) is dominant to white (*b*):

   Parents: *B/b* × *B/b*
   Progeny: 3 black:1 white (1 *B/B* : 2 *B/b* : 1 *b/b*)

4. **a.** This is simply a matter of counting genotypes; there are 9 genotypes in the Punnett square. Alternatively, you know there are three genotypes possible per gene, for example *R/R*, *R/r*, and *r/r*, and since both genes assort independently, there are 3 × 3 = 9 total genotypes.

   b. Again, simply count. The genotypes are

1 *R/R* ; *Y/Y*	1 *r/r* ; *Y/Y*	1 *R/R* ; *y/y*	1 *r/r* ; *y/y*
2 *R/r* ; *Y/Y*	2 *r/r* ; *Y/y*	2 *R/r* ; *y/y*	
2 *R/R* ; *Y/y*			
4 *R/r* ; *Y/y*			

c. To find a formula for the number of genotypes, first consider the following:

Number of genes	Number of genotypes	Number of phenotypes
1	$3 = 3^1$	$2 = 2^1$
2	$9 = 3^2$	$4 = 2^2$
3	$27 = 3^3$	$8 = 2^3$

Note that the number of genotypes is 3 raised to some power in each case. In other words, a general formula for the number of genotypes is 3^n, where "n" equals the number of genes.

For allelic relationships that show complete dominance, the number of phenotypes is 2 raised to some power. The general formula for the number of phenotypes observed is 2^n, where "n" equals the number of genes.

d. The round, yellow phenotype is *R/–* ; *Y/–*. Two ways to determine the exact genotype of a specific plant are through selfing or conducting a testcross.

With selfing, complete heterozygosity will yield a 9:3:3:1 phenotypic ratio. Homozygosity at one locus will yield a 3:1 phenotypic ratio, while homozygosity at both loci will yield only one phenotypic class.

With a testcross, complete heterozygosity will yield a 1:1:1:1 phenotypic ratio. Homozygosity at one locus will yield a 1:1 phenotypic ratio, while homozygosity at both loci will yield only one phenotypic class.

5. Each die has six sides, so the probability of any one side (number) is $^1/_6$.

To get specific red, green, and blue numbers involves "and" statements that are independent. So each independent probability is multiplied together.

a. $(^1/_6)(^1/_6)(^1/_6) = (^1/_6)^3 = {^1/_{216}}$

b. $(^1/_6)(^1/_6)(^1/_6) = (^1/_6)^3 = {^1/_{216}}$

c. $(^1/_6)(^1/_6)(^1/_6) = (^1/_6)^3 = {^1/_{216}}$

d. To get no sixes is the same as getting anything but sixes:

$$(1 - {^1/_6})(1 - {^1/_6})(1 - {^1/_6}) = (^5/_6)^3 = {^{125}/_{216}}.$$

e. There are three ways to get two sixes and one five:

6R, 6G, 5B $(^1/_6)(^1/_6)(^1/_6)$

or +

6R, 5G, 6B $(^1/_6)(^1/_6)(^1/_6)$

or +

5R, 6G, 6B $(^1/_6)(^1/_6)(^1/_6)$

$$= 3(^1/_6)^3 = {^3/_{216}} = {^1/_{72}}$$

f. Here there are "and" and "or" statements: p(three sixes "or" three fives)

= p(6R "and" 6G "and" 6B "or" 5R "and" 5G "and" 5B)

$= (1/6)^3 + (1/6)^3 = 2(1/6)^3 = 2/216 = 1/108$

g. There are six ways to fulfill this:

$(1/6)^3 + (1/6)^3 + (1/6)^3 + (1/6)^3 + (1/6)^3 + (1/6)^3 = 6(1/6)^3 = 1/36$

h. The easiest way to approach this problem is to consider each die separately. The first die thrown can be any number. Therefore, the probability for it is 1.

The second die can be any number except the number obtained on the first die. Therefore, the probability of not duplicating the first die is 1 – p(first die duplicated) $= 1 - 1/6 = 5/6$.

The third die can be any number except the numbers obtained on the first two dice. Therefore, the probability is 1 – p(first two dice duplicated) $= 1 - 2/6 = 2/3$.

Therefore, the probability of all different dice is $(1)(5/6)(2/3) = 10/18 = 5/9$.

6. a. Before beginning the specific problems, calculate the probabilities associated with each jar.

jar 1 p(R) 600/(600 + 400) = 0.6
p(W) = 400/(600 + 400) = 0.4

jar 2 p(B) = 900/(900 + 100) = 0.9
p(W) 100/(900 + 100) = 0.1

jar 3 p(G) 10/(10 + 990) = 0.01
p(W) 990/(10 + 990) = 0.99

(1) p(R, B, G) = (0.6)(0.9)(0.01) = 0.0054

(2) p(W, W, W) = (0.4)(0.1)(0.99) = 0.0396

(3) Before plugging into the formula, you should realize that, while white can come from any jar, red and green must come from specific jars (jar 1 and jar 3). Therefore, white must come from jar 2:

p(R, W, G) = (0.6)(0.1)(0.01) = 0.0006

(4) p(R, W, W) = (0.6)(0.1)(0.99) = 0.0594

(5) There are three ways to satisfy this:

R, W, W or W, B, W or W, W, G

= (0.6)(0.1)(0.99) + (0.4)(0.9)(0.99) + (0.4)(0.1)(0.01)

= 0.0594 + 0.3564 + 0.0004 = 0.4162

(6) At least one white is the same as 1 minus no whites:

p(at least 1 W) = 1 – p(no W) = 1 – p(R, B, G)

= 1 – (0.6)(0.9)(0.01) = 1 – 0.0054 = 0.9946

b. The cross is *R/r* × *R/r*. The probability of red (*R/–*) is $^3/_4$, and the probability of white (*r/r*) is $^1/_4$. Because only one white is needed, the only unacceptable result is all red.

In n trials, the probability of all red is $(^3/_4)^n$. Because the probability of failure must be no greater than 5 percent:

$(^3/_4)^n < 0.05$

$n > 10.41$, or 11 seeds.

c. The p(failure) = 0.8 for each egg. Since all eggs are implanted simultaneously, the p(5 failures) = $(0.8)^5$. The p(at least one success) = 1 – $(0.8)^5$ = 1 – 0.328 = 0.672

7. a. By considering the pedigree (see below), you will discover that the cross in question is *T/t* × *T/t*. Therefore, the probability of being a taster is $^3/_4$, and the probability of being a nontaster is $^1/_4$.

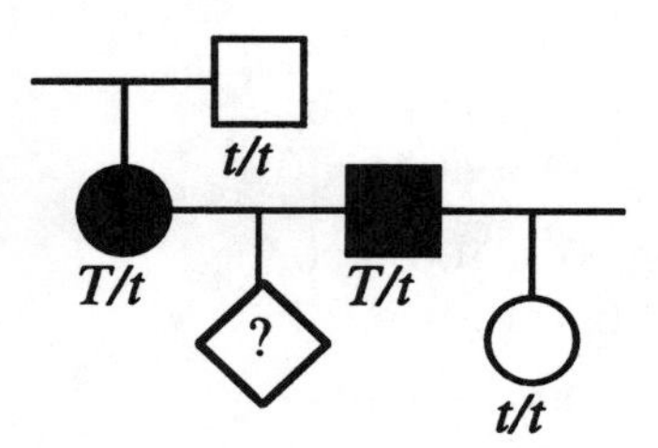

Also, the probability of having a boy equals the probability of having a girl equals $^1/_2$.

(1) p(nontaster girl) = p(nontaster) × p(girl) = $^1/_4 \times ^1/_2 = ^1/_8$

(2) p(taster girl) = p(taster) × p(girl) = $^3/_4 \times ^1/_2 = ^3/_8$

(3) p(taster boy) = p(taster) × p(boy) = $^3/_4 \times ^1/_2 = ^3/_8$

b. p(taster for first two children) = p(taster for first child) × p(taster for second child) = $^3/_4 \times ^3/_4 = ^9/_{16}$

8. *Unpacking the Problem*

1. Yes. The pedigree is given below.

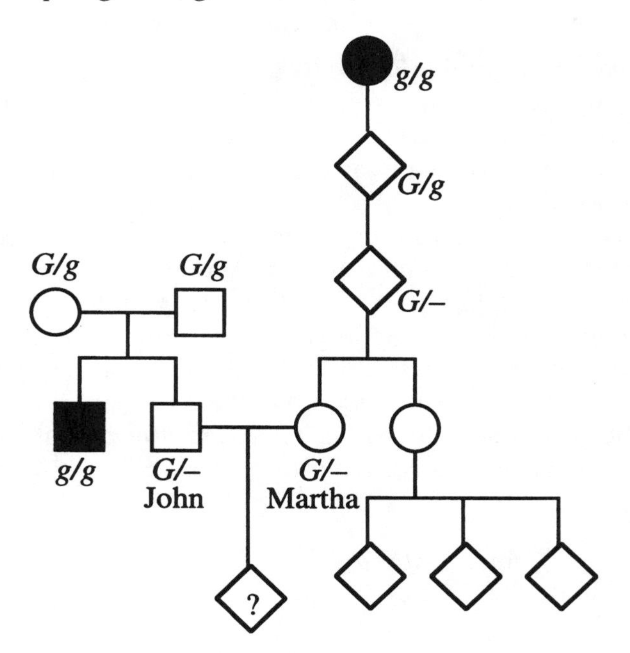

2. In order to state this problem as a Punnett square, you must first know the genotypes of John and Martha. The genotypes can be determined only through considering the pedigree. Even with the pedigree, however, the genotypes can be stated only as *G/–* for both John and Martha.

   The probability that John is carrying the allele for galactosemia is $^2/_3$, rather than the $^1/_2$ that you might guess. To understand this, recall that John's parents must be heterozygous in order to have a child with the recessive disorder while still being normal themselves (the assumption of normalcy is based on the information given in the problem). John's parents were both *G/g*. A Punnett square for their mating would be:

   The cross is

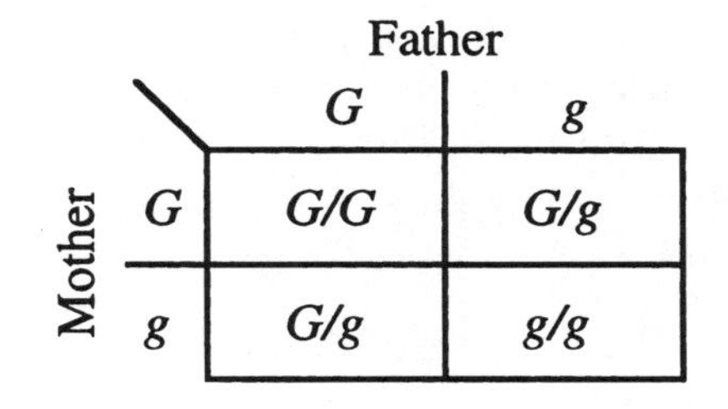

Mother \ Father	*G*	*g*
G	*G/G*	*G/g*
g	*G/g*	*g/g*

P $G/g \times G/g$

F_1 g/g John's brother

$G/–$ John (either G/G or G/g)

The expected ratio of the F_1 is 1 G/G : 2 G/g : 1 g/g. Because John does not have galactosemia (an assumption based on the information given in the problem), he can be either G/G or G/g, which occurs at a ratio of 1:2. Therefore, his probability of carrying the g allele is $^2/_3$.

The probability that Martha is carrying the g allele is based on the following chain of logic. Her great-grandmother had galactosemia, which means that she had to pass the allele to Martha's grandparent. Because the problem states nothing with regard to the grandparent's phenotype, it must be assumed that the grandparent was normal, or G/g. The probability that the grandparent passed it to Martha's parent is $^1/_2$. Next, the probability that Martha's parent passed the allele to Martha is also $^1/_2$, assuming that the parent actually has it. Therefore, the probability that Martha's parent has the allele and passed it to Martha is $^1/_2 \times {}^1/_2$, or $^1/_4$.

In summary:

John $p(G/G) = {}^1/_3$
$p(G/g) = {}^2/_3$

Martha $p(G/G) = {}^3/_4$
$p(G/g) = {}^1/_4$

This information does not fit easily into a Punnett square.

3. While the above information could be put into a branch diagram, it does not easily fit into one and overcomplicates the problem, just as a Punnett square would.

4. The mating between John's parents illustrates Mendel's first law.

5. The scientific words in this problem are *galactosemia*, *autosomal*, and *recessive*.

   Galactosemia is a metabolic disorder characterized by the absence of the enzyme galactose 1-phosphate uridyl transferase, which results in an accumulation of galactose. In the vast majority of cases, galactosemia results in an enlarged liver, jaundice, vomiting, anorexia, lethargy, and very early death if galactose is not omitted from the diet (initially, the child obtains galactose from milk).

   Autosomal refers to genes that are on the autosomes.

   Recessive means that in order for an allele to be expressed, it must be the only form of the gene present in the organism.

6. The major assumption is that if nothing is stated about a person's phenotype, the person is of normal phenotype. Another assumption that may be of value, but is not actually needed, is that all people marrying into these two families are normal and do not carry the allele for galactosemia.

7. The people not mentioned in the problem, but who must be considered, are John's parents and Martha's grandparent and parent descended from her affected great-grandmother.
8. The major statistical rule needed to solve the problem is the product rule ("and" rule).
9. Autosomal recessive disorders are assumed to be rare and assumed to occur equally frequently in males and females. They are also assumed to be expressed if the person is homozygous for the recessive genotype.
10. Rareness leads to the assumption that people who marry into a family that is being studied do not carry the allele, which was assumed in entry (6).
11. The only certain genotypes in the pedigree are John's parents, John's brother, and Martha's great-grandmother and grandmother. All other individuals have uncertain genotypes.
12. John's family can be treated simply as a heterozygous-by-heterozygous cross, with John having a $^2/_3$ probability of being a carrier, while it is unknown if either of Martha's parents carry the allele.
13. The information regarding Martha's sister and her children turns out to be irrelevant to the problem.
14. The problem contains a number of assumptions that have not been necessary in problem solving until now.
15. I can think of a number. Can you?

Solution to the Problem

p(child has galactosemia) = p(John is G/g) × p(Martha is G/g) × p(both parents passed g to the child) = $(^2/_3)(^1/_4)(^1/_4) = {^2/_{48}} = {^1/_{24}}$

9. Charlie, his mate, or both, obviously were not pure-breeding, because his F_2 progeny were of two phenotypes. Let A = black and white, and a = red and white. If both parents were heterozygous, then red and white would have been expected in the F_1 generation. Red and white were not observed in the F_1 generation, so only one of the parents was heterozygous. The cross is

P $A/a \times A/A$

F_1 $1\ A/a : 1\ A/A$

Two F_1 heterozygotes (A/a) when crossed would give 1 A/A (black and white) : 2 A/a (black and white) : 1 a/a (red and white). If the red and white F_2 progeny were from more than one mate of Charlie's, then the farmer acted correctly. However, if the F_2 progeny came only from one mate, the farmer may have acted too quickly.

10. Because the parents are heterozygous, both are A/a. Both twins could be albino or both twins could be normal ("and," "or," "and" = multiply, add, multiply). The probability of being normal ($A/-$) is $^3/_4$, and the probability of being albino (a/a) is $^1/_4$.

$$p(\text{both normal}) + p(\text{both albino})$$
$$p(\text{first normal}) \times p(\text{second normal}) + p(\text{first albino}) \times p(\text{second albino})$$
$$(3/4)(3/4) + (1/4)(1/4) = 9/16 + 1/16 = 5/8$$

11. The plants are approximately 3 blotched:1 unblotched. This suggests that blotched is dominant to unblotched and that the original plant, which was selfed, was a heterozygote.

a. Let *A* = blotched, *a* = unblotched.
P *A/a* (blotched) × *A/a* (blotched)
F_1 1 *A/A* : 2 *A/a* : 1 *a/a*
3 *A/–* (blotched):1 *a/a* (unblotched)

b. All unblotched plants should be pure-breeding in a testcross with an unblotched plant (*a/a*), and one-third of the blotched plants should be pure-breeding.

12. In theory, it cannot be proved that an animal is not a carrier for a recessive allele. However, in an *A/–* × *a/a* cross, the more dominant-phenotype progeny produced, the less likely it is that the parent is *A/a*. In such a cross half the progeny would be *a/a* and half would be *A/a*. With *n* dominant phenotype progeny, the probability that the parent is *A/a* is $(1/2)^n$.

13. The results suggest that winged (*A/–*) is dominant to wingless (*a/a*) (cross 2 gives a 3:1 ratio). If that is correct, the crosses become

Pollination	Genotypes	Number of progeny plants: Winged	Wingless
Winged (selfed)	*A/A* × *A/A*	91	1*
Winged (selfed)	*A/a* × *A/a*	90	30
Wingless (selfed)	*a/a* × *a/a*	4*	80
Winged × wingless	*A/A* × *a/a*	161	0
Winged × wingless	*A/a* × *a/a*	29	31
Winged × wingless	*A/A* × *a/a*	46	0
Winged × winged	*A/A* × *A/–*	44	0
Winged × winged	*A/A* × *A/–*	24	0

The five unusual plants are most likely due either to human error in classification or to contamination. Alternatively, they could result from environmental effects on development. For example, too little water may have prevented the seed pods from becoming winged even though they are genetically winged.

14. a. The disorder appears to be dominant because all affected individuals have an affected parent. If the trait was recessive, then I-1, II-2, III-1, and III-8 would all have to be carriers (heterozygous for the rare allele).

b. Assuming dominance, the genotypes are

I: *d/d, D/d*

II: *D/d, d/d, D/d, d/d*

III: *d/d, D/d, d/d, D/d, d/d, d/d, D/d, d/d*

IV: *D/d, d/d, D/d, d/d, d/d, d/d, d/d, D/d, d/d*

c. The mating is *D/d* × *d/d*. The probability of an affected child (*D/d*) equals $^1/_2$, and the probability of an unaffected child (*d/d*) equals $^1/_2$. Therefore, the chance of having four unaffected children (since each is an independent event) is $^1/_2 \times {}^1/_2 \times {}^1/_2 \times {}^1/_2 = {}^1/_{16}$.

15. a. *Pedigree 1:* The best answer is recessive because two unaffected individuals had affected progeny. Also, the disorder skips generations and appears in a mating between two related individuals.

Pedigree 2: The best answer is dominant because two affected parents have an unaffected child. Also, it appears in each generation, roughly half the progeny are affected, and all affected individuals have an affected parent.

Pedigree 3: The best answer is dominant for many of the reasons stated for pedigree 2. Inbreeding, while present in the pedigree, does not allow an explanation of recessive because it cannot account for individuals in the second generation.

Pedigree 4: The best answer is recessive. Two unaffected individuals had affected progeny.

b. Genotypes of pedigree 1:
Generation I: *A/–, a/a*
Generation II: *A/a, A/a, A/a, A/–, A/–, A/a*
Generation III: *A/a, A/a*
Generation IV: *a/a*

Genotypes of pedigree 2:
Generation I: *A/a, a/a, A/a, a/a*
Generation II: *a/a, a/a, A/a, A/a, a/a, a/a, A/a, A/a, a/a*
Generation III: *a/a, a/a, a/a, a/a, a/a, A/–, A/–, A/–, A/a, a/a*
Generation IV: *a/a, a/a, a/a*

Genotypes of pedigree 3:
Generation I: *A/–, a/a*
Generation II: *A/a, a/a, a/a, A/a*
Generation III: *a/a, A/a, a/a, a/a, A/a, a/a*
Generation IV: *a/a, A/a, A/a, A/a, a/a, a/a*

Genotypes of pedigree 4:
Generation I: *a/a, A/–, A/a, A/a*
Generation II: *A/a, A/a, A/a, a/a, A/–, a/a, A/–, A/–, A/–, A/–, A/–*
Generation III: *A/a, a/a, A/a, A/a, a/a, A/a*

16. a. The pedigree is:

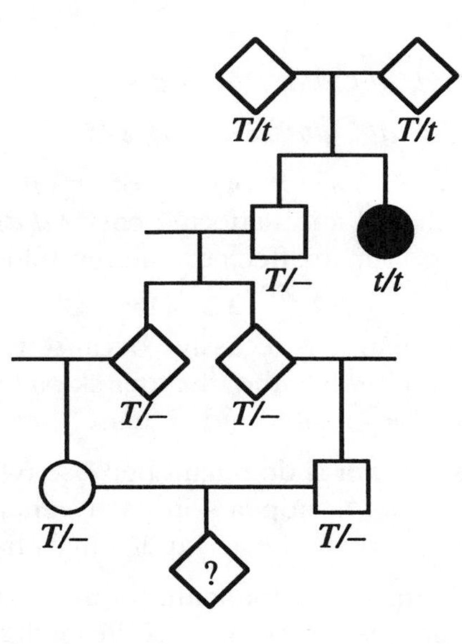

b. The probability that the child of the two first cousins will have Tay-Sachs disease is a function of three probabilities: p(the woman is T/t) × p(the man is T/t) × p(both donate t);

$= (^2/_3)(^1/_2)(^1/_2) \times (^2/_3)(^1/_2)(^1/_2) \times {^1/_4} = {^1/_{144}}$

To understand the probabilities of the first two events, see the discussion for problem 8 part (2) of this chapter.

17. a. Autosomal recessive: affected individuals inherited the trait from unaffected parents and a daughter inherited the trait from an unaffected father.

b. Both parents must be heterozygous to have a $^1/_4$ chance of having an affected child. Parent 2 is heterozygous, since her father is homozygous for the recessive allele and parent 1 has a $^1/_2$ chance of being heterozygous, since his father is heterozygous because 1's paternal grandmother was affected. Overall, $1 \times {^1/_2} \times {^1/_4} = {^1/_8}$.

18. a. Yes. It is inherited as an autosomal dominant trait.

b. Susan is highly unlikely to have Huntington disease. Her great-grandmother (individual II-2) is 75 years old and has yet to develop it, when nearly 100% of people carrying the allele will have developed the disease by that age. If her great-grandmother does not have it, Susan cannot inherit it.

Alan is somewhat more likely than Susan to develop Huntington disease. His grandfather (individual III-7) is only 50 years old, and approximately 20% of the people with the allele have yet to develop the disease by that age. Therefore it can be estimated that the grandfather has a 10% chance of being a carrier (50% chance he inherited the allele from his father × 20% chance he has not yet developed symptoms). If Alan's grandfather eventually develops Huntington disease, then there is a probability of 50% that Alan's father inherited it from him, and a probability of 50% that Alan received that allele from his father. Therefore, Alan has a $1/10 \times 1/2 \times 1/2 = 1/40$ current probability of developing Huntington disease and $1/2 \times 1/2 = 1/4$ probability if his grandfather eventually develops it.

19. a. Assuming the trait is rare, expect that all individuals marrying into the pedigree do not carry the disease-causing allele.

I: *P/P*, *p/p*, *p/p*, *P/P*

II: *P/P*, *P/p*, *P/p*, *P/p*, *P/p*

III: *P/P*, *P/–*, *P/–*, *P/P*

IV: *P/–*, *P/–*

b. For their child to have PKU, both A and B must be carriers and both must donate the recessive allele.

The probability that individual A has the PKU allele is derived from individual II-2. II-2 must be *P/p* since her father must be *p/p*. Therefore, the probability that II-2 passed the PKU allele to individual III-2 is $1/2$. If III-2 received the allele, the probability that he passed it to individual IV-1 (A) is $1/2$. Therefore, the probability that A is a carrier is $1/2 \times 1/2 = 1/4$.

The probability that individual B has the allele goes back to the mating of II-3 and II-4, both of whom are heterozygous. Their child, III-3, has a $2/3$ probability of having received the PKU allele and a probability of $1/2$ of passing it to IV-2 (B). Therefore, the probability that B has the PKU allele is $2/3 \times 1/2 = 1/3$.

If both parents are heterozygous, they have a $1/4$ chance of both passing the *p* allele to their child.

p(child has PKU) = *p*(A is *P/p*) × *p*(B is *P/p*) × *p*(both parents donate *p*)

$$1/4 \quad \times \quad 1/3 \quad \times \quad 1/4 \quad = 1/48$$

c. If the first child is normal, no additional information has been gained and the probability that the second child will have PKU is the same as the probability that the first child will have PKU, or $1/48$.

d. If the first child has PKU, both parents are heterozygous. The probability of having an affected child is now $1/4$, and the probability of having an unaffected child is $3/4$.

20. a. In order to draw this pedigree, you should realize that if an individual's status is not mentioned, then there is no way to assign a genotype to that person. The parents of the boy in question had a phenotype (and genotype) that differed from his. Therefore, both parents were heterozygous and the boy, who is a nonroller, homozygous recessive. Let *R* stand for the ability to roll the tongue and *r* stand for the inability to roll the tongue. The pedigree becomes

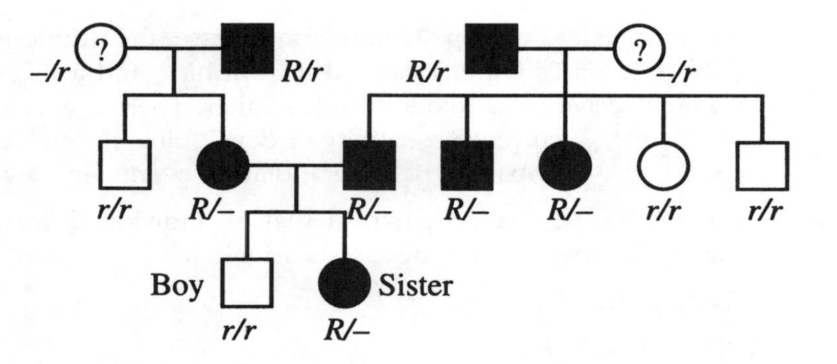

b. Assuming the twins are identical, there might be either an environmental component to the expression of that gene or developmental noise (see Chapter 1). Another possibility is that the *R* allele is not fully penetrant and that some genotypic "rollers" do not express the phenotype.

21. a. Let *C* stand for the normal allele and *c* stand for the allele that causes cystic fibrosis.

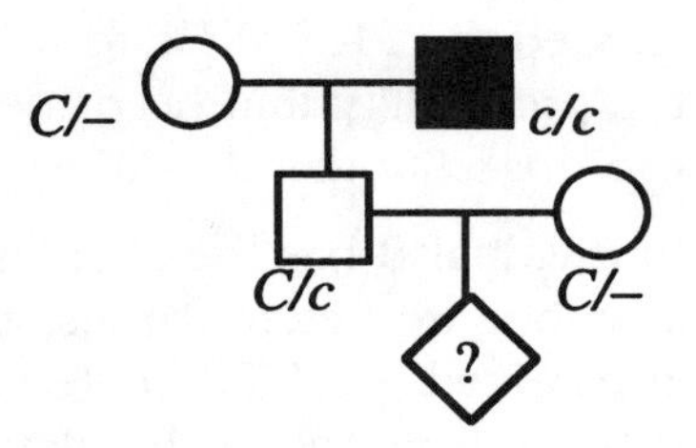

b. The man has a 100% probability of having the *c* allele. His wife, who is from the general population, has a $^1/_{50}$ chance of having the *c* allele. If both have the allele, then $^1/_4$ of their children will have cystic fibrosis. The probability that their first child will have cystic fibrosis is

p(man has *c*) × p(woman has *c*) × p(both pass *c* to the child)

$1.0 \times {}^1/_{50} \times {}^1/_4 = {}^1/_{200} = 0.005$

c. If the first child does have cystic fibrosis, then the woman is a carrier of the *c* allele. Since both parents are *C/c*, the chance that the second child will be normal is the probability of a normal child in a heterozygous × heterozygous mating, or $^3/_4$.

22. a. The inheritance pattern for red hair suggested by this pedigree is recessive, since most red-haired individuals are from parents without this trait.

b. By observing those around us, the allele appears to be somewhat rare.

23. *Taster by taster cross.* Tasters can be either *T/T* or *T/t*, and the genotypic status cannot be determined until a large number of progeny are observed. A failure to obtain a 3:1 ratio in the matings of two tasters would be expected because there are three types of matings:

Mating	Genotypes	Phenotypes
T/T × *T/T*	all *T/T*	all tasters
T/T × *T/t*	1 *T/T* : 1 *T/t*	all tasters
T/t × *T/t*	1 *T/T* : 2 *T/t* : 1 *t/t*	3 tasters : 1 nontaster

Taster by nontaster cross. There are two types of matings that resulted in the observed progeny:

Mating	Genotypes	Phenotypes
T/T × *t/t*	all *T/t*	all tasters
T/t × *t/t*	1 *T/t* : 1 *t/t*	1 tasters : 1 nontasters

Again, the failure to obtain either a 1:0 ratio or a 1:1 ratio would be expected because of the two mating types.

Nontaster by nontaster cross. There is only one mating that is nontaster by nontaster (*t/t* × *t/t*), and 100% of the progeny would be expected to be nontasters. Of 223 children 5 were classified as tasters. Some could be the result of mutation (unlikely), some could be the result of misclassification (likely), some could be the result of a second gene that affects the expression of the gene in question (possible), some could be the result of developmental noise (possible), and some could be due to illegitimacy (possible).

24. Use the following symbols:

Gene function	Dominant allele	Recessive allele
Color	*R* = red	*r* = yellow
Loculed	*L* = two	*l* = many
Height	*H* = tall	*h* = dwarf

The starting plants are pure-breeding, so their genotypes are

red, two-loculed, dwarf *R/R* ; *L/L* ; *h/h* and yellow, many-loculed, tall *r/r* ; *l/l* ; *H/H.*

The farmer wants to produce a pure-breeding line that is yellow, two-loculed, and tall, which would have the genotype *r/r* ; *L/L* ; *H/H.*

The two pure-breeding starting lines will produce an F_1 that will be *R/r* ; *L/l* ; *H/h*. By doing an $F_1 \times F_1$ cross and selecting yellow, two-loculed, and tall plants, the known genotype will be *r/r* ; *L/–* ; *H/–*. The task then will be to do sequential testcrosses for both the *L/–* and *H/–* genes among these yellow, two-loculed, and tall plants. Because the two genes in question are possibly homozygous recessive in different plants, each plant that is yellow, two-loculed, and tall will have to be testcrossed twice.

For each testcross, the plant will obviously be discarded if the testcross reveals a heterozygous state for the gene in question. If no recessive allele is detected, then the minimum number of progeny that must be examined to be 95% confident that the plant is homozygous is based on the frequency of the dominant phenotype if heterozygous, which is $^1/_2$. In n progeny, the probability of obtaining all dominant progeny in a test cross given that the plant is heterozygous is $(^1/_2)^n$. To be 95% confident of homozygosity, the following formula is used, where 5% is the probability that it is not homozygous:

$$(^1/_2)^n = 0.05$$

n = 4.3, or 5 phenotypically dominant progeny must be obtained from each testcross to be 95% confident that the plant is homozygous.

25. Let *A* represent achondroplasia and *a* represent normal height. Let *N* represent neurofibromatosis and *n* represent the normal allele. Because both conditions are extremely rare, the affected individuals are assumed to be heterozygous. The genes are also assumed to assort independently. The cross is

P *A/a* ; *n/n* × *a/a* ; *N/n*

F_1 1 *A/a* ; *n/n* : 1 *a/a* ; *n/n* : 1 *a/a* ; *N/n* : 1 *A/a* ; *N/n*

1 anchondroplasia : 1 normal : 1 neurofibromatosis: 1 anchondroplasia , neurofibromatosis

26. **a.** *C/c* ; *S/s* × *C/c* ; *S/s* There are 3 short:1 long, and 3 dark:1 albino.

Therefore, each gene is heterozygous in the parents.

b. *C/C*; *S/s* × *C/C*; *s/s* There are no albino, and there are 1 long: 1 short indicating a testcross for this trait.

c. *C/c* ; *S/S* × *c/c* ; *S/S* There are no long, and there are 1 dark: 1 albino.

d. *c/c* ; *S/s* × *c/c* ; *S/s* All are albino, and there are 3 short:1 long.

e. *C/c* ; *s/s* × *C/c* ; *s/s* All are long, and there are 3 dark:1 albino.

f. *C/C*; *S/s* × *C/C*; *S/s* There are no albino, and there are 3 short: 1 long.

g. *C/c* ; *S/s* × *C/c* ; *s/s* There are 3 dark:1 albino, and 1 short:1 long.

27. **a. and b.** Cross 2 indicates that purple (*G*) is dominant to green (*g*), and cross 1 indicates cut (*P*) is dominant to potato (*p*).

Cross 1: *G/g* ; *P/p* × *g/g* ; *P/p* There are 3 cut:1 potato, and 1 purple:1 green.

Cross 2: G/g ; *P/p* × *G/g* ; *p/p* There are 3 purple:1 green, and 1 cut:1 potato.

Cross 3: *G/G* ; *P/p* × *g/g* ; *P/p* There are no green, and there are 3 cut:1 potato.

Cross 4: *G/g* ; *P/P* × *g/g* ; *p/p* There are no potato, and there are 1 purple:1 green.

Cross 5: *G/g* ; *p/p* × *g/g* ; *P/p* There are 1 cut:1 potato, and there are 1 purple:1 green.

28. **a.** Since each gene assorts independently, each probability should be considered separately and then multiplied together for the answer.

For (1) $^1/_2$ will be A, $^3/_4$ will be B, $^1/_2$ will be C, $^3/_4$ will be D, and $^1/_2$ will be E.

$^1/_2 \times ^3/_4 \times ^1/_2 \times ^3/_4 \times ^1/_2 = ^9/_{128}$

For (2) $^1/_2$ will be a, $^3/_4$ will be B, $^1/_2$ will be c, $^3/_4$ will be D, and $^1/_2$ will be e.

$^1/_2 \times ^3/_4 \times ^1/_2 \times ^3/_4 \times ^1/_2 = ^9/_{128}$

For (3) it is the sum of (1) and (2) = $^9/_{128} + ^9/_{128} = ^9/_{64}$

For (4) it is 1 – (part 3) = $1 - ^9/_{64} = ^{55}/_{64}$

b. For (1) $^1/_2$ will be *A/a*, $^1/_2$ will be *B/b*, $^1/_2$ will be *C/c*, $^1/_2$ will be *D/d*, and $^1/_2$ will be *E/e*.

$^1/_2 \times ^1/_2 \times ^1/_2 \times ^1/_2 \times ^1/_2 = ^1/_{32}$

For (2) $^1/_2$ will be *a/a*, $^1/_2$ will be *B/b*, $^1/_2$ will be *c/c*, $^1/_2$ will be *D/d*, and $^1/_2$ will be *e/e*.

$^1/_2 \times ^1/_2 \times ^1/_2 \times ^1/_2 \times ^1/_2 = ^1/_{32}$

For (3) it is the sum of (1) and (2) = $^1/_{16}$.

For (4) it is 1 – (part 3) = $1 - ^1/_{16} = ^{15}/_{16}$

29. The single yellow fly may either be recessive *y/y* or be yellow because of a diet of silver salts. Cross that fly with a known recessive yellow fly and raise half the larvae on a diet with no silver salts and half the larvae on a diet with silver salts. A true recessive will result in flies with yellow bodies on both diets, while a phenocopy that is genetically *Y/–* will produce flies with brown or yellow bodies, depending on the diet.

Phenocopies are caused by environmental factors. Many drugs used by pregnant women result in genetic defects that are phenocopies. One example is cleft lip or palate caused by Valium taken before the fetal face is completely formed. Retardation caused by the consumption of alcohol during pregnancy is another phenocopy effect.

30. **a.** Cataracts appear to be caused by a dominant allele because affected people have affected parents. Dwarfism appears to be caused by a recessive allele because affected people have unaffected parents.

b. Using *A* for cataracts, *a* for no cataracts, *B* for normal height, and *b* for dwarfism, the genotypes are

III: *a/a* ; *B/b*, *a/a* ; *B/b*, *A/a* ; *B/–*, *a/a* ; *B/–*, *A/a* ; *B/b*, *A/a* ; *B/b*, *a/a* ; *B/–*, *a/a* ; *B/–*, *A/a* ; *b/b*

c. The mating is *a/a* ; *b/b* (IV-1) × *A/–* ; *B/–* (IV-5). Recall that the probability of a child's being affected by any disease is a function of the probability of each parent's carrying the allele in question and the probability that one parent (for a dominant disorder) or both parents (for a recessive disorder) donate it to the child. Individual IV-1 is homozygous for these two genes; therefore, the only task is to determine the probabilities associated with individual IV-5.

The probability that individual IV-5 is heterozygous for dwarfism is $^2/_3$. Thus the probability that she has the *b* allele and will pass it to her child is $^2/_3 \times {}^1/_2 = {}^1/_3$.

The probability that individual IV-5 is homozygous for cataracts is $^1/_3$, and the probability that she is heterozygous is $^2/_3$. If she is homozygous for the allele that causes cataracts, she must pass it to her child; if she is heterozygous for cataracts, she has a probability of $^1/_2$ of passing it to her child.

The probability that the first child is a dwarf with cataracts is the probability that the child inherits the *A* and *b* alleles from its mother which is $(^1/_3 \times 1)(^2/_3 \times {}^1/_2) + (^2/_3 \times {}^1/_2)(^2/_3 \times {}^1/_2) = {}^2/_9$. Alternatively, you can calculate the chance of inheriting the *b* allele $(^2/_3 \times {}^1/_2)$ and not inheriting the *a* allele $(1 - {}^1/_3)$ or $^1/_3 \times {}^2/_3 = {}^2/_9$.

The probability of having a phenotypically normal child is the probability that the mother donates the *a* and *B* (or not *b*) alleles which is $(^2/_3 \times {}^1/_2)(1 - {}^1/_3) = {}^2/_9$.

31. a. and b. Begin with any two of the three lines and cross them. If, for example, you began with *a/a* ; *B/B* ; *C/C* × *A/A* ; *b/b* ; *C/C*, the progeny would all be *A/a* ; *B/b* ; *C/C*. Crossing two of these would yield

9 *A/–* ; *B/–* ; *C/C*
3 *a/a* ; *B/–* ; *C/C*
3 *A/–* ; *b/b* ; *C/C*
1 *a/a* ; *b/b* ; *C/C*

The *a/a* ; *b/b* ; *C/C* genotype has two of the genes in a homozygous recessive state and occurs in $^1/_{16}$ of the offspring. If that were crossed with *A/A* ; *B/B* ; *c/c*, the progeny would all be *A/a* ; *B/b* ; *C/c*. Crossing two of them (or "selfing") would lead to a 27:9:9:9:3:3:3:1 ratio, and the plant occurring in $^1/_{64}$ of the progeny would be the desired *a/a* ; *b/b* ; *c/c*.

There are several different routes to obtaining *a/a* ; *b/b* ; *c/c*, but the one outlined above requires only four crosses.

32. a. At year 1, the genotypic ratio is

A/A 0.55
A/a 0.40
a/a 0.05

Because all plants self-pollinate, the three crosses are

A/A × *A/A* → 100% *A/A*
A/a × *A/a* → 25% *A/A*, 50% *A/a*, 25% *a/a*
a/a × *a/a* → 100% *a/a*

In other words, the two homozygous classes grow at the expense of the heterozygous class. At year 2, the genotypic ratio would be

A/A $0.55 + {}^{1}/_{4}(0.40) = 0.65$
A/a ${}^{1}/_{2}(0.40) = 0.20$
a/a $0.5 + {}^{1}/_{4}(0.40) = 0.15$

Again, when this population self-pollinates, the homozygous classes will grow, by the same percentages, at the expense of the heterozygous class. At year 3, the genotypic ratio would be

A/A $0.65 + {}^{1}/_{4}(0.20) = 0.70$
A/a ${}^{1}/_{2}(0.20) = 0.10$
a/a $0.15 + {}^{1}/_{4}(0.20) = 0.20$

A third year of selfing will produce the following genotypic ratio:

A/A $0.70 + {}^{1}/_{4}(0.10) = 0.725$
A/a ${}^{1}/_{2}(0.10) = 0.05$
a/a $0.20 + {}^{1}/_{4}(0.10) = 0.225$

b. Consider the cross *A/a* × *A/a*. Only ${}^{1}/_{2}$ the progeny would be expected to be heterozygous. This will hold for each locus that began as heterozygous. Therefore, overall heterozygosity will be reduced by 50% with each generation of selfing. After the third generation of selfing, only $({}^{1}/_{2})^3 \times 40\%$ $= {}^{1}/_{8}(40\%) = 5\%$ of the loci would remain heterozygous.

c. The same mathematical concepts apply in each situation. The generalized formula for a reduction of heterozygosity with selfing is $({}^{1}/_{2})^n X$, where n = the number of generations and X = the initial percentage of heterozygosity.

33. P s^+/s^+ × s/Y
↓
F_1 ${}^{1}/_{2}$ s^+/s normal female
${}^{1}/_{2}$ s^+/Y normal male

s^+/s × s^+/Y
↓
F_2 ${}^{1}/_{4}$ s^+/s^+ normal female
${}^{1}/_{4}$ s^+/s normal female
${}^{1}/_{4}$ s^+/Y normal male
${}^{1}/_{4}$ s/Y small male

In the cross: P s^+/s × s/Y
↓
Progeny ${}^{1}/_{4}$ s^+/s normal female
${}^{1}/_{4}$ s/s small female
${}^{1}/_{4}$ s^+/Y normal male
${}^{1}/_{4}$ s/Y small male

34. **a.** Your mother gives you 23 chromosomes, half of all that she has and half of all that you have. Therefore, you have half of all your genes in common with your mother.

b. For any heterozygous gene (for example, *A/a*) in a parent, it is possible that

You	Brother
A	*A*
a	*a*
A	*a*
a	*A*

Thus, there are two of four combinations between you and your brother that will be a match. Therefore, you and your brother will have half your genes in common for each gene that is heterozygous in your parents. You and your brother will have identical alleles from your parents for each gene that is homozygous in your parents. The degree of homozygosity in a parent is unknown. For this reason, the final answer can be stated only for heterozygosity in the parents, and homozygosity must be ignored.

35. First assume that the cross is autosomal:

P *G/–* (graceful female) × *g/g* (gruesome male)

F_1
1 *G/g* graceful female
1 *–/g* ? female
1 *G/g* graceful male
1 *–/g* ? male

This cross does not meet the observation, so it must be wrong. The cross also cannot be *G/–* × *g*/Y (X-linked), because graceful females will result.

Next assume that the female is the heterogametic sex and the gene is sex-linked:

P *G*/O (graceful female) × *g/g* (gruesome male)

F_1 1 *g*/O gruesome female

1 *G/g* graceful male

Notice that this outcome matches the observed results. Therefore, in the schmoo the female is the heterogametic sex and this gene is sex-linked. The O can be a female-determining chromosome or no chromosome.

36. **a. and b.** The disease does not appear to be autosomal because the trait is inherited in a sex-specific way.

c. It cannot be Y-linked because females are affected.

d. Yes. All daughters and none of the sons will be affected if the father has an X-linked dominant trait.

e. Not likely. If the mother is a carrier, it is highly unlikely she would pass the gene to all her daughters and none of her sons.

37. Let H = hypophosphatemia and h = normal. The cross is $H/\text{Y} \times h/h$, yielding H/h (females) and h/Y (males). The answer is 0%.

38. If the historical record is accurate, the data suggest Y linkage. Another explanation is an autosomal gene that is dominant in males and recessive in females. This has been observed for other genes in both humans and other species.

39. You should draw pedigrees for this question.

a.

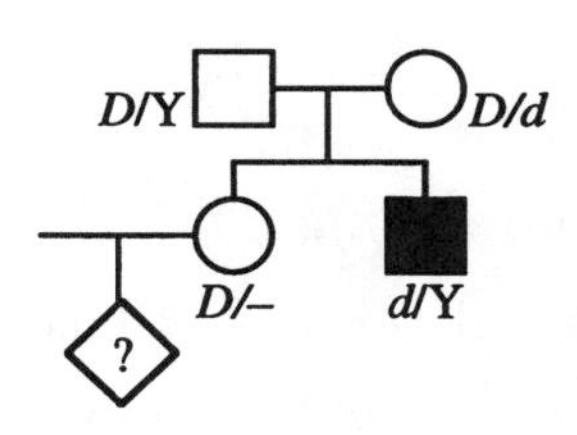

The "maternal grandmother" had to be a carrier, *D/d*. The probability that the woman inherited the *d* allele from her is $^1/_2$. The probability that she passes it to her child is $^1/_2$. The probability that the child is male is $^1/_2$. The total probability of the woman's having an affected child is $^1/_2 \times {}^1/_2 \times {}^1/_2 = {}^1/_8$.

b. The same pedigree as part a applies. The "maternal grandmother" had to be a carrier, *D/d*. The probability that your mother received the allele is $^1/_2$. The probability that your mother passed it to you is $^1/_2$. The total probability is $^1/_2 \times {}^1/_2 = {}^1/_4$.

c.

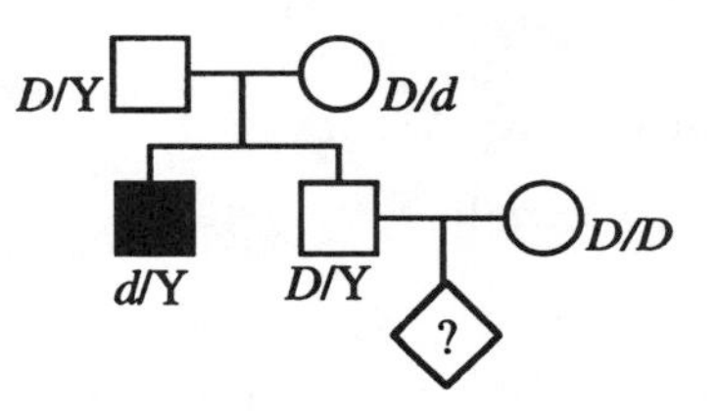

Because your father does not have the disease, you cannot inherit the allele from him. Therefore, the probability of inheriting an allele will be based on the chance that your mother is heterozygous. Since she is "unrelated" to the pedigree, assume that this is zero.

40. a. Because none of the parents are affected, the disease must be recessive. Because the inheritance of this trait appears to be sex-specific, it is most likely X-linked. If it were autosomal, all three parents would have to be carriers and, by chance, only sons and none of the daughters inherited the trait (which is quite unlikely).

b. I A/Y, A/a, A/Y

II A/Y, $A/–$, a/Y, $A/–$, A/Y, a/Y, a/Y, $A/–$, a/Y, $A/–$

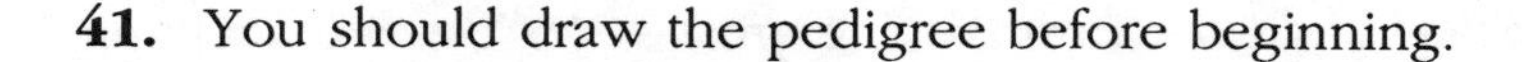

41. You should draw the pedigree before beginning.

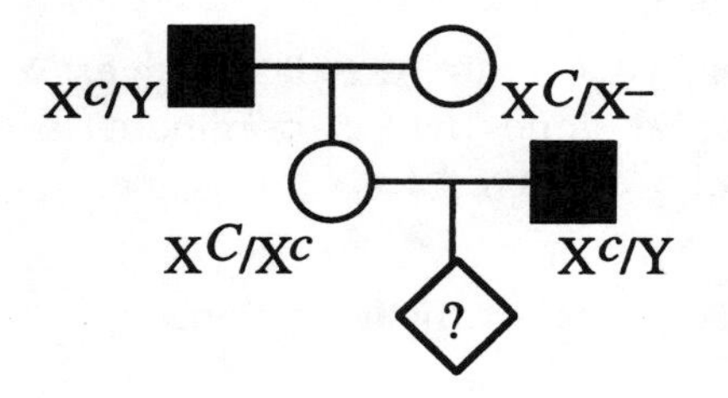

a. X^C/X^c, X^c/X^c

b. p(colorblind) × p(male) = $(^1/_2)(^1/_2) = {}^1/_4$

c. The girls will be 1 normal (X^C/X^c) : 1 colorblind (X^c/X^c)

d. The cross is $X^C/X^c \times X^c/Y$, yielding 1 normal : 1 colorblind for both sexes

42. a. This problem involves X-inactivation. Let B = black and b = *orange.*

Females	Males
X^B/X^B = black	X^B/Y = black
X^b/X^b = orange	X^b/Y = orange
X^B/X^b = calico	

b. P X^b/X^b (orange) × X^B/Y (black)

F_1 X^B/X^b (calico female)

X^b/Y (orange male)

c. P X^B/X^B (black) × X^b/Y (orange)

F_1 X^B/X^b (calico female)

X^B/Y (black male)

d. Because the males are black or orange, the mother had to have been calico. Half the daughters are black, indicating that their father was black.

e. Males were orange or black, indicating that the mothers were calico. Orange females indicate that the father was orange.

43. The man is B/b ; X^D/X^d so $^1/_2 \times {}^1/_2 = {}^1/_4$ of the sperm will be b; X^d

44. a. Recessive (unaffected parents have affected progeny) and X-linked (only assumption is that the grandmother, I-2, is a carrier). If autosomal, then I-1, I-2, and II-6 would all have to be carriers.

b. Generation I: X^A/Y, X^A/X^a

Generation II: X^A/X^A, X^a/Y, X^AY, X^A/X^-, X^A/X^a, X^A/Y

Generation III: X^A/X^A, X^A/Y, X^A/X^a, X^A/X^a, X^A/Y, X^AX^A, X^a/Y, X^A/Y, X^A/X^-

c. Because it is stated that the trait is rare, the assumption is that no one marrying into the pedigree carries the recessive allele. Therefore, the first couple has no chance of an affected child, because the son received a Y chromosome from his father. The second couple has a 50 percent chance of having affected sons and no chance of having affected daughters. The third couple has no chance of having an affected child, but all of their daughters will be carriers.

45. **a.** Autosomal recessive: excluded by unaffected female in third generation.

b. Autosomal dominant: consistent.

c. X-linked recessive: excluded by affected female with unaffected father.

d. X-linked dominant: excluded by unaffected female in third generation.

e. Y-linked: excluded by affected females.

46. **a.** From cross 6, Bent (*B*) is dominant to normal (*b*). Both parents are "bent," yet some progeny are "normal."

b. From cross 1, it is X-linked. The trait is inherited in a sex-specific manner — all sons have the mother's phenotype.

c. In the following table, the Y chromosome is stated; the X is implied.

	Parents		Progeny	
Cross	Female	Male	Female	Male
1	*b/b*	*B*/Y	*B/b*	*b*/Y
2	*B/b*	*b*/Y	*B/b*, *b/b*	*B*/Y, *b*/Y
3	*B/B*	*b*/Y	*B/b*	*B*/Y
4	*b/b*	*b*/Y	*b/b*	*b*/Y
5	*B/B*	*B*/Y	*B/B*	*B*/Y
6	*B/b*	*B*/Y	*B/B*, *B/b*	*B*/Y, *b*/Y

47. ***Unpacking the Problem***

1. *Normal* is used to mean wild type, or red eye color and long wings.
2. Both "*line*" and "*strain*" are used to denote pure-breeding fly stocks, and the words are interchangeable.
3. Your choice.
4. Three characters are being followed: eye color, wing length, and sex.
5. For eye color, there are two phenotypes: red and brown. For wing length there are two phenotypes: long and short. For sex there are two phenotypes: male and female.
6. The F_1 females designated normal have red eyes and long wings.
7. The F_1 males that are called short-winged have red eyes and short wings.
8. The F_2 ratio is

   $^3/_8$ red eyes, long wings

   $^3/_8$ red eyes, short wings

   $^1/_8$ brown eyes, long wings

   $^1/_8$ brown eyes, short wings

9. Because there is not the expected 9:3:3:1 ratio, one of the factors that distorts the expected dihybrid ratio must be present. Such factors can be sex linkage, epistasis, genes on the same chromosome, environmental effect, reduced penetrance, or a lack of complete dominance in one or both genes.
10. With sex linkage, traits are inherited in a sex-specific way. With autosomal inheritance, males and females have the same probabilities of inheriting the trait.
11. The F_2 does not indicate sex-specific inheritance.
12. The F_1 data do show sex-specific inheritance — all males are brown-eyed, like their mothers, whereas all females are red-eyed, like their fathers.
13. The F_1 suggests that long is dominant to short and red is dominant to brown. The F_2 data show a 3 red: 1 brown ratio, indicating the dominance of red but a 1:1 long: short ratio indicative of a testcross. Without the F_1 data, it is not possible to determine whether long or short is dominant.
14. If Mendelian notation is used, then the red and long alleles need to be designated with uppercase letters, for example *R* and *L*, while the brown (*r*) and short (*l*) alleles need to be designated with lowercase letters. If *Drosophila* notation is used, then the brown allele may be designated with a lowercase *b* and the wild-type (red) allele with a b^+; the short wing-length gene with an *s* and the wild-type (long) allele with an s^+. (Genes are often named after their mutant phenotype.)
15. To deduce the inheritance of these phenotypes means to provide all genotypes for all animals in the three generations discussed and account for the ratios observed.

Solution to the Problem

Start this problem by writing the crosses and results so that all the details are clear.

P brown, short female × red, long male

F_1 red, long females
red, short males

These results tell you that red-eyed is dominant to brown-eyed and, since both females and males are red-eyed, that this gene is autosomal. Since males differ from females in their genotype with regard to wing length, this trait is sex-linked. Knowing that *Drosophila* females are XX and males are XY, the long-winged females tell us that long is dominant to short and that the gene is X-linked. Let *B* = red, *b* = brown, *S* = long, and *s* = short. The cross can be rewritten as follows:

P *b/b* ; *s/s* × *B/B* ; *S/Y*

F_1 $^1/_2$ *B/b* ; *S/s* females
$^1/_2$ *B/b* ; *s/*Y males

F_2	$^1/_{16}$ *B/B* ; *S/s*	red, long, female
	$^1/_{16}$ *B/B* ; *s/s*	red, short, female
	$^1/_8$ *B/b* ; *S/s*	red, long, female
	$^1/_8$ *B/b* ; *s/s*	red, short, female
	$^1/_{16}$ *b/b* ; *S/s*	brown, long, female
	$^1/_{16}$ *b/b* ; *s/s*	brown, short, female
	$^1/_{16}$ *B/B* ; *S*/Y	red, long, male
	$^1/_{16}$ *B/B* ; *s*/Y	red, short, male
	$^1/_8$ *B/b* ; *S*/Y	red, long, male
	$^1/_8$ *B/b* ; *s*/Y	red, short, male
	$^1/_{16}$ *b/b* ; *S*/Y	brown, long, male
	$^1/_{16}$ *b/b* ; *s*/Y	brown, short, male

The final phenotypic ratio is

$^3/_8$ red, long
$^3/_8$ red, short
$^1/_8$ brown, long
$^1/_8$ brown, short

with equal numbers of males and females in all classes.

48. The different sex-specific phenotypes found in the F_1 indicate sex-linkage — the females inherit the trait of their fathers. The first cross also indicates that the wild-type large spots is dominant over the lacticolor small spots. Let A = wild type and a = lacticolor.

Cross 1: If the male is assumed to be the hemizygous sex, then it soon becomes clear that the predictions do not match what was observed:

P *a/a* female × *A*/Y male

F_1 *A/a* wild-type females
a/Y lacticolor males

Therefore, assume that the female is the hemizygous sex. Let Z stand for the sex-determining chromosome in females. The cross becomes

P *a*/Z female × *A/A* male

F_1	*A/a*	wild-type male
	A/Z	wild-type female
F_2	$^1/_4$ *A*/Z	wild-type females
	$^1/_2$ *A*/–	wild-type males
	$^1/_4$ *a*/Z	lacticolor females

Cross 2:

P	*A/Z* female × *a/a* male	
F_1	*a/Z*	lacticolor females
	A/a	wild-type males
F_2	$^1/_4$ *A/Z*	wild-type females
	$^1/_4$ *A/a*	wild-type males
	$^1/_4$ *a/Z*	lacticolor females
	$^1/_4$ *a/a*	lacticolor males

49. On the basis of phenotype, the woman appears to have two different cell lines for G6PD activity in her red blood cells. If *G* = normal enzyme activity and *g* = reduced enzyme activity and malaria resistance, then the woman appears to have *G*/– and *g/g* cells. This can best be explained by X-inactivation in the woman's cells. Assume that she is *G/g*. In approximately half her cells the G allele–containing X chromosome will be inactivated, leaving only a functional *g* allele. Those cells will be resistant to the malarial parasite. In the other half of her cells the *g* allele–containing X chromosome will be inactivated. Those cells will have a functional *G* allele and will be susceptible to the parasite.

50. The suggestion is that the woman was a carrier for testicular feminization. Testicular feminization is an X-linked recessive disorder that renders individuals unresponsive to androgens. Chromosomal males with the disorder exhibit an almost idealized feminine appearance, with large breasts, little to no body hair, extremely smooth skin; they are sterile. Females usually show no effect in the heterozygous state.

The gene for testicular feminization is on the X chromosome, and therefore would be subjected to inactivation an expected 50 percent of the time. If testicular feminization is responsible for the woman's unusual phenotype, then the allele that results in this disorder in males would be functioning in approximately 50 percent of her cells. Those cells would be unresponsive to androgens, which would account for the increased breast size and the lack of pubic hair. That these symptoms are confined to one side of her body suggests that she had an unusual midline split as to which X chromosome was inactivated. Her right side expressed the testicular feminization allele, while her left side expressed the normal allele. The menstrual irregularities are completely compatible with her being a carrier for testicular feminization.

If she were a carrier, then 50 percent of her sons would be expected to suffer from testicular feminization. Her brothers would also have a 50 percent risk of having the disorder. Also, her daughters would be expected to have a 50 percent chance of being carriers, which could lead to this disorder in 50 percent of her grandsons. In other words, the pedigree is completely compatible with the suggestion that she was a carrier for testicular feminization:

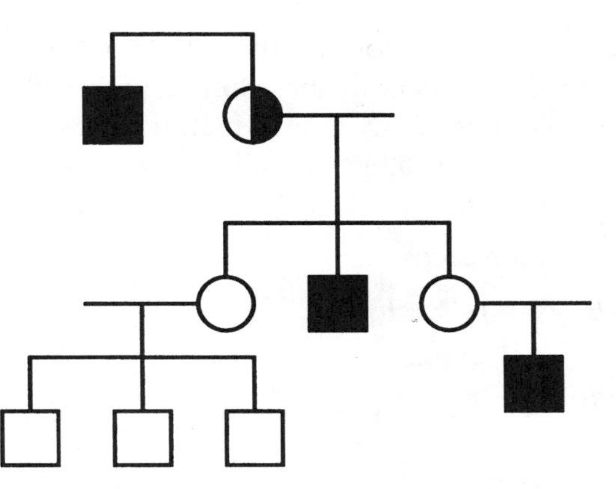

51. **a.** Note that only males are affected and that they had unaffected parents. For rare traits, this is what is expected for an X-linked recessive disorder.

b. The mothers of all affected sons must be heterozygous for the disorder. In addition, the daughters of all affected men must be heterozygous. Finally, barring mutation, individual I-2 must have been heterozygous or I-1 must have had the trait.

52. Note that only males are affected and that in all but one case, the trait can be traced through the female side. However, there is one example of an affected male having affected sons. If the trait is X-linked, this male's wife must be a carrier. Depending on how rare this trait is in the general population, this suggests that the disorder is caused by an autosomal dominant with expression limited to males.

53. First, draw the pedigree.

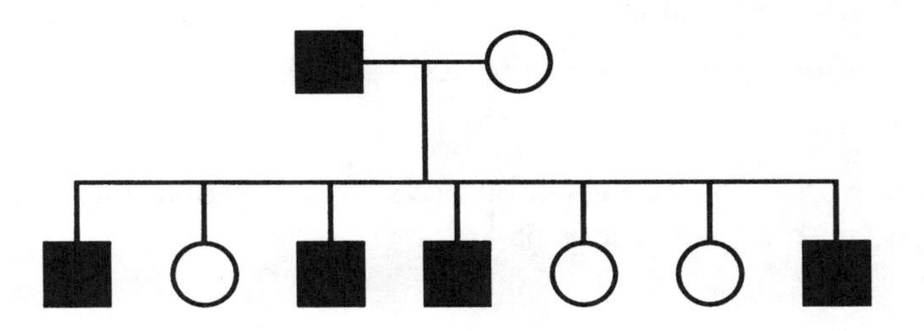

Let the genes be designated by the pigment produced by the normal allele: red pigment, *R*; green pigment, *G*; and blue pigment, *B*.

Recall that the sole X in males comes from the mother, while females obtain an X from each parent. Also recall that a difference in phenotype between sons and daughters is usually due to an X-linked gene. Because all the sons are colorblind and neither the mother nor the daughters are, the mother must carry a different allele for colorblindness on each X chromosome. In other words, she is heterozygous for both X-linked genes, and they are in repulsion: *R g/r G*. With regard to the autosomal gene, she must be *B/–*.

Because all the daughters are normal, the father, who is colorblind, must be able to complement the defects in the mother with regard to his X chromosome. Because he has only one X with which to do so, his genotype must be *R G*/Y ; *b/b*. Likewise, the mother must be able to complement the father's defect, so she must be *B/B*.

The original cross is therefore

P *R g/r G* ; *B/B* × *R G/Y* ; *b/b*

F_1

Females	Males
R g/R G ; *B/b*	*R g*/Y ; *B/b*
r G/R G ; *B/b*	*r G*/Y ; *B/b*

54. Because the disorder is X-linked recessive, the affected male had to have received the allele, *a*, from the female common ancestor in the first generation. The probability that the affected man's wife also carries the *a* allele is the probability that she also received it from the female common ancestor. That probability is $^1/_8$.

The probability that the couple will have an affected boy is

p(father donates Y) × *p*(the mother has *a*) × *p*(mother donates *a*)

$^1/_2 \times {}^1/_8 \times {}^1/_2 = {}^1/_{32}$

The probability that the couple will have an affected girl is

p(father donates X^a) × *p*(the mother has *a*) × *p*(mother donates *a*)

$^1/_2 \times {}^1/_8 \times {}^1/_2 = {}^1/_{32}$

The probability of normal children

= 1 – *p*(affected children)

= 1 – *p*(affected male) – *p*(affected female)

$= 1 - {}^1/_{32} - {}^1/_{32} = {}^{30}/_{32} = {}^{15}/_{16}$

Half the normal children will be boys, with a probability of $^{15}/_{32}$, and half will be girls, with a probability of $^{15}/_{32}$.

3 Chromosomal Basis of Heredity

1. The key function of mitosis is to generate two daughter cells genetically identical with the original parent cell.

2. Two key functions of meiosis are to halve the DNA content and to reshuffle the genetic content of the organism to generate genetic diversity among the progeny.

3. It's pretty hard to beat several billions of years of evolution, but it might be simpler if DNA did not replicate prior to meiosis. The same events responsible for halving the DNA and producing genetic diversity could be achieved in a single cell division if homologous chromosomes paired, recombined, randomly aligned during metaphase and separated during anaphase, etc. However, you would lose the chance to check and repair DNA that replication allows.

4. In large part, this question is asking, "Why sex?" Parthogenesis (the ability to reproduce without fertilization — in essence, cloning) is not common among multicellular organisms. Nearly all plants and animals reproduce sexually, and despite the numerical advantages of cloning, most multicellular species that have adopted it as their only method of reproducing have become extinct. However, there is no agreed upon explanation of why the loss of sexual reproduction usually leads to early extinction or conversely, why sexual reproduction is associated with evolutionary success.

   On the other hand, the immediate effects of such a scenario are obvious. All offspring will be genetically identical with their mothers, and males would be extinct within one generation.

5. As cells divide mitotically, each chromosome consists of identical sister chromatids that are separated to form genetically identical daughter cells. Although the second division of meiosis appears to be a similar process, the "sister" chromatids are likely to be different. Recombination during earlier meiotic stages has swapped regions of DNA between sister and nonsister chromosomes such that the two daughter cells of this division are typically not genetically identical.

6. The resulting cells will have the identical genotype: *A/a* ; *B/b*.

7. Yes, it could work, but certain DNA repair mechanisms (such as postreplication recombination repair, see Chapter 16 of the companion text) could not be invoked prior to cell division. There would be just two cells as products of this meiosis, rather than four.

8. The general formula for the number of different male/female centromeric combinations possible is 2^n, where n = number of different chromosome pairs. In this case, $2^5 = 32$.

9. Since the DNA levels vary sixfold, the range covers cells that are haploid (spores or cells of the gametophyte stage) to cells that are triploid (the endosperm) and dividing (after DNA has replicated but prior to cell division). The following cells would fit the DNA measurements:

0.7	haploid cells
1.4	diploid cells in G_1 or haploid cells after S but prior to cell division
2.1	triploid cells of the endosperm
2.8	diploid cells after S but prior to cell division
4.2	triploid cells after S but prior to cell division

10. Assume both plants are diploid sporophytes, each homozygous for the alleles being studied (*r/r* for red and *b/b* for brown). Both can undergo meiosis to generate spores that then divide mitotically into adult haploid gametophytes. Assuming that each original parent generated equal numbers of spores, equal numbers of *r* (red) and *b* (brown) plants would result.

11.

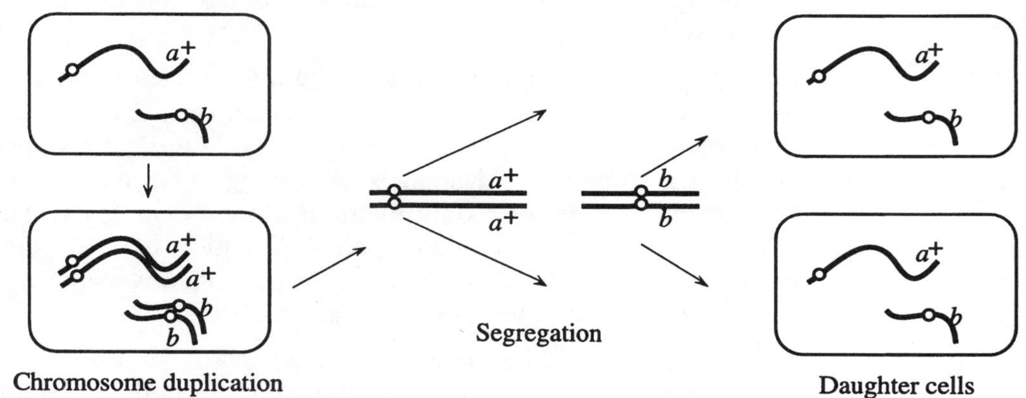

12. In the following schematic drawings, chromosomes (or chromatids) that are radioactive are indicated by the "grains" that would be observed after autoradiography. After the second mitotic division, a number of outcomes are possible owing to the random alignment and separation of the radioactive and nonradioactive chromatids.

13.

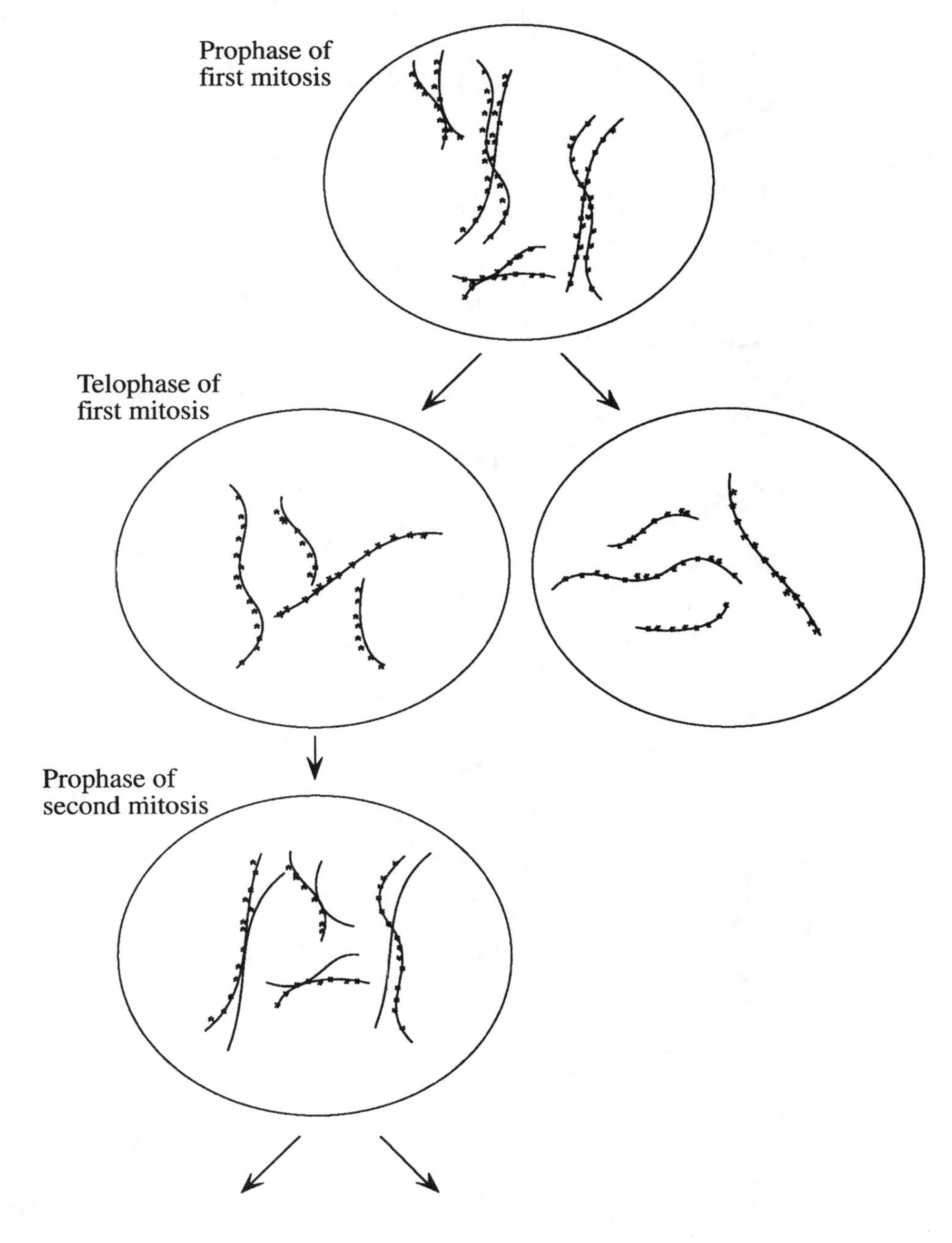

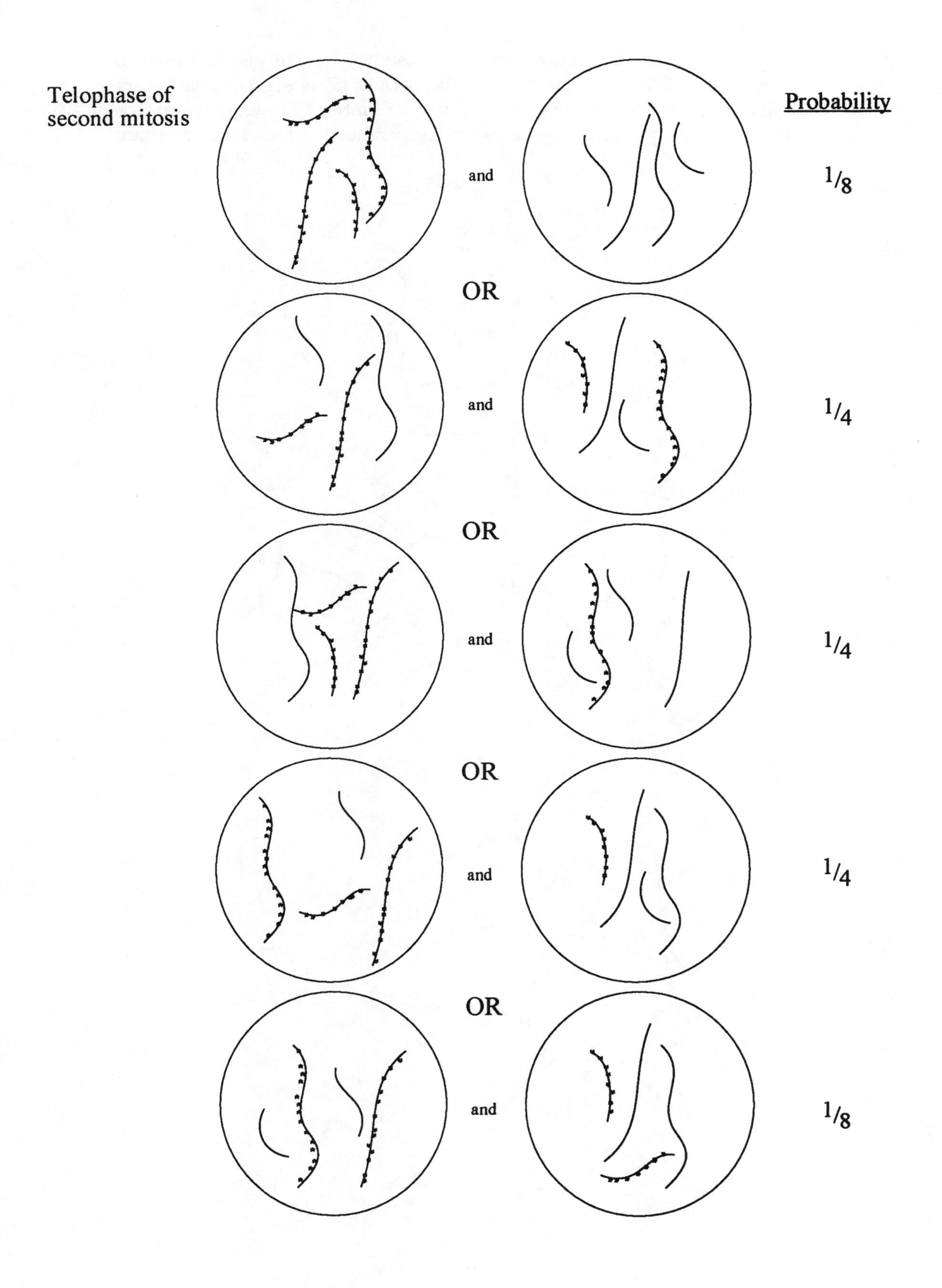
Telophase of second mitosis
Probability
and
$^1/_8$
OR
and
$^1/_4$
OR
and
$^1/_4$
OR
and
$^1/_4$
OR
and
$^1/_8$

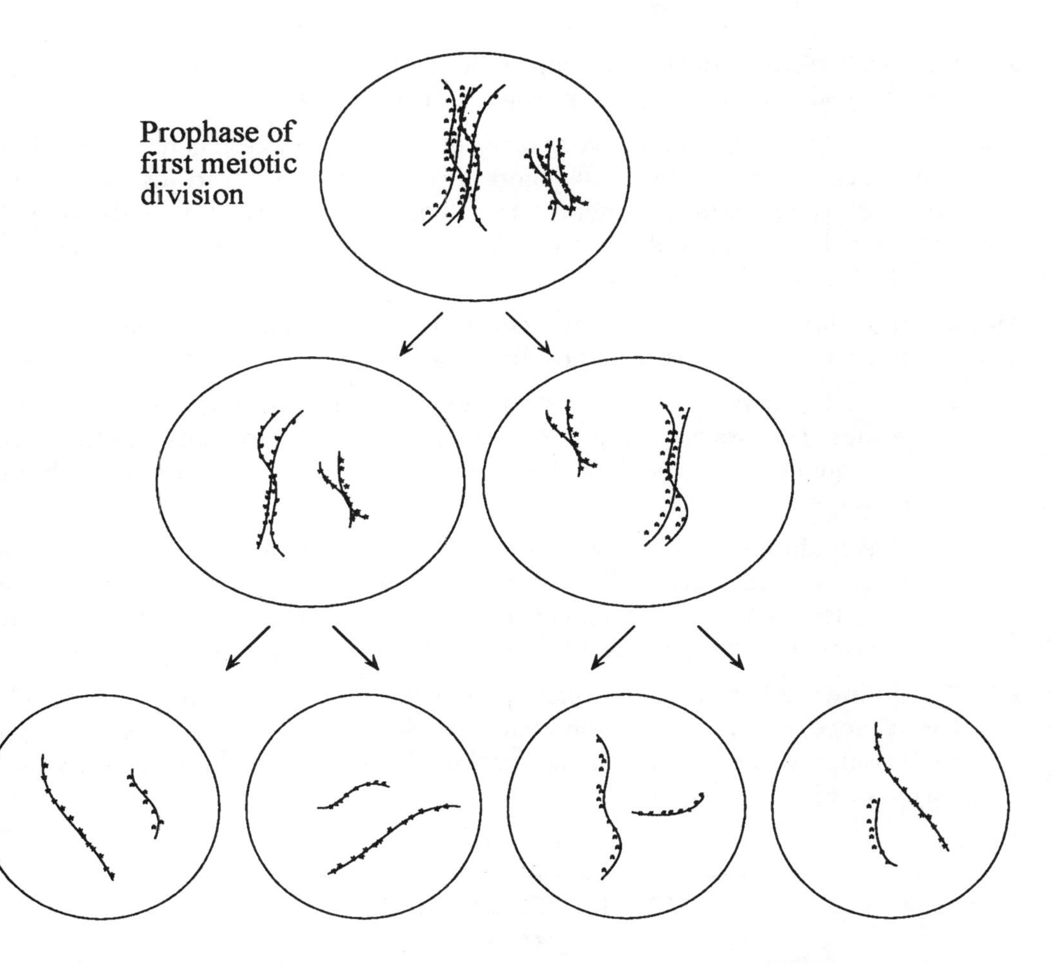

14. PFGE separates DNA molecules by size. When DNA is carefully isolated from *Neurospora* (which has seven different chromosomes), seven bands should be produced using this technique. Similarly, the pea has seven different chromosomes and will produce seven bands (homologous chromosomes will co-migrate as a single band).

15. There is a total of 4 m of DNA and nine chromosomes per haploid set. On average, each is $^4/_9$ m long. At metaphase, their average length is 13 μm, so the average packing ratio is 13×10^{-6} m : 4.4×10^{-1} m, or roughly 1:34,000! This remarkable achievement is accomplished through the interaction of the DNA with proteins. At its most basic, eukaryotic DNA is associated with histones in units called *nucleosomes* and during mitosis coils into a solenoid and, as loops, associates with and winds into a central core of nonhistone protein called the *scaffold*.

16. The nucleus contains the genome and separates it from the cytoplasm. However, during cell division, the nuclear envelope dissociates (breaks down); it is the job of the microtubule-based spindle to separate the chromosomes (divide the genetic material), around which nuclei reform during telophase. In this sense, it can be viewed as a passive structure that is divided by the cell's cytoskeleton.

17. Yes. Half of our genetic makeup is derived from each parent, each parent's genetic makeup is derived half from each of their parents, etc.

18. Since the "half" inherited is very random, the chances of receiving exactly the same half is vanishingly small. Ignoring recombination and just focusing on which chromosomes are inherited from one parent (for example, the one they inherited from their father or the one from their mother?), there are 2^{23} = 8,388,608 possible combinations!

19. a. In a diploid cell, expect two chromosomes (a pair of homologs) to each have a single locus of radioactivity.

b. Expect many regions of radioactivity scattered throughout the chromosomes. The exact number and pattern would be dependent on the specific sequence in question and where and how often it is present within the genome.

c. The multiple copies of the genes for ribosomal RNA are organized into large tandem arrays called *nucleolar organizers* (NO_s). Therefore, expect broader areas of radioactivity compared with (a). The number of these regions would equal the number of NO_s present in the organism.

20. The following is meant to be examples of what is possible. It is also possible, for instance, that more than one band would be present in (a), depending on the position of the restriction sites within the sequence complementary to the probe used.

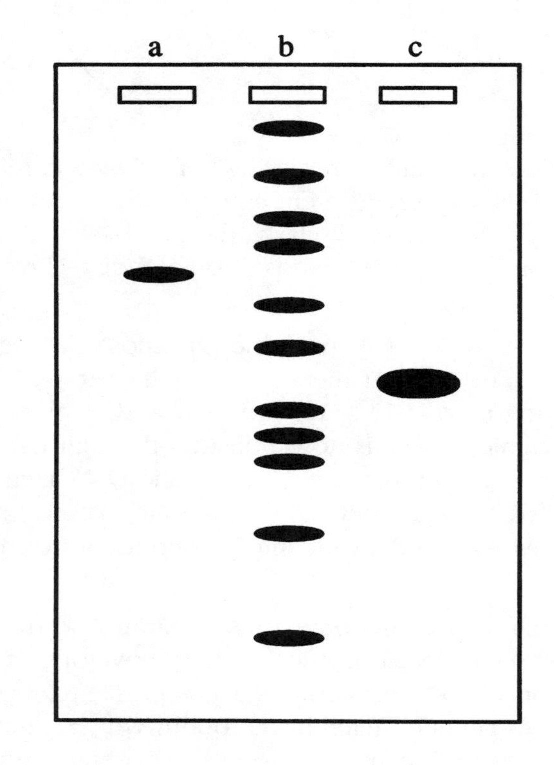

21. Remember, the endosperm is formed from two polar nuclei (which are genetically identical) and one sperm nucleus.

Female	Male	Polar nuclei	Sperm	Endosperm
s/s	*S/S*	*s* and *s*	*S*	*S/s/s*
S/S	*s/s*	*S* and *S*	*s*	*S/S/s*
S/s	*S/s*	$^1/_2$ *S* and *S*	$^1/_2$ *S*	$^1/_4$ *S/S/S*
				$^1/_4$ *S/S/s*
		$^1/_2$ *s* and *s*	$^1/_2$ *s*	$^1/_4$ *S/s/s*
				$^1/_4$ *s/s/s*

22. First, examine the crosses and the resulting genotypes of the endosperm:

Female	Male	Polar nuclei	Sperm	Endosperm
f'/f'	f''/f''	f' and f'	f''/f''	$f'/f'/f''$ (floury)
f''/f''	f'/f'	f'' and f''	f'/f'	$f''/f''/f'$ (flinty)

As can be seen, the phenotype of the endosperm correlates with the predominant allele present.

23. **a.** A sporophyte of *A/a* ; *B/b* genotype will produce gametophytes in the following proportions:

$^1/_4$ *A* ; *B*
$^1/_4$ *A* ; *b*
$^1/_4$ *a* ; *B*
$^1/_4$ *a* ; *b*

b. Random fertilization of the spores from the above gametophytes can occur 4 × 4 = 16 possible ways. Four of these combinations (*A* ; *B* × *a* ; *b*, *a* ; *b* × *A* ; *B*, *A* ; *b* × *a* ; *B*, *a* ; *B* × *A* ; *b*) will result in the desired *A/a* ; *B/b* sporophyte genotype. Therefore $^1/_4$ of the next generation should be of this genotype.

24. All daughter cells will still be *A/a* ; *B/b* ; *C/c*. Mitosis generates daughter cells genetically identical with the original cell.

25.

P	ad^- ; *a* × ad^+ ; α
Transient diploid	ad^+/ad^- ; *a*/α
F_1	$^1/_4$ ad^+ ; *a*, white
	$^1/_4$ ad^- ; *a*, purple
	$^1/_4$ ad^+ ; α, white
	$^1/_4$ ad^- ; α, purple

26.

	Mitosis	Meiosis
fern	sporophyte gametophyte	sporophyte (sporangium)
moss	sporophyte gametophyte	sporophyte (atheridium and archegonium)
flowering plant	sporophyte gametophyte	sporophyte (anther and ovule)
pine tree	sporophyte gametophyte	sporophyte (pine cone)
mushroom	sporophyte gametophyte	sporophyte (ascus or basidium)
frog	somatic cells	gonads
butterfly	somatic cells	gonads
snail	somatic cells	gonads

27. This problem is tricky because the answers depend on how a cell is defined. In general, geneticists consider the transition from one cell to two cells to occur with the onset of anaphase in both mitosis and meiosis even though cytoplasmic division occurs at a later stage.

a. 46 chromosomes, each with 2 chromatids = 92 chromatids

b. 46 chromosomes, each with 2 chromatids = 92 chromatids

c. 46 physically separate chromosomes in each of two about-to-be-formed cells

d. 23 chromosomes in each of two about-to-be-formed cells, each with 2 chromatids = 46 chromatids

e. 23 chromosomes in each of two about-to-be-formed cells

28. (5) chromosome pairing

29. His children will have to inherit the satellite-containing 4 (probability = $^1/_2$), the abnormally staining 7 (probability = $^1/_2$), and the Y chromosome (probability = $^1/_2$). To get all three, the probability is $(^1/_2)(^1/_2)(^1/_2) = {^1/_8}$.

30. The parental set of centromeres can match either parent, which means there are two ways to satisfy the problem. For any one pair, the probability of a centromere from one parent going into a specific gamete is $^1/_2$. For n pairs, the probability of all the centromeres being from one parent is $(^1/_2)^n$. Therefore, the total probability of having a haploid complement of centromeres from either parent is $2(^1/_2)^n = (^1/_2)^{n-1}$.

31. Dear Monk Mendel:

I have recently read your most engrossing manuscript detailing the results of your most wise experiments with garden peas. I salute both your curiosity and your ingenuity in conducting said experiments, thereby opening up for scientific exploration an entire new area of our Maker's universe. Dear Sir, your findings are extraordinary!

While I do not pretend to compare myself to you in any fashion, I beg to bring to your attention certain findings I have made with the aid of that most fascinating and revealing instrument, the microscope. I have been turning my attention to the smallest of worlds with an instrument that I myself have built, and I have noticed some structures that may parallel in behavior the factors that you have postulated in the pea.

I have worked with grasshoppers, however, not your garden peas. Although you are a man of the cloth, you are also a man of science, and I pray that you will not be offended when I state that I have specifically studied the reproductive organs of male grasshoppers. Indeed, I did not limit myself to studying the organs themselves; instead, I also studied the smaller units that make up the male organs and have beheld structures most amazing within them.

These structures are contained within numerous small bags within the male organs. Each bag has a number of these structures, which are long and threadlike at some times and short and compact at other times. They come together in the middle of a bag, and then they appear to divide equally. Shortly thereafter, the bag itself divides, and what looks like half of the threadlike structures goes into each new bag. Could it be, Sir, that these threadlike structures are the very same as your factors? I know, of course, that garden peas do not have male organs in the same way that grasshoppers do, but it seems to me that you found it necessary to emasculate the garden peas in order to do some crosses, so I do not think it too farfetched to postulate a similarity between grasshoppers and garden peas in this respect.

Pray, Sir, do not laugh at me and dismiss my thoughts on this subject even though I have neither your excellent training nor your astounding wisdom in the Sciences. I remain your humble servant to eternity!

32. (1) Impossible: the alleles of the same genes are on nonhomologous chromosomes

(2) Meiosis II.

(3) Meiosis II.

(4) Meiosis II.

(5) Mitosis.

(6) Impossible: appears to be mitotic anaphase but alleles of sister chromatids are not identical

(7) Impossible: too many chromosomes

(8) Impossible: too many chromosomes

(9) Impossible: too many chromosomes

(10) Meiosis I.

(11) Impossible: appears to be meiosis of homozygous *a/a* ; *B/B*

(12) Impossible: the alleles of the same genes are on nonhomologous chromosomes

4 GENE INTERACTION

1. The woman must be *A/O*, so the mating is *A/O* × *A/B*. Their children will be

Genotype	Phenotype
$^1/_4$ *A/A*	A
$^1/_4$ *A/B*	AB
$^1/_4$ *A/O*	A
$^1/_4$ *B/O*	B

2. You are told that the cross of two erminette fowls results in 22 erminette, 14 black, and 12 pure white. Two facts are important: (1) the parents consist of only one phenotype, yet the offspring have three phenotypes, and (2) the progeny appear in an approximate ratio of 1:2:1. These facts should tell you immediately that you are dealing with a heterozygous × heterozygous cross involving one gene and that the erminette phenotype must be the heterozygous phenotype.

When the heterozygote shows a different phenotype from either of the two homozygotes, the heterozygous phenotype results from incomplete dominance or codominance. Because two of the three phenotypes contain black, either fully or in an occasional feather, you might classify erminette as an instance of incomplete dominance because it is intermediate between fully black and fully white. Alternatively, because erminette has both black and white feathers, you might classify the phenotype as codominant. Your decision will rest on whether you look at the whole animal (incomplete dominance) or at individual feathers (codominance). This is yet another instance where what you conclude is determined by how you observe.

To test the hypothesis that the erminette phenotype is a heterozygous phenotype, you could cross an erminette with either, or both, of the homozygotes. You should observe a 1:1 ratio in the progeny of both crosses.

3. **a.** The original cross and results were

P long, white × round, red

F_1 oval, purple

F_2			
	9 long, red	19 oval, red	8 round, white
	15 long, purple	32 oval, purple	16 round, purple
	8 long,. white	16 oval, white	9 round, red
	32 long	67 oval	33 round

The data show that, when the results are rearranged by shape, a 1:2:1 ratio is observed for color within each shape category. Likewise, when the data are rearranged by color, a 1:2:1 ratio is observed for shape within each color category:

9 long, red	15 long, purple	8 round, white
19 oval, red	32 oval, purple	16 oval, white
9 round, red	16 round, purple	8 long, white
37 red	63 purple	32 white

A 1:2:1 ratio is observed when there is a heterozygous × heterozygous cross. Therefore, the original cross was a dihybrid cross. Both oval and purple must represent an incomplete dominant phenotype.

Let L = long, L' = round, R = red and R' = white. The cross becomes

P L/L ; R'/R' × L'/L' ; R/R

F_1 L/L' ; R/R' × L/L' ; R/R'

			$1/4$ R/R =	$1/16$	long, red
F_2	$1/4$ L/L	×	$1/2$ R/R' =	$1/8$	long, purple
			$1/4$ R'/R' =	$1/16$	long, white
			$1/4$ R/R =	$1/8$	oval, red
	$1/2$ L/L'	×	$1/2$ R/R' =	$1/4$	oval, purple
			$1/4$ R'/R' =	$1/8$	oval, white
			$1/4$ R/R =	$1/16$	round, red
	$1/4$ L'/L'	×	$1/2$ R/R' =	$1/8$	round, purple
			$1/4$ R'/R' =	$1/16$	round, white

b. A long, purple × oval, purple cross is as follows:

P $L/L\ ;\ R/R' \times L/L'\ ;\ R/R'$

			$^1/_4\ R/R$ =	$^1/_8$ long, red
F_1	$^1/_2\ L/L$	×	$^1/_2\ R/R'$ =	$^1/_4$ long, purple
			$^1/_4\ R'/R'$ =	$^1/_8$ long, white
			$^1/_4\ R/R$ =	$^1/_8$ oval, red
	$^1/_2\ L/L'$	×	$^1/_2\ R/R'$ =	$^1/_4$ oval, purple
			$^1/_4\ R'/R'$ =	$^1/_8$ oval, white

4. From the cross $c^+/c^{ch} \times c^{ch}/c^h$ the progeny are

$^1/_4$	c^+/c^{ch}	full color
$^1/_4$	c^+/c^h	full color
$^1/_4$	c^{ch}/c^{ch}	chinchilla
$^1/_4$	c^{ch}/c^h	chinchilla

Thus, 0% of the progeny will be Himalayan.

5. **a.** The data indicate that there is a single gene with multiple alleles. The order of dominance is

black > sepia > cream > albino

Cross	Parents	Progeny	Conclusion
Cross 1:	*b/a* × *b/a*	3 *b/–* : 1 *a/a*	Black is dominant to albino.
Cross 2:	*b/s* × *a/a*	1 *b/a* : 1 *s/a*	Black is dominant to sepia; sepia is dominant to albino.
Cross 3:	*c/a* × *c/a*	3 *c/–* : 1 *a/a*	Cream is dominant to albino.
Cross 4:	*s/a* × *c/a*	1 *c/a* : 2 *s/–* : 1 *a/a*	Sepia is dominant to cream.
Cross 5:	*b/c* × *a/a*	1 *b/a* : 1 *c/a*	Black is dominant to cream.
Cross 6:	*b/s* × *c/–*	1 *b/–* : 1 *s/–*	"–" can be *c* or *a*.
Cross 7:	*b/s* × *s/–*	1 *b/s* : 1 *s/–*	"–" can be *s*, *c*, or *a*.
Cross 8:	*b/c* × *s/c*	1 *s/c* : 2 *b/–* : 1 *c/c*	
Cross 9:	*s/c* × *s/c*	3 *s/–* : 1 *c/c*	
Cross 10:	*c/a* × *a/a*	1 *c/a*:1 *a/a*	

b. The progeny of the cross *b/s* × *b/c* will be $^3/_4$ black ($^1/_4$ *b/b*, $^1/_4$ *b/c*, $^1/_4$ *b/s*) : $^1/_4$ sepia (*s/c*).

6. Both codominance (=) and classical dominance (>) are present in the multiple allelic series for blood type: $A = B$, $A > O$, $B > O$.

Parents' phenotype	Parents' possible genotypes	Parents' possible children
a. AB × O	*A/B* × *O/O*	*A/O*, *B/O*
b. A × O	*A/A* or *A/O* × *O/O*	*A/O*, *O/O*
c. A × AB	*A/A* or *A/O* × *A/B*	*A/A*, *A/B*, *A/O*, *B/O*
d. O × O	*O/O* × *O/O*	*O/O*

The possible genotypes of the children are

Phenotype	Possible genotypes
O	*O/O*
A	*A/A*, *A/O*
B	*B/B*, *B/O*
AB	*A/B*

Using the assumption that each set of parents had one child, the following combinations are the only ones that will work as a solution:

Parents	Child
a. AB × O	B
b. A × O	A
c. A × AB	AB
d. O × O	O

7. *M* and *N* are codominant alleles. The rhesus group is determined by classically dominant alleles. The *ABO* alleles are mixed codominance and classical dominance (see Problem 6).

Person		Blood type			Possible paternal contribution	
Husband	O	M	Rh^+	*O*	*M*	*R* or *r*
Wife's lover	AB	MN	Rh^-	*A* or *B*	*M* or *N*	*r*
Wife	A	N	Rh^+	*A* or *O*	*N*	*R* or *r*
Child 1	O	MN	Rh^+	*O*	*M*	*R* or *r*
Child 2	A	N	Rh^+	*A* or *O*	*N*	*R* or *r*
Child 3	A	MN	Rh^-	*A* or *O*	*M*	*r*

The wife must be *A/O* ; *N/N* ; *R/r*. (She has a child with type O blood and another child that is Rh^- so she must carry both of these recessive alleles.) Only the husband could donate O to child 1. Only the lover could donate N to child 2. Both the husband and the lover could have donated the necessary alleles to child 3.

8. The key to solving this problem is in the statement that breeders cannot develop a pure-breeding stock and that a cross of two platinum foxes results in some normal progeny. Platinum must be dominant to normal color and heterozygous (*A/a*). An 82:38 ratio is very close to 2:1. Because a 1:2:1 ratio is expected in a heterozygous cross, one genotype is nonviable. It must be the *A/A*, homozygous platinum, genotype that is nonviable, because the homozygous recessive genotype is normal color (*a/a*). Therefore, the platinum allele is a pleiotropic allele that governs coat color in the heterozygous state and is lethal when homozygous.

9. **a.** Because Pelger crossed with normal stock results in two phenotypes in a 1:1 ratio, either Pelger or normal is heterozygous (*A/a*) and the other is homozygous (*a/a*) recessive. The problem states that normal is true-breeding, or *a/a*. Pelger must be *A/a*.

b. The cross of two Pelger rabbits results in three phenotypes. This means that the Pelger anomaly is dominant to normal. This cross is *A/a* × *A/a*, with an expected ratio of 1:2:1. Because the normal must be *a/a*, the extremely abnormal progeny must be *A/A*. There were only 39 extremely abnormal progeny because the others died before birth.

c. The Pelger allele is pleiotropic. In the heterozygous state it is dominant for nuclear segmentation of white blood cells. In the homozygous state it is a recessive lethal.

You could look for the nonsurviving fetuses in utero. Because the hypothesis of embryonic death when the Pelger allele is homozygous predicts a one-fourth reduction in litter size, you could also do an extensive statistical analysis of litter size, comparing normal × normal with Pelger × Pelger.

d. By analogy with rabbits, the absence of a homozygous Pelger anomaly in humans can be explained as recessive lethality. Also, because 1 in 1000 people have the Pelger anomaly, a heterozygous × heterozygous mating would be expected in only 1 of 1 million ($^1/_{1000} \times {}^1/_{1000}$) random matings, and then only 1 in 4 of the progeny would be expected to be homozygous. Thus, the homozygous Pelger anomaly is expected in only 1 of 4 million births. This is extremely rare and might not be recognized.

e. By analogy with rabbits, among the children of a man and a woman with the Pelger anomaly two-thirds of the surviving progeny would be expected to show the Pelger anomaly and one–third would be expected to be normal. The developing fetus that is homozygous for the Pelger allele would not be expected to survive until birth.

10. **a.** The sex ratio is expected to be 1:1.

b. The female parent was heterozygous for an X-linked recessive lethal allele. This would result in 50% fewer males than females.

c. Half the female progeny should be heterozygous for the lethal allele and half should be homozygous for the nonlethal allele. Individually mate the F_1 females and determine the sex ratio of their progeny.

11. Note that a cross of the short-bristled female with a normal male results in two phenotypes with regard to bristles and an abnormal sex ratio of 2 females:1 male. Furthermore, all the males are normal, while the females are normal and short in equal numbers. Whenever the sexes differ with respect to phenotype among the progeny, an X-linked gene is implicated. Because only the normal phenotype is observed in. males, the short-bristled phenotype must be heterozygous, and the allele must be a recessive lethal. Thus the first cross was *A/a* × *a*/Y.

Long-bristled females (*a/a*) were crossed with long-bristled males (*a*/Y). All their progeny would be expected to be long-bristled (*a/a* or *a*/Y). Short-bristled females (*A/a*) were crossed with long-bristled males (*a*/Y). The progeny expected are

$^1/_4$ *A/a* short-bristled females	$^1/_4$ *a*/Y long-bristled males
$^1/_4$ *a/a* long-bristled females	$^1/_4$ *A*/Y nonviable

12. In order to do this problem, you should first restate the information provided. The following two genes are independently assorting:

h/h = hairy	*s/s* = no effect
H/h = hairless	*S/s* suppresses H/h, giving hairy
H/H = lethal	*S/S* = lethal

a. The cross is *H/h* ; *S/s* × *H/h* ; *S/s*. Because this is a typical dihybrid cross, the expected ratio is 9:3:3:1. However, the problem cannot be worked in this simple fashion because of the epistatic relationship of these two genes. Therefore, the following approach should be used.

For the *H* gene, you expect $^1/_4$ *H/H* : $^1/_2$ *H/h* : $^1/_4$ *h/h*. For the *S* gene, you expect $^1/_4$ *S/S* : $^1/_2$ *S/s* : $^1/_4$ *s/s*. To get the final ratios, multiply the frequency of the first genotype by the frequency of the second genotype.

$^1/_4$ *H/H*		all progeny die regardless of the *S* gene		
		$^1/_4$ *S/S* =	$^1/_8$ *H/h* ; *S/S*	die
$^1/_2$ *H/h*	×	$^1/_2$ *S/s* =	$^1/_4$ *H/h* ; *S/s*	hairy
		$^1/_4$ *s/s* =	$^1/_8$ *H/h* ; *s/s*	hairless
		$^1/_4$ *S/S* =	$^1/_{16}$ *h/h* ; *S/S*	die
$^1/_4$ *h/h*	×	$^1/_2$ *S/s* =	$^1/_8$ *h/h* ; *S/s*	hairy
		$^1/_4$ *s/s* =	$^1/_{16}$ *h/h* ; *s/s*	hairy

Of the $^9/_{16}$ living progeny, the ratio of hairy to hairless is 7:2.

b. This cross is *H/h* ; *s/s* × *H/h* ; *S/s*. A 1:2:1 ratio is expected for the *H* gene and a 1:1 ratio is expected for the *S* gene.

$^1/_4$ *H/H*		all progeny die regardless of the *S* gene		
$^1/_2$ *H/h*	×	$^1/_2$ *S/s* =	$^1/_4$ *H/h* ; *S/s*	hairy
		$^1/_2$ *s/s* =	$^1/_4$ H/h ; *s/s*	hairless
$^1/_4$ *h/h*	×	$^1/_2$ *S/s* =	$^1/_8$ *h/h* ; *S/s*	hairy
		$^1/_2$ *s/s* =	$^1/_8$ *h/h* ; *s/s*	hairy

Of the $^3/_4$ living progeny, the ratio of hairy to hairless is 2:1.

13. Note that the F_2 are in an approximate 9:6:1 ratio. This suggests a dihybrid cross in which *A/–* ; *b/b* has the same appearance as *a/a* ; *B/–*. Let the disk phenotype be the result of *A/–* ; *B/–* and the long phenotype be the result of *a/a* ; *b/b*. The crosses are

P	*A/A* ; *B/B* (disk) × *a/a* ; *b/b* (long)	
F_1	*A/a* ; *B/b* (disk)	
F_2	9	*A/–* ; *B/–* (disk)
	3	*a/a* ; *B/–* (sphere)
	3	*A/–* ; *b/b* (sphere)
	1	*a/a* ; *b/b* (long)

14. The suggestion from the data is that the two albino lines had mutations in two different genes. When the extracts from the two lines were placed in the same test tube, they were capable of producing color because the gene product of one line was capable of compensating for the absence of a gene product from the second line.

a. The most obvious control is to cross the two pure-breeding lines. The cross would be *A/A* ; *b/b* × *a/a* ; *B/B*. The progeny will be *A/a* ; *B/b*, and all should be reddish purple.

b. The most likely explanation is that the red pigment is produced by the action of at least two different gene products. When petals of the two plants were ground together, the different defective enzyme of each plant was complemented by the normal enzyme of the other.

c. The genotypes of the two lines would be *A/A* ; *b/b* and *a/a* ; *B/B*.

d. The F_1 would all be pigmented, *A/a* ; *B/b*. This is an example of complementation. The mutants are defective for different genes. The F_2 would be

9	*A/–* ; *B/–* Pigmented
3	*a/a* ; *B/–* White
3	*A/–* ; *b/b* White
1	*a/a* ; *b/b* White

15. a. This is an example where one phenotype in the parents gives rise to three phenotypes in the offspring. The "frizzle" is the heterozygous phenotype and shows incomplete dominance.

P A/a (frizzle) × A/a (frizzle)

F_1 1 A/A (normal) : 2 A/a (frizzle) : 1 a/a (woolly)

b. If A/A (normal) is crossed to a/a (woolly), all offspring will be A/a (frizzle).

16. a. The best explanation is that Marfan's syndrome is inherited as a dominant autosomal trait, since roughly half the children of all affected individuals also express the trait. If it were recessive, then all individuals marrying affected spouses would have to be heterozygous for an allele that when homozygous causes Marfan's.

b. The pedigree shows both pleiotropy (multiple affected traits) and variable expressivity (variable degree of expressed phenotype). Penetrance is the percentage of individuals with a specific genotype who express the associated phenotype. There is no evidence of decreased penetrance in this pedigree.

c. Pleiotropy indicates that the gene product is required in a number of different tissues, organs, or processes. When the gene is mutant, all tissues needing the gene product will be affected. Variable expressivity of a phenotype for a given genotype indicates modification by one or more other genes, random noise, or environmental effects.

17. *Unpacking the Problem*

1. The character being studied is petal color.
2. The wild-type phenotype is blue.
3. A *variant* is a phenotypic difference from wild type that is observed.
4. There are two variants: pink and white.
5. "In nature" means that the variants did not appear in laboratory stock and, instead, were found growing wild.
6. Possibly the variants appeared as a small patch or even a single plant within a larger patch of wild type.
7. Seeds would be grown to check the outcome from each cross.
8. Given that no sex linkage appears to exist (sex is not specified in parents or offspring), "blue × white" means the same as "white × blue." Similar results would be expected because the trait being studied appears to be autosomal.
9. The first two crosses show a 3:1 ratio in the F_2, suggesting the segregation of one gene. The third cross has a 9:4:3 ratio for the F_2, suggesting that two genes are segregating.
10. Blue is dominant to both white and pink.
11. *Complementation* refers to generation of wild-type progeny from the cross of two strains that are mutant in different genes.

12. The ability to make blue pigment requires two enzymes that are individually defective in the pink or white strains. The F_1 progeny of this cross is blue, since each has inherited one nonmutant allele for both genes and can therefore produce both functional enzymes.

13. Blueness from a pink × white cross arises through complementation.

14. The following ratios are observed: 3:1, 9:4:3.

15. There are monohybrid ratios observed in the first two crosses.

16. There is a modified 9:3:3:1 ratio in the third cross.

17. A monohybrid ratio indicates that one gene is segregating, while a dihybrid ratio indicates that two genes are segregating.

18. 15:1, 12:3:1, 9:6:1, 9:4:3, 9:7, 1:2:1, 2:1

19. There is a modified dihybrid ratio in the third cross.

20. A modified dihybrid ratio most frequently indicates the interaction of two or more genes.

21. Recessive epistasis is indicated by the modified dihybrid ratio.

22.

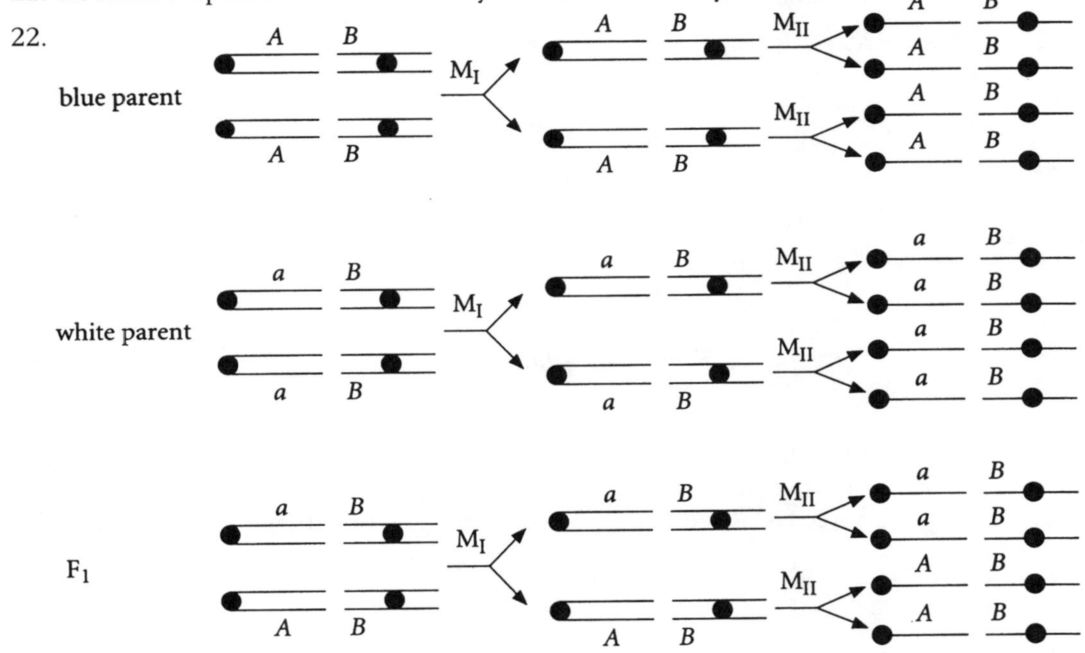

Solution to the Problem

a. Let A = wild-type, a = white, B = wild-type, and b pink.

Cross 1:	P	blue × white	A/A ; $B/B \times a/a$; B/B
	F_1	all blue	All A/a ; B/B
	F_2	3 blue:1 white	3 $A/-$; B/B:1 a/a ; B/B

Cross 2:	P	blue × pink	*A/A* ; *B/B* ´ *A/A* ; *b/b*
	F_1	all blue	All *A/A* ; *B/b*
	F_2	3 blue:1 pink	3 *A/A* ; *B/–* : 1 *A/A* ; *b/b*
Cross 3:	P	pink × white	*A/A* ; *b/b* × *a/a* ; *B/B*
	F_1	all blue	All *A/a* ; *B/b*
	F_2	9 blue	9 *A/–* ; *B/–*
		4 white	3 *a/a* ; *B/–* : 1 *a/a* ; *b/b*
		3 pink	3 *A/–* ; *b/b*

When the allele *a* is homozygous, the expression of alleles *B* or *b* is blocked or masked. The white phenotype is epistatic to the pigmented phenotypes. It is likely that the product of the *A* gene produces an intermediate that is then modified by the product of the *B* gene. If the plant is *a/a*, this intermediate is not made and the phenotype of the plant is the same regardless of the ability to produce functional *B* product.

b. The cross is

F_2 blue × white

F_3 3/8 blue

1/8 pink

4/8 white

Begin by writing as much of each genotype as can be assumed

F_2 *A/–* ; *B/–* × *a/a* ; *–/–*

F_3 3/8 *A/–* ; *B/–*

1/8 *A/–* ; *b/b*

4/8 *a/a* ; *–/–*

Notice that both *a/a* and *b/b* appear in the F_3 progeny. In order for these homozygous recessives to occur, each parent must have at least one *a* and one *b*. Using this information, the cross becomes

F_2 *A/a* ; *B/b* × *a/a* ; *–/b*

F_3 3/8 A/a ; *B/b*

1/8 *A/a* ; *b/b*

4/8 *a/a* ; *b/–*

The only remaining question is whether the white parent was homozygous recessive, *b/b*, or heterozygous, *B/b*. If the white parent had been homozygous recessive, then the cross would have been a testcross of the blue parent, and the progeny ratio would have been 1 blue:1 pink:2 white, or 1 *A/a* ; *B/b* : 1 *A/a* ; *b/b* : 1 *a/a* ; *B/b* : 1 *a/a* ; *b/b*. This was not observed.

Therefore, the white parent had to have been heterozygous, and the F_2 cross was A/a ; $B/b \times a/a$; B/b.

18. a. Note that the third cross has a 13:3 F_2 ratio. This is a modified dihybrid ratio, indicating two genes. Let the deviation from wild type in line 1 be symbolized by a and the deviation from wild type in line 2 be symbolized by B. Whether an uppercase or a lowercase symbol was chosen for the deviation was dictated by the F_1 result in the first two crosses. That is, if the F_1 is wild-type, then the deviation is recessive. However, if the F_1 is not wild-type, then the deviation is dominant.

Obviously, epistasis is involved in the production of pisatin. In line 1, the wild-type allele makes pisatin, and the recessive deviation does not produce pisatin. In line 2, the wild-type allele allows the expression of the pisatin that is normally made by the first gene, while the dominant deviation blocks expression of that wild-type product.

b. Cross 1:

P	a/a ; $b/b \times A/A$; b/b	
F_1	A/a ; b/b	
F_2	3 $A/–$; b/b : 1 a/a ; b/b	

Cross 2:

P	A/A ; $B/B \times A/A$; b/b	
F_1	A/A ; B/b	
F_2	3 A/A ; $B/–$: 1 A/A ; b/b	

Cross 3:

P	a/a ; $b/b \times A/A$; B/B	
F_1	A/a ; B/b	
F_2	9 $A/–$; $B/–$	no pisatin
	3 $A/–$; b/b	pisatin
	3 a/a ; $B/–$	no pisatin
	1 a/a ; b/b	no pisatin

c. Line 1 does not make pisatin, while line 2 blocks the expression of pisatin.

19. It is possible to produce black offspring from two pure-breeding recessive albino parents if albinism results from mutations in two different genes. If the cross is designated

$$A/A \ ; \ b/b \times a/a \ ; \ B/B$$

all offspring would be

$$A/a \ ; \ B/b$$

and they would have a black phenotype because of complementation.

20. The data indicate that white is dominant to solid purple. Note that the F_2 are in an approximate 12:3:1 ratio. In order to achieve such a ratio, epistasis must be involved.

a. Because a modified 9:3:3:1 ratio was obtained in the F_2, the F_1 had to be a double heterozygote. Solid purple occurred at one-third the rate of white, which means that it will be in the form of either *D/– ; e/e* or *d/d ; E/–*. In order to achieve a double heterozygote in the F_1, the original white parent also has to be either *D/– ; e/e* or *d/d ; E/–*.

Arbitrarily assume that the original cross was *D/D ; e/e* (white) × *d/d ; E/E* (purple). The F_1 would all be *D/d ; E/e*. The F_2 would be

9 *D/– ; E/–*	white, by definition
3 *d/d ; E/–*	purple, by definition
3 *D/– ; e/e*	white, by deduction
1 *d/d ; e/e*	spotted purple, by deduction

Under these assumptions, *D* blocks the expression of both *E* and *e*. The *d* allele has no effect on the expression of *E* and *e*. *E* results in solid purple, while *e* results in spotted purple. It would also be correct, of course, to assume the opposite set of epistatic relationships (*E* blocks the expression of *D* or *d*, *D* results in solid purple, and *d* results in spotted purple).

b. The cross is white × solid purple. While the solid purple genotype must be *d/d ; E/–*, as defined in part (a), the white genotype can be one of several possibilities. Note that the progeny phenotypes are in a 1:2:1 ratio and that one of the phenotypes, spotted, must be *d/d ; e/e*. In order to achieve such an outcome, the purple genotype must be *d/d ; E/e*. The white genotype of the parent must contain both a *D* and a *d* allele in order to produce both white (*D/–*) and spotted plants (*d/d*). At this point, the cross has been deduced to be *D/d ; –/–* (white) × *d/d ; E/e* (purple).

If the white plant is *E/E*, the progeny will be

$^1/_2$ *D/d ; E/–*	white
$^1/_2$ *d/d ; E/–*	solid purple

This was not observed. If the white plant is *E/e*, the progeny will be

$^3/_8$ *D/d ; E/–*	white
$^1/_8$ *D/d ; e/e*	white
$^3/_8$ *d/d ; E/–*	solid purple
$^1/_8$ *d/d ; e/e*	spotted purple

The phenotypes were observed, but in a different ratio. If the white plant is *e/e*, the progeny will be

$^1/_4$ *D/d ; E/e*	white
$^1/_4$ *D/d ; e/e*	white
$^1/_4$ *d/d ; E/e*	solid purple
$^1/_4$ *d/d ; e/e*	spotted purple

This was observed in the progeny. Therefore, the parents were *D/d* ; *e/e* (white) × *d/d* ; *E/e* (purple).

21. **a. and b.** Crosses 1–3 show a 3:1 ratio, indicating that brown, black, and yellow are all alleles of one gene. Crosses 4–6 show a modified 9:3:3:1 ratio, indicating that at least two genes are involved. Those crosses also indicate that the presence of color is dominant to its absence. Furthermore, epistasis must be involved for there to be a modified 9:3:3:1 ratio.

By looking at the F_1 of crosses 1–3, the following allelic dominance relationships can be determined: black > brown > yellow. Arbitrarily assign the following genotypes for homozygotes: B^l/B^l = black, B^r/B^r = brown, B^y/B^y = yellow.

By looking at the F_2 of crosses 4–6, a white phenotype is composed of two categories: the double homozygote and one class of the mixed homozygote/heterozygote. Let lack of color be caused by *c/c*. Color will therefore be *C/–*.

Parents	F_1	F_2
1 B^r/B^r ; C/C × B^y/B^y ; C/C	B^r/B^y ; C/C	3 $B^r/–$; C/C:1 B^y/B^y ; C/C
2 B^l/B^l ; C/C × B^r/B^r ; C/C	B^l/B^r ; C/C	3 $B^l/–$; C/C:1 B^r/B^r ; C/C
3 B^l/B^l ; C/C × B^y/B^y ; C/C	B^l/B^y ; C/C	3 $B^l/–$; C/C:1 B^y/B^y ; C/C
4 B^l/B^l ; c/c × B^y/B^y ; C/C	B^l/B^y ; C/c	9 $B^l/–$; $C/–$:3 B^y/B^y ; $C/–$:3 $B^l/–$; c/c:1 B^y/B^y ; c/c
5 B^l/B^l ; c/c × B^r/B^r ; C/C	B^l/B^r ; C/c	9 $B^l/–$; $C/–$:3 B^r/B^r ; $C/–$:3 $B^l/–$; c/c:1 B^r/B^r ; c/c
6 B^l/B^l ; C/C × B^y/B^y ; c/c	B^l/B^y ; C/c	9 $B^l/–$; $C/–$:3 B^y/B^y ; $C/–$:3 $B^l/–$; c/c:1 B^y/B^y ; c/c

22. It is possible to produce normally pigmented offspring from albino parents if albinism results from mutations in either of two different genes. If the cross is designated

$$A/A \cdot b/b \times a/a \cdot B/B$$

then all the offspring would be

$$A/a \cdot B/b$$

and they would have a pigmented phenotype because of complementation.

23. **a.** To solve this problem you will have to use a trial-and-error approach. The first decision regards the number of genes involved. Cross 2 tells you that there are at least two genes because the cross of two pure-breeding strains shows a new phenotype (complementation). Cross 5, however, indicates that there are at least three genes involved because a 1:7 ratio is not observed with two genes.

Cross 1 indicates that lines 1 and 2 cannot compensate for each other, suggesting a shared homozygous defective gene. Cross 2 indicates that lines 1 and 3 can compensate for each other's defects, and cross 3 indicates that lines 2 and 3 cannot. This suggests that line 2 shares a defective homozygous gene with line 3 and that line 1 is normal for the defective gene seen in line 3 but not in line 2.

At this point, you must arbitrarily make some assumptions and test them against the results. You could assume, for instance, that lines 1 and 2 share a defect in gene A, and that lines 2 and 3 share a defect in B.

line 1:	*a/a*	*B/B*
line 2:	*a/a*	*b/b*
line 3:	*A/A*	*b/b*

However, as stated above, the segregation of just two genes cannot explain all the data.

Let the three genes involved be *A*, *B*, and *D*. Because progeny of lines 1 and 3 are red, only one can be mutant for *D*, and the data from cross 4 indicates that it is line 1. Cross 5 indicates that line 2 is also *d/d*. Now the crosses can be explained. The genotypes of the lines are

line 1:	*a/a*	*B/B*	*d/d*
line 2:	*a/a*	*b/b*	*d/d*
line 3:	*A/A*	*b/b*	*D/D*

Cross 1: *a/a* ; *B/B* ; *d/d* × *a/a* ; *b/b* ; *d/d* →	all	*a/a* ; *B/b* ; *d/d*	white
Cross 2: *a/a* ; *B/B* ; *d/d* × *A/A* ; *b/b* ; *D/D* →	all	*A/a* ; *B/b* ; *D/d*	red
Cross 3: *a/a* ; *b/b* ; *d/d* × *A/A* ; *b/b* ; *D/D* →	all	*A/a* ; *b/b* ; *D/d*	white
Cross 4: *A/a* ; *B/b* ; *D/d* × *a/a* ; *B/B* ; *d/d* →	1/8	*A/a* ; *B/b* ; *D/d*	red
	1/8	*A/a* ; *B/b* ; *d/d*	white
	1/8	*a/a* ; *B/b* ; *D/d*	white
	1/8	*a/a* ; *B/b* ; *d/d*	white
	1/8	*A/a* ; *B/B* ; *D/d*	red
	1/8	*A/a* ; *B/B* ; *d/d*	white
	1/8	*a/a* ; *B/B* ; *D/d*	white
	1/8	*a/a* ; *B/B* ; *d/d*	white
Cross 5: *A/a* ; *B/b* ; *D/d* × *a/a* ; *b/b* ; *d/d* →	1/8	*A/a* ; *B/b* ; *D/d*	red
	1/8	*A/*a ; *B/b* ; *d/d*	white

$^1/_8$ *A/a* ; *b/b* ; *D/d* white
$^1/_8$ *A/a* ; *b/b* ; *d/d* white
$^1/_8$ *a/a* ; *B/b* ; *D/d* white
$^1/_8$ *a/a* ; *B/b* ; *d/d* white
$^1/_8$ *a/a* ; *b/b* ; *D/d* white
$^1/_8$ *a/a* ; *b/b* ; *d/d* white

Cross 6: *A/a* ; *B/b* ; *D/d* × *A/A* ; *b/b* ; *D/D* → $^1/_8$ *A/A* ; *B/b* ; *D/d* red
$^1/_8$ *A/a* ; *B/b* ; *D/d* red
$^1/_8$ *A/A* ; *B/b* ; *D/D* red
$^1/_8$ *A/a* ; *B/b* ; *D/D* red
$^1/_8$ *A/A* ; *b/b* ; *D/D* white
$^1/_8$ *A/a* ; *b/b* ; *D/D* white
$^1/_8$ *A/A* ; *b/b* ; *D/d* white
$^1/_8$ *A/a* ; *b/b* ; *D/d* white

b. The cross is *A/a* ; *B/b* ; *D/d* × *A/a* ; *b/b* ; *D/d*. The red progeny will have to be

A/– ; *B/–* ; *D/–* , which equals $(^3/_4)\ (^1/_2)\ (^3/_4) = {}^9/_{32}$.

24. The first step in each cross is to write as much of the genotype as possible from the phenotype.

Cross 1: *A/–* ; *B/–* × *a/a* ; *b/b* → 1 *A/–* ; *B/–*:2 *?/a*; *?/b* : 1 *a/a* ; *b/b*

Because the double recessive appears, the blue parent must be *A/a* ; *B/b*. The $^1/_2$ purple then must be *A/a* ; *b/b* and *a/a* ; *B/b*.

Cross 2: *?/?* ; *?/?* × *?/?* ; *?/?* → 1 *A/–* ; *B/–* : 2 *?/?* ; *?/?* : 1 *a/a* ; *b/b*

The two parents must be, in either order, *A/a* ; *b/b* and *a/a* ; *B/b*. The two purple progeny must be the same. The blue progeny are *A/a* ; *B/b*.

Cross 3: *A/–* ; *B/–* × *A/–* ; *B/–* → 3 *A/–* ; *B/–* : 1 *?/?* ; *?/?*

The only conclusions possible here are that one parent is either *A/A* or *B/B* and the other parent is *B/b* if the first is *A/A* or *A/a* if the first is *B/B*.

Cross 4: *A/–* ; *B/–* × *?/?* ; *?/?* → 3 *A/–* ; *B/–* : 4 *?/?* ; *?/?* : 1 *a/a* ; *b/b*

The purple parent can be either *A/a* ; *b/b* or *a/a* ; *B/b* for this answer. Assume the purple parent is *A/a* ; *b/b*. The blue parent must be *A/a* ; *B/b*. The progeny are

$^3/_4$ *A/–*	×	$^1/_2$ *B/b*	= $^3/_8$ *A/–* ; *B/b*	blue
		$^1/_2$ *b/b*	= $^3/_8$ *A/–* ; *b/b*	purple

$^1/_4$ *a/a*	×	$^1/_2$ *B/b*	= $^1/_8$ *a/a* ; *B/b*	purple
		$^1/_2$ *b/b*	= $^1/_8$ *a/a* ; *b/b*	scarlet

Cross 5: *A/–* ; *b/b* × *a/a* ; *b/b* → 1 *A/–* ; *b/b* : 1 *a/a* ; *b/b*

As written this is a testcross for gene *A*. The purple parent and progeny are *A/a* ; *b/b*. Alternatively, the purple parent and progeny could be *a/a* ; *B/b*.

25. The F_1 progeny of cross 1 indicate that sun red is dominant to pink. The F_2 progeny, which are approximately in a 3:1 ratio, support this. The same pattern is seen in crosses 2 and 3, with sun red dominant to orange and orange dominant to pink. Thus, we have a multiple allelic series with sun red > orange > pink. In all three crosses, the parents must be homozygous.

If c^{sr} = sun red, c^o = orange, and c^p = pink, then the crosses and the results are

Cross	Parents	F_1	F_2
1	$c^{sr}/c^{sr} \times c^p/c^p$	c^{sr}/c^p	$3\ c^{sr}/- : 1\ c^p/c^p$
2	$c^o/c^o \times c^{sr}/c^{sr}$	c^{sr}/c^o	$3\ c^{sr}/- : 1\ c^o/c^o$
3	$c^o/c^o \times c^p/c^p$	c^o/cc^p	$3\ c^o/- : 1\ c^p/c^p$

Cross 4 presents a new situation. The color of the F_1 differs from that of either parent, suggesting that two separate genes are involved. An alternative explanation is either codominance or incomplete dominance. If either codominance or incomplete dominance is involved, then the F_2 will appear in a 1:2:1 ratio. If two genes are involved, then a 9:3:3:1 ratio, or some variant of it, will be observed. Because the wild-type phenotype appears in the F_1 and F_2, it appears that complementation is occurring. This requires two genes. The progeny actually are in a 9:4:3 ratio. This means that two genes are involved and that there is epistasis. Furthermore, for three phenotypes to be present in the F_2, the two F_1 parents must have been heterozygous.

Let *a* stand for the scarlet gene and *A* for its colorless allele, and assume that there is a dominant allele, *C*, that blocks the expression of the alleles that we have been studying to this point.

Cross 4:	P	c^o/c^o ; *A/A* × *C/C* ; *a/a*	
	F_1	C/c^o ; *A/a*	
	F_2	9 *C/–* ; *A/–*	yellow
		3 *C/–* ; *a/a*	scarlet
		3 c^o/c^o ; *A/–*	orange
		1 c^o/c^o ; *a/a*	orange (epistasis, with c^o blocking the expression of *a/a*)

26. a. P *A/A* (agouti) × *a/a* (nonagouti)

gametes *A* and *a*

F_1 *A/a* (agouti)
gametes *A* and *a*

F_2 1 *A/A* (agouti):2 *A/a* (agouti):1 *a/a* (nonagouti)

b. P *B/B* (wild type) × *b/b* (cinnamon)
gametes *B* and *b*

F_1 *B/b* (wild type)
gametes *B* and *b*

F_2 1 *B/B* (wild type) : 2 *B/b* (wild type) : 1 *b/b* (cinnamon)

c. P *A/A* ; *b/b* (cinnamon or brown agouti) × *a/a* ; *B/B* (black nonagouti)
gametes *A* ; *b* and *a* ; *B*

F_1 *A/a* ; *B/b* (wild type, or black agouti)

d.

9 *A/–* ; *B/–*	black agouti
3 *a/a* ; *B/–*	black nonagouti
3 *A/–*; *b/b*	cinnamon
1 *a/a* ; *b/b*	chocolate

e. P *A/A* ; *b/b* (cinnamon) × *a/a* ; *B/B* (black nonagouti)
gametes *A* ; *b* and *a* ; *B*

F_1 *A/a* ; *B/b* (wild type)
gametes *A* ; *B*, *A* ; *b*, *a* ; *B*, and *a* ; *b*

F_2	9	*A/–* ; *B/–*	wild type
		1 *A/A* ; *B/B*	
		2 *A/a* ; *B/B*	
		2 *A/A* ; *B/b*	
		4 *A/a* ; *B/b*	
	3	*a/a* ; *B/–*	black nonagouti
		1 *a/a* ; *B/B*	
		2 *a/a* ; *B/b*	
	3	*A/–* ; *b/b*	cinnamon
		1 *A/A* ; *b/b*	
		2 *A/a* ; *b/b*	
	1	*a/a* ; *b/b*	chocolate

f. P A/a ; $B/b \times A/A$; b/b (wild type) (cinnamon) — A/a ; $B/b \times a/a$; B/B (wild type) (black nonagouti)

F_1	1/4 A/A ; B/b wild type	1/4 A/a ; B/B wild type
	1/4 A/a ; B/b wild type	1/4 A/a ; B/b wild type
	1/4 A/A ; b/b cinnamon	1/4 a/a ; B/B black nonagouti
	1/4 A/a ; b/b cinnamon	1/4 a/a ; B/b black nonagouti

g. P A/a ; $B/b \times a/a$; b/b
(wild type) (chocolate)

F_1
1/4 A/a ; B/b wild type
1/4 A/a ; b/b cinnamon
1/4 a/a ; B/b black nonagouti
1/4 a/a ; b/b chocolate

h. To be albino, the mice must be *c/c*, but the genotype with regard to the *A* and *B* genes can be determined only by looking at the F_2 progeny.

Cross 1: P c/c ; ?/? ; ?/? $\times$ C/C ; A/A ; B/B

F_1 C/c ; $A/–$; $B/–$

F_2	87 wild type	$C/–$; $A/–$; $B/–$
	32 cinnamon	$C/–$; $A/–$; b/b
	39 albino	c/c ; ?/? ; ?/?

For cinnamon to appear in the F_2, the F_1 parents must be *B/b*. Because the wild type is *B/B*, the albino parent must have been *b/b*. Now the F_1 parent can be written *C/c* ; *A/–* ; *B/b*. With such a cross, one-fourth of the progeny would be expected to be albino (*c/c*), which is what is observed. Three-fourths of the remaining progeny would be black, either agouti or nonagouti, and one-fourth would be either cinnamon, if agouti, or chocolate, if nonagouti. Because chocolate is not observed, the F_1 parent must not carry the allele for nonagouti. Therefore, the F_1 parent is *A/A* and the original albino must have been *c/c* ; *A/A* ; *b/b*.

Cross 2: P c/c ; ?/? ; ?/? $\times$ C/C ; A/A ; B/B

F_1 C/c ; $A/–$; $B/–$

F_2	62 wild type	*C/– ; A/– ; B/–*
	18 albino	*c/c ; ?/? ; ?/?*

This is a 3:1 ratio, indicating that only one gene is heterozygous in the F_1. That gene must be *C/c*. Therefore, the albino parent must be *c/c ; A/A ; B/B*.

Cross 3: P *c/c ; ?/? ; ?/?* × *C/C ; A/A ; B/B*

F_1 *C/c ; A/– ; B/–*

F_2	96 wild type	*C/– ; A/– ; B/–*
	30 black	*C/– ; a/a ; B/–*
	41 albino	*c/c ; ?/? ; ?/?*

For a black nonagouti phenotype to appear in the F_2, the F_1 must have been heterozygous for the *A* gene. Therefore, its genotype can be written *C/c ; A/a ; B/–* and the albino parent must be *c/c ; a/a ; ?/?*. Among the colored F_2 a 3:1 ratio is observed, indicating that only one of the two genes is heterozygous in the F_1. Therefore, the F_1 must be *C/c ; A/a ; B/B* and the albino parent must be *c/c ; a/a ; B/B*.

Cross 4: P *c/c ; ?/? ; ?/?* × *C/C ; A/A ; B/B*

F_1 *C/c ; A/– ; B/–*

F_2	287 wild type	*C/– ; A/– ; B/–*
	86 black	*C/– ; a/a ; B/–*
	92 cinnamon	*C/– ; A/– ; b/b*
	29 chocolate	*C/– ; a/a ; b/b*
	164 albino	*c/c ; ?/? ; ?/?*

To get chocolate F_2 progeny the F_1 parent must be heterozygous for all genes and the albino parent must be *c/c ; a/a ; b/b*.

27. To solve this problem, first restate the information.

A/–	yellow	*A/– ; R/–*	gray
R/–	black	*a/a ; r/r*	white

The cross is gray × yellow, or *A/– ; R/–* × *A/– ; r/r*. The F_1 progeny are

$^3/_8$ yellow	$^1/_8$ black
$^3/_8$ gray	$^1/_8$ white

For white progeny, both parents must carry an *r* and an *a* allele. Now the cross can be rewritten as *A/a ; R/r* × *A/a ; r/r*

28. a. The stated cross is

P single-combed (*r/r* ; *p/p*) × walnut-combed (*R/R* ; *P/P*)

F_1 *R/r* ; *P/p* walnut

F_2 9 *R/–* ; *P/–* walnut
3 *r/r* ; *P/–* pea
3 R/– ; *p/p* rose
1 *r/r* ; *p/p* single

b. The stated cross is

P walnut-combed × rose-combed

and the F_1 progeny are

	Phenotypes	Possible genotypes
3/8	rose	*R/–* ; *p/p*
3/8	walnut	*R/–* ; *P/–*
1/8	pea	*r/r* ; *P/–*
1/8	single	*r/r* ; *p/p*

The 3 *R/–* : 1 *r/r* ratio indicates that the parents were heterozygous for the *R* gene. The 1 *P/–* : 1 *p/p* ratio indicates a testcross for this gene. Therefore, the parents were *R/r* ; *P/p* and *R/r* ; *p/p*.

c. The stated cross is

P walnut-combed × rose-combed

F_1 walnut (*R/–* ; *P/p*)

To get this result, one of the parents must be homozygous *R*, but both need not be, and the walnut parent must be homozygous *P/P*.

d. The following genotypes produce the walnut phenotype:

R/R ; *P/P*, *R/r* ; *P/P*, *R/R* ; *P/p*, *R/r* ; *P/p*

29. Notice that the F_1 shows a difference in phenotype correlated with sex. At least one of the two genes is X-linked. The F_2 ratio suggests independent assortment between the two genes. Because purple is present in the F_1, the parental white-eyed male must have at least one *P* allele. The presence of white eyes in the F_2 suggests that the F_1 was heterozygous for pigment production, which means that the male also must carry the *a* allele. A start on the parental genotypes can now be made:

P *A/A* ; *P/P* × *a/–* ; *P/–*, where "– " could be either a Y chromosome or a second allele.

The question now is, which gene is X-linked? If the *A* gene is X-linked, the cross is

P *A/A* ; *p/p* × *a/*Y ; *P/P*

F_1 *A/a* ; *P/p* × *A/*Y ; *P/p*

All F_2 females will inherit the *A* allele from their father. Under this circumstance, no white-eyed females would be observed. Therefore, the *A* gene cannot be X-linked. The cross is

P	A/A ; $p/p \times a/a$; P/Y	
F_1	A/a ; P/p	purple-eyed females
	A/a ; p/Y	red-eyed males
F_2	Females	Males
	$^3/_8$ $A/–$; P/p purple	$^3/_8$ $A/–$; P/Y purple
	$^3/_8$ $A/–$; p/p red	$^3/_8$ $A/–$; p/Y red
	$^1/_8$ a/a ; P/p white	$^1/_8$ a/a ; P/Y white
	$^1/_8$ a/a ; p/p white	$^1/_8$ a/a ; p/Y white

30. The results indicate that two genes are involved (modified 9:3:3:1 ratio), with white blocking the expression of color by the other gene. The ratio of white:color is 3:1, indicating that the F_1 is heterozygous (W/w). Among colored dogs, the ratio is 3 black:1 brown, indicating that black is dominant to brown and the F_1 is heterozygous (B/b). The original brown dog is w/w ; b/b and the original white dog is W/W ; B/B. The F_1 progeny are W/w ; B/b and the F_2 progeny are

9 $W/–$; $B/–$	white
3 w/w ; $B/–$	black
3 $W/–$; b/b	white
1 w/w ; b/b	brown

31.

Cross	Results	Conclusion
$A/–$; $C/–$; $R/– \times a/a$; c/c ; R/R	50% colored	Colored or white will depend on the *A* and *C* genes. Because half the seeds are colored, one of the two genes is heterozygous.
$A/–$; $C/–$; $R/– \times a/a$; C/C ; r/r	25% colored	Color depends on *A* and *R* in this cross. If only one gene were heterozygous, 50% would be colored. Therefore, both *A* and *R* are heterozygous. The seed is A/a ; C/C ; R/r.
$A/–$; $C/–$; $R/– \times A/A$; c/c ; r/r	50% colored	This supports the above conclusion.

32. a. The *A/A* ; *C/C* ; *R/R* ; *pr/pr* parent produces pigment that is not converted to purple. The phenotype is red. The *a/a* ; *c/c* ; *r/r* ; *Pr/Pr* does not produce pigment. The phenotype is yellow.

b. The F_1 will be *A/a* ; *C/c* ; *R/r* ; *Pr/pr*, which will produce pigment. The pigment will be converted to purple.

c. The difficult way to determine the phenotypic ratios is to do a branch diagram, yielding the following results.

$^{81}/_{256}$	*A/– ; C/– ; R/– ; Pr/–*	purple	$^{9}/_{256}$	*a/a ; C/– ; R/– ; Pr/–*	yellow
$^{27}/_{256}$	*A/– ; C/– ; R/– ; pr/pr*	red	$^{9}/_{256}$	*a/a ; C/– ; R/– ; pr/pr*	yellow
$^{27}/_{256}$	*A/– ; C/– ; r/r ; Pr/–*	yellow	$^{9}/_{256}$	*a/a ; C/– ; r/r ; Pr/–*	yellow
$^{9}/_{256}$	*A/– ; C/– ; r/r ; pr/pr*	yellow	$^{3}/_{256}$	*a/a ; C/– ; r/r ; pr/pr*	yellow
$^{27}/_{256}$	*A/– ; c/c ; R/– ; Pr/–*	yellow	$^{27}/_{256}$	*a/a ; c/c ; R/– ; Pr/–*	yellow
$^{9}/_{256}$	*A/– ; c/c ; R/– ; pr/pr*	yellow	$^{3}/_{256}$	*a/a ; c/c ; R/– ; pr/pr*	yellow
$^{9}/_{256}$	*A/– ; c/c ; r/r ; Pr/–*	yellow	$^{3}/_{256}$	*a/a ; c/c ; r/r ; Pr/–*	yellow
$^{3}/_{256}$	*A/– ; c/c ; r/r ; pr/pr*	yellow	$^{1}/_{256}$	*a/a ; c/c ; r/r ; pr/pr*	yellow

The final phenotypic ratio is 81 purple:27 red:148 yellow.

The easier method of determining phenotypic ratios is to recognize that four genes are involved in a heterozygous × heterozygous cross. Purple requires a dominant allele for each gene. The probability of all dominant alleles is $(^3/_4)^4 = {}^{81}/_{256}$. Red results from all dominant alleles except for the *Pr* gene. The probability of that outcome is $(^3/_4)^3(^1/_4) = {}^{27}/_{256}$. The remainder of the outcomes will produce no pigment, resulting in yellow. That probability is $1 - {}^{81}/_{256} - {}^{27}/_{256} = {}^{148}/_{256}$.

d. The cross is *A/a* ; *C/c* ; *R/r* ; *Pr/pr* × *a/a* ; *c/c* ; *r/r* ; *pr/pr*. Again, either a branch diagram or the easier method can be used. The final probabilities are

purple $= (^1/_2)^4 = {}^1/_{16}$
red $= (^1/_2)^3(^1/_2) = {}^1/_{16}$
yellow $= 1 - {}^1/_{16} - {}^1/_{16} = {}^7/_8$

33. a. The cross is

P *td* ; *su* (wild type) × td^+ ; su^+ (wild type)

F_1	1 *td* ; *su*	wild type
	1 *td* ; su^+	requires tryptophan
	1 td^+ ; su^+	wild type
	1 td^+ ; *su*	wild type

b. 1 tryptophan-dependent:3 tryptophan-independent

34. a. This type of gene interaction is called *epistasis*. The phenotype of *e/e* is epistatic to the phenotypes of *B/–* or *b/b*.

b. The progeny of generation I have all possible phenotypes. Progeny II-3 is beige (*e/e*), so both parents must be heterozygous *E/e*. Progeny II-4 is brown (*b/b*), so both parents must also be heterozygous *B/b*. Progeny III-3 and III-5 are brown, so II-2 and II-5 must be *B/b*. Progeny III-2 and III-7 are beige (*e/e*), so all their parents must be *E/e*.

The following are the inferred genotypes:

I 1 (*B/b* ; *E/e*) 2 (*B/b* ; *E/e*)

II 1 (*b/b* ; *E/e*) 2 (*B/b* ; *E/e*) 3 (*–/–* ; *e/e*) 4 (*b/b* ; *E/–*)
5 (*B/b* ; *E/e*) 6 (*b/b* ; *E/e*)

III 1 (*B/b* ; *E/–*) 2 (*–/b* ; *e/e*) 3 (*b/b* ; *E/–*) 4 (*B/b* ; *E/–*)
5 (*b/b* ; *E/–*) 6 (*B/b* ;*E/–*) 7 (*–/b* ; *e/e*)

35. P *A/A* ; *B/B* ; *C/C* ; *D/D* ; *S/S* × *a/a* ; *b/b* ; *c/c* ; *d/d* ; *s/s*

F_1 *A/a* ; *B/b* ; *C/c* ; *D/d* ; *S/s*

F_2 *A/a* ; *B/b* ; *C/c* ; *D/d* ; *S/s* × *A/a* ; *B/b* ; *C/c* ; *D/d* ; *S/s*

($^3/_4$ *A/–*)($^3/_4$ *B/–*)($^3/_4$ *C/–*)($^3/_4$ *D/–*)($^3/_4$ *S/–*) = $^{243}/_{1024}$	agouti
($^3/_4$ *A/–*)($^3/_4$ *B/–*)($^3/_4$ *C/–*)($^3/_4$ *D/–*)($^1/_4$ *s/s*) = $^{81}/_{1024}$	spotted agouti
($^3/_4$ *A/–*)($^3/_4$ *B/–*)($^3/_4$ *C/–*)($^1/_4$ *d/d*)($^3/_4$ *S/–*) = $^{81}/_{1024}$	dilute agouti
($^3/_4$ *A/–*)($^3/_4$ *B/–*)($^3/_4$ *C/–*)($^1/_4$ *d/d*)($^1/_4$ *s/s*) = $^{27}/_{1024}$	dilute spotted agouti
($^3/_4$ *A/–*)($^3/_4$ *B/–*)($^1/_4$ *c/c*)($^3/_4$ *D/–*)($^3/_4$ *S/–*) = $^{81}/_{1024}$	albino
($^3/_4$ *A/–*)($^3/_4$ *B/–*)($^1/_4$ *c/c*)($^3/_4$ *D/–*)($^1/_4$ *s/s*) = $^{27}/_{1024}$	albino
($^3/_4$ *A/–*)($^3/_4$ *B/–*)($^1/_4$ *c/c*)($^1/_4$ *d/d*)($^3/_4$ *S/–*) = $^{27}/_{1024}$	albino
($^3/_4$ *A/–*)($^3/_4$ *B/–*)($^1/_4$ *c/c*)($^1/_4$ *d/d*)($^1/_4$ *s/s*) = $^{9}/_{1024}$	albino
($^3/_4$ *A/–*)($^1/_4$ *b/b*)($^3/_4$ *C/–*)($^3/_4$ *D/–*)($^3/_4$ *S/–*) = $^{81}/_{1024}$	cinnamon
($^3/_4$ *A/–*)($^1/_4$ *b/b*)($^3/_4$ *C/–*)($^3/_4$ *D/–*)($^1/_4$ *s/s*) = $^{27}/_{1024}$	cinnamon spotted
($^3/_4$ *A/–*)($^1/_4$ *b/b*)($^3/_4$ *C/–*)($^1/_4$ *d/d*)($^3/_4$ *S/–*) = $^{27}/_{1024}$	dilute cinnamon
($^3/_4$ *A/–*)($^1/_4$ *b/b*)($^3/_4$ *C/–*)($^1/_4$ *d/d*)($^1/_4$ *s/s*) = $^{9}/_{1024}$	dilute spotted cinnamon
($^3/_4$ *A/–*)($^1/_4$ *b/b*)($^1/_4$ *c/c*)($^3/_4$ *D/–*)($^3/_4$ *S/–*) = $^{27}/_{1024}$	albino
($^3/_4$ *A/–*)($^1/_4$ *b/b*)($^1/_4$ *c/c*)($^3/_4$ *D/–*)($^1/_4$ *s/s*) = $^{9}/_{1024}$	albino
($^3/_4$ *A/–*)($^1/_4$ *b/b*)($^1/_4$ *c/c*)($^1/_4$ *d/d*)($^3/_4$ *S/–*) = $^{9}/_{1024}$	albino
($^3/_4$ *A/–*)($^1/_4$ *b/b*)($^1/_4$ *c/c*)($^1/_4$ *d/d*)($^1/_4$ *s/s*) = $^{3}/_{1024}$	albino

$(^1/_4\ a/a)(^3/_4\ B/–)(^3/_4\ C/–)(^3/_4\ D/–)(^3/_4\ S/–) = {}^{81}/_{1024}$	black
$(^1/_4\ a/a)(^3/_4\ B/–)(^3/_4\ C/–)(^3/_4\ D/–)(^1/_4\ s/s) = {}^{27}/_{1024}$	black spotted
$(^1/_4\ a/a)(^3/_4\ B/–)(^3/_4\ C/–)(^1/_4\ d/d)(^3/_4\ S/–) = {}^{27}/_{1024}$	dilute black
$(^1/_4\ a/a)(^3/_4\ B/–)(^3/_4\ C/–)(^1/_4\ d/d)(^1/_4\ s/s) = {}^{9}/_{1024}$	dilute spotted black
$(^1/_4\ a/a)(^3/_4\ B/–)(^1/_4\ c/c)(^3/_4\ D/–)(^3/_4\ S/–) = {}^{27}/_{1024}$	albino
$(^1/_4\ a/a)(^3/_4\ B/–)(^1/_4\ c/c)(^3/_4\ D/–)(^1/_4\ s/s) = {}^{9}/_{1024}$	albino
$(^1/_4\ a/a)(^3/_4\ B/–)(^1/_4\ c/c)(^1/_4\ d/d)(^3/_4\ S/–) = {}^{9}/_{1024}$	albino
$(^1/_4\ a/a)(^3/_4\ B/–)(^1/_4\ c/c)(^1/_4\ d/d)(^1/_4\ s/s) = {}^{3}/_{1024}$	albino
$(^1/_4\ a/a)(^1/_4\ b/b)(^3/_4\ C/–)(^3/_4\ D/–)(^3/_4\ S/–) = {}^{27}/_{1024}$	brown
$(^1/_4\ a/a)(^1/_4\ b/b)(^3/_4\ C/–)(^3/_4\ D/–)(^1/_4\ s/s) = {}^{9}/_{1024}$	brown spotted
$(^1/_4\ a/a)(^1/_4\ b/b)(^3/_4\ C/–)(^1/_4\ d/d)(^3/_4\ S/–) = {}^{9}/_{1024}$	dilute brown
$(^1/_4\ a/a)(^1/_4\ b/b)(^3/_4\ C/–)(^1/_4\ d/d)(^1/_4\ s/s) = {}^{3}/_{1024}$	dilute spotted brown
$(^1/_4\ a/a)(^1/_4\ b/b)(^1/_4\ c/c)(^3/_4\ D/–)(^3/_4\ S/–) = {}^{9}/_{1024}$	albino
$(^1/_4\ a/a)(^1/_4\ b/b)(^1/_4\ c/c)(^3/_4\ D/–)(^1/_4\ s/s) = {}^{3}/_{1024}$	albino
$(^1/_4\ a/a)(^1/_4\ b/b)(^1/_4\ c/c)(^1/_4\ d/d)(^3/_4\ S/–) = {}^{3}/_{1024}$	albino
$(^1/_4\ a/a)(^1/_4\ b/b)(^1/_4\ c/c)(^1/_4\ d/d)(^1/_4\ s/s) = {}^{1}/_{1024}$	albino

36. **a. and b.** The two starting lines are *i/i* ; *D/D* ; *M/M* ; *W/W* and *I/I* ; *d/d* ; *m/m* ; *w/w*, and you are seeking *i/i* ; *d/d* ; *m/m* ; *w/w*. There are many ways to proceed, one of which follows below.

I. *i/i* ; *D/D* ; *M/M* ; *W/W* × *I/I* ; *d/d* ; *m/m* ; *w/w*

II. *I/i* ; *D/d* ; *M/m* ; *W/w* × *I/i* ; *D/d* ; *M/m* ; *W/w*

III. select *i/i* ; *d/d* ; *m/m* ; *w/w*, which has a probability of $(^1/_4)^4 = {}^1/_{256}$.

c. and d. In the first cross, all progeny chickens will be of the desired genotype to proceed to the second cross. Therefore, the only problems are to be sure that progeny of both sexes are obtained for the second cross, which is relatively easy, and that enough females are obtained to make the time required for the desired genotype to appear feasible. Because chickens lay one to two eggs a day, the more egg-laying females there are, the faster the desired genotype will be obtained. In addition, it will be necessary to obtain a male and a female of the desired genotype in order to establish a pure-breeding line.

Assume that each female will lay two eggs a day, that money is no problem, and that excess males cause no problems. By hatching 200 eggs from the first cross, approximately 100 females will be available for the second cross. These 100 females will produce 200 eggs each day. Thus, in one week a total of 1400 eggs will be produced. Of these 1400 eggs, there will be approximately 700 of each sex. At a probability of $^1/_{256}$, the desired genotype should be achieved 2.7 times for each sex within that first week.

37. Pedigrees like this are quite common. They indicate lack of penetrance due to epistasis or environmental effects. Individual A must have the dominant autosomal gene.

38. In cross 1, the following can be written immediately:

P *M/– ; D/– ; w/w* (dark purple) × *m/m ; ?/? ; ?/?* (white with yellowish spots)

F_1 $^1/_2$ *M/– ; D/– ; w/w* dark purple
$^1/_2$ *M/– ; d/d ; w/w* light purple

All progeny are colored, indicating that no *W* allele is present in the parents. Because the progeny are in a 1:1 ratio, only one of the genes in the parents is heterozygous. Also, the light purple progeny, *d/d*, indicates which gene that is. Therefore, the genotypes must be

P *M/M ; D/d ; w/w* × *m/m ; d/d ; w/w*

F_1 $^1/_2$ *M/m ; D/d ; w/w*
$^1/_2$ *M/m ; d/d ; w/w*

In cross 2, the following can be written immediately:

P *m/m ; ?/? ; ?/?* (white with yellowish spots) × *M/– ; d/d ; w/w* (light purple)

F_1 $^1/_2$ *M/– ; ?/? ; W/–* white with purple spots
$^1/_4$ *M/– ; D/– ; w/w* dark purple
$^1/_4$ *M/– ; d/d ; w/w* light purple

For light and dark purple to appear in a 1:1 ratio among the colored plants, one of the parents must be heterozygous *D/d*. The ratio of white to colored is 1:1, a testcross, so one of the parents is heterozygous *W/w*. All plants are purple, indicating that one parent is homozygous *M/M*. Therefore, the genotypes are

P *m/m ; D/d ; W/w* × *M/M ; d/d ; w/w*

F_1 $^1/_2$ *M/m ; –/d ; W/w*
$^1/_4$ *M/m ; D/d ; w/w*
$^1/_4$ *M/m ; d/d ; w/w*

39. a. This is a dihybrid cross resulting in a 13 white:3 red ratio of progeny in the F_2. This ratio of white to red indicates that the double recessive is not the red phenotype. Instead, the general formula for color is represented by *a/a ; B/–*.

Let line 1 be *A/A* ; *B/B* and line 2 be *a/a* ; *b/b*. The F_1 is *A/a* ; *B/b*. Assume that *A* blocks color in line 1 and *b/b* blocks color in line 2. The F_1 will be white because of the presence of *A*. The F_2 would be

9 *A/–* ; *B/–*	white
3 *A/–* ; *b/b*	white
3 *a/a* ; *B/–*	red
1 *a/a* ; *b/b*	white

b. Cross 1: *A/A* ; *B/B* × *A/a* ; *B/b* → all *A/–* ; *B/–* white

Cross 2: *a/a* ; *b/b* × *A/a* ; *B/b*

$^1/_4$ *A/a* ; *B/b*	white
$^1/_4$ *A/a* ; *b/b*	white
$^1/_4$ *a/a* ; *b/b*	white
$^1/_4$ *a/a* ; *B/b*	red

40. a. Note that blue is always present, indicating *E/E* (blue) in both parents. Because of the ratios that are observed neither *C* nor *D* is varying. In this case, the gene pairs that are involved are *A/a* and *B/b*. The parents are *A/A* ; *b/b* × *a/a* ; *B/B* or *A/A* ; *B/B* × *a/a* ; *b/b*.

The F_1 are *A/a* ; *B/b* and the F_2 are

9	*A/–* ; *B/–*	blue + red, or purple
3	*A/–* ; *b/b*	blue + yellow, or green
3	*a/a* ; *B/–*	blue + $white_2$, or blue
1	*a/a* ; *b/b*	blue + $white_2$, or blue

b. Blue is not always present, indicating *E/e* in the F_1. Because green never appears, the F_1 must be *B/B* ; *C/C* ; *D/D*. The parents are *A/A* ; *e/e* × *a/a* ; *E/E* or *A/A* ; *E/E* × *a/a* ; *e/e*.

The F_1 are *A/a* ; *E/e*, and the F_2 are

9	*A/–* ; *E/–*	red + blue, or purple
3	*A/–* ; *e/e*	red + $white_1$, or red
3	*a/a* ; *E/–*	$white_2$ + blue, or blue
1	*a/a* ; *e/e*	$white_2$ + $white_1$, or white

c. Blue is always present, indicating that the F_1 is *E/E*. No green appears, indicating that the F_1 is also *B/B*. The two genes involved are *A* and *D*. The parents are *A/A* ; *d/d* × *a/a* ; *D/D* or *A/A* ; *D/D* × *a/a* ; *d/d*.

The F_1 are *A/a* ; *D/d* and the F_2 are

9	*A/–* ; *D/–*	blue + red + $white_4$, or purple
3	*A/–* ; *d/d*	blue + red, or purple
3	*a/a* ; *D/–*	blue + $white_2$ + $white_4$, or blue
1	*a/a* ; *d/d*	$white_2$ + blue + red, or purple

d. The presence of yellow indicates *b/b* ; *e/e* in the F_2. Therefore, the parents are *B/B* ; *e/e* × *b/b* ; *E/E* or *B/B* ; *E/E* × *b/b* ; *e/e*.

The F_1 are *B/b* ; *E/e* and the F_2 are

9	*B/–* ; *E/–*	red + blue, or purple
3	*B/–* ; *e/e*	red + $white_1$, or red
3	*b/b* ; *E/–*	yellow + blue, or green
1	*b/b* ; *e/e*	yellow + $white_1$, or yellow

41. a. Note that cross 1 suggests that one gene is involved and that single is dominant to double. Cross 2 supports this conclusion. Now note that in crosses 3 and 4, a 1:1 ratio is seen in the progeny, suggesting that superdouble is an allele of both single and double. Superdouble must be heterozygous, however, and it must be dominant to both single and double. Because the heterozygous superdouble yields both single and double when crossed with the appropriate plant, it cannot be heterozygous for the single allele. Therefore, it must be heterozygous for the double allele. A multiple allelic series has been detected: superdouble > single > double.

For now, assume that only one gene is involved and attempt to rationalize the crosses with the assumptions made above.

Cross	Parents	Progeny	Conclusion
1	$A^S/A^S \times A^D/A^D$	A^S/A^D	A^S is dominant to A^D
2	$A^S/A^D \times A^S A^D$	$3\ A^S/–:1\ A^D/A^D$	supports Cross 1 conclusion
3	$A^D/A^D \times A^{Sd}/A^D$	$1\ A^{Sd}/A^D:1\ A^D/A^D$	A^{Sd} is dominant to A^D
4	$A^S/A^S \times A^{Sd}/A^D$	$1\ A^{Sd}/A^S:1\ A^S/A^D$	A^{Sd} is dominant to A^S
5	$A^D/A^D \times A^{Sd}/A^S$	$1\ A^{Sd}/A^D:1\ A^D/A^S$	supports conclusion of heterozygous superdouble
6	$A^D/A^D \times A^S/A^D$	$1\ A^D/A^D:1\ A^D/A^S$	supports conclusion of heterozygous superdouble

b. While this explanation does rationalize all the crosses, it does not take into account either the female sterility or the origin of the superdouble plant from a double-flowered variety.

A number of genetic mechanisms could be proposed to explain the origin of superdouble from the double-flowered variety. Most of the mechanisms will be discussed in later chapters and so will not be mentioned here. However, it can safely be assumed at this point that, whatever the mechanism, it was aberrant enough to block the proper formation of the complex structure of the female flower. Because of female sterility, no homozygote for superdouble can be observed.

42. a. All the crosses suggest two independently assorting genes. However, that does not mean that there are a total of only two genes governing eye color. In fact, there are four genes controlling eye color that are being studied here. Let

a/a = defect in the yellow line 1
b/b = defect in the yellow line 2
d/d = defect in the brown line
e/e = defect in the orange line

The genotypes of each line are as follows:

yellow 1: *a/a* ; *B/B* ; *D/D* ; *E/E*
yellow 2: *A/A* ; *b/b* ; *D/D* ; *E/E*
brown: *A/A* ; *B/B* ; *d/d* ; *E/E*
orange: *A/A* ; *B/B* ; *D/D* ; *e/e*

b. P *a/a* ; *B/B* ; *D/D* ; *E/E* × *A/A* ; *b/b* ; *D/D* ; *E/E* — yellow 1 × yellow 2

F_1 *A/a* ; *B/b* ; *D/D* ; *E/E* — red

F_2
9 *A/–* ; B/– ; *D/D* ; *E/E* — red
3 *a/a* ; *B/–* ; *D/D* ; *E/E* — yellow
3 *A/–* ; *b/b* ; *D/D* ; *E/E* — yellow
1 *a/a* ; *b/b* ; *D/D* ; *E/E* — yellow

P *a/a* ; *B/B* ; *D/D* ; *E/E* × *A/A* ; *B/B* ; *d/d* ; *E/E* — yellow 1 × brown

F_1 *A/a* ; *B/B* ; *D/d* ; *E/E* — red

F_2
9 *A/–* ; *B/B* ; *D/–*; *E/E* — red
3 *a/a* ; *B/B* ; *D/–* ; *E/E* — yellow
3 *A/–* ; *B/B* ; *d/d* ; *E/E* — brown
1 *a/a* ; *B/B* ; *d/d* ; *E/E* — yellow

P *a/a* ; *B/B* ; *D/D* ; *E/E* × *A/A* ; *B/B* ; *D/D* ; *e/e* — yellow 1 × orange

F_1 *A/a* ; *B/B* ; *D/D* ; *E/e* — red

F_2
9 *A/–* ; *B/B* ; *D/D* ; *E/–* — red
3 *a/a* ; *B/B* ; *D/D* ; *E/–* — yellow
3 *A/–* ; *B/B* ; *D/D* ; *e/e* — orange
1 *a/a* ; *B/B* ; *D/D* ; *e/e* — orange

P *A/A* ; *b/b* ; *D/D* ; *E/E* × *A/A* ; *B/B* ; *d/d* ; *E/E* — yellow 2 × brown

F_1 *A/A* ; *B/b* ; *D/d* ; *E/E* — red

F_2
9 *A/A* ; *B/–* ; *D/–* ; *E/E* — red
3 *A/A* ; *b/b* ; *D/–* ; *E/E* — yellow
3 *A/A* ; *B/–* ; *d/d* ; *E/E* — brown
1 *A/A* ; *b/b* ; *d/d* ; *E/E* — yellow

P *A/A* ; *b/b* ; *D/D* ; *E/E* × *A/A* ; *B/B* ; *D/D* ; *e/e* — yellow 2 × orange

F_1 *A/A* ; *B/b* ; *D/D* ; *E/e* — red

F_2
9 *A/A* ; B/– ; *D/D* ; *E/–* — red
3 *A/A* ; *b/b* ; *D/D* ; *E/–* — yellow
3 *A/A* ; *B/–* ; *D/D* ; *e/e* — orange
1 *A/A* ; *b/b* ; *D/D* ; *e/e* — yellow

P	A/A ; B/B ; d/d ; E/E × A/A ; B/B ; D/D ; e/e	brown × orange
F_1	A/A ; B/B ; D/d ; E/e	red
F_2	9 A/A ; B/B ; $D/–$; $E/–$	red
	3 A/A ; B/B ; d/d ; $E/–$	brown
	3 A/A ; B/B ; $D/–$; e/e	orange
	1 A/A ; B/B ; d/d ; e/e	orange

c. When constructing a biochemical pathway, remember that the earliest gene that is defective in a pathway will determine the phenotype of a doubly defective genotype. Look at the following doubly recessive homozygotes from the crosses. Notice that the double defect d/d ; e/e has the same phenotype as the defect a/a ; e/e. This suggests that the E gene functions earlier than do the A and D genes. Using this logic, the following table can be constructed:

Genotype	Phenotype	Conclusion
a/a ; B/B ; D/D ; e/e	orange	E functions before A
A/A ; B/B ; d/d ; e/e	orange	E functions before D
a/a ; b/b ; D/D ; E/E	yellow	B functions before A
A/A ; b/b ; D/D ; e/e	yellow	B functions before E
a/a ; B/B ; d/d ; E/E	yellow	A functions before D
A/A ; b/b ; d/d ; E/E	yellow	B functions before D

The genes function in the following sequence: B, E, A, D. The metabolic path is

$$\text{yellow 2} \xrightarrow{B} \text{orange} \xrightarrow{E} \text{yellow} \xrightarrow{A} \text{brown} \xrightarrow{D} \text{red}$$

43. a. A trihybrid cross would give a 63:1 ratio. Therefore, there are three R loci segregating in this cross.

b. P R_1/R_1 ; R_2/R_2 ; R_3/R_3 × r_1/r_1 ; r_2/r_2 ; r_3/r_3

F_1 R_1/r_1 ; R_2/r_2 ; R_3/r_3

F_2	27	$R_1/–$; $R_2/–$; $R_3/–$	red
	9	$R_1/–$; $R_2/–$; r_3/r_3	red
	9	$R_1/–$; r_2/r_2 ; $R_3/–$	red
	9	r_1/r_1 ; $R_2/–$; $R_3/–$	red
	3	$R_1/–$; r_2/r_2 ; r_3/r_3	red
	3	r_1/r_1 ; $R_2/–$; r_3/r_3	red
	3	r_1/r_1 ; r_2/r_2 ; $R_3/–$	red
	1	r_1/r_1 ; r_2/r_2 ; r_3/r_3	white

c. (1) In order to obtain a 1:1 ratio, only one of the genes can be heterozygous. A representative cross would be R_1/r_1 ; r_2/r_2 ; r_3/r_3 × r_1/r_1 ; r_2/r_2 ; r_3/r_3.

(2) In order to obtain a 3 red:1 white ratio, two alleles must be segregating and they cannot be within the same gene. A representative cross would be R_1/r_1 ; R_2/r_2 ; r_3/r_3 × r_1/r_1 ; r_2/r_2 ; r_3/r_3.

(3) In order to obtain a 7 red: 1 white ratio, three alleles must be segregating, and they cannot be within the same gene. The cross would be R_1/r_1 ; R_2/r_2 ; R_3/r_3 × r_1/r_1 ; r_2/r_2 ; r_3/r_3.

d. The formula is $1 - (^1/_2)^n$, where n = the number of loci that are segregating in the representative crosses above.

44. a. The trait is recessive (parents without the trait have children with the trait) and autosomal (daughters can inherit the trait from unaffected fathers). Looking to generation III, there is also evidence that there are two different genes that when defective result in deaf-mutism.

Assuming that one gene has alleles *A* and *a* and the other has *B* and *b*. The following genotypes can be inferred:

I-1 and I-2	*A/a* ; *B/B*	I-3 and I-4	*A/A* ; *B/b*
II-(1, 3, 4, 5, 6)	*A/–* ; *B/B*	II- (9, 10, 12, 13, 14, 15)	*A/A* ; *B/–*
II-2 and II-7	*a/a* ; *B/B*	II-8 and II-11	*A/A* ; *b/b*

b. Generation III shows complementation. All are *A/a* ; *B/b*.

45. a. The first impression from the pedigree is that the gene causing blue sclera and brittle bones is pleiotropic with variable expressivity. If two genes were involved, it would be highly unlikely that all people with brittle bones also had blue sclera.

b. Sons and daughters inherit from affected fathers, so the allele appears to be autosomal.

c. The trait appears to be inherited as a dominant but with incomplete penetrance. For the trait to be recessive, many of the nonrelated individuals marrying into the pedigree would have to be heterozygous (e.g., I-1, I-3, II-8, II-11). Individuals II4, II14, III2, and III14 have descendants with the disorder, although they do not themselves express the disorder. Therefore, $^4/_{20}$ people that can be inferred to carry the gene do not express the trait. That is 80% penetrance. (Penetrance could be significantly less than that, since many possible carriers have no shown progeny.) The pedigree also exhibits variable expressivity. Of the 16 individuals who have blue sclera, 10 do not have brittle bones. Usually, expressivity is put in terms of none, variable, and highly variable, rather than expressed as percentages.

46. a. and b. Assuming that both the *Brown* and the *Van Scoy* lines were homozygous, the parental cross suggests that nonhygienic behavior is dominant to hygienic. A consideration of the specific behavior of the F_1 × *Brown* progeny, however, suggests that the behavior is separable into two processes; one involves uncapping and one involves removal of dead pupae. If this is true, uncapping and removal of dead pupae, two behaviors that normally go together in the *Brown* line, have been separated in the F_2 progeny, suggesting that they are controlled by different and unlinked genes. Those bees that lack uncapping behavior are still able to express removal of dead pupae if environmental conditions are such that they do not need to uncap a compartment first. Also, the uncapping behavior is epistatic to removal of dead pupae.

Let: U = no uncapping, u = uncapping, R = no removal, r = removal

P	u/u ; r/r × U/U ; R/R	
	(*Brown*) (*Van Scoy*)	
	U/u ; R/r × u/u ; r/r	(F_1 cross *Brown*)
Progeny	$^1/_4$ U/u ; R/r	nonhygienic
	$^1/_4$ U/u ; r/r	nonhygienic (no uncapping but removal if uncapped)
	$^1/_4$ u/u ; R/r	uncapping but no removal
	$^1/_4$ u/u ; r/r	hygienic

47. a. Note that the first two crosses are reciprocal and that the male offspring differ in phenotype between the two crosses. This indicates that the gene is on the X chromosome.

Also note that the F_1 females in the first two crosses are sickle. This indicates that sickle is dominant to round. The third cross also indicates that oval is dominant to sickle. Therefore, this is a multiple allelic series with oval > sickle > round.

Let W^o = oval, W^s = sickle, and W^r = round. The three crosses are

Cross 1: W^s/W^s × W^r/Y → W^s/W^r and W^s/Y

Cross 2: W^r/W^r × W^s/Y → W^s/W^r and W^r/Y

Cross 3: W^s/W^s × W^o/Y → W^o/W^s and W^s/Y

b. W^o/W^s × W^r/Y

$^1/_4$ W^o/W^r	female oval
$^1/_4$ W^s/W^r	female sickle
$^1/_4$ W^o/Y	male oval
$^1/_4$ W^s/Y	male sickle

48. a. First note that there is a phenotypic difference between males and females, indicating X-linkage. This means that the male progeny express both alleles of the female parent. A beginning statement of the genotypes could be as follows, where H indicates the gene and the numbers 1, 2, and 3 indicate the variants:

P	$H^1/H^2 \times H^3/Y$	
F_1	$^1/_4\ H^1/H^3$	one-banded female
	$^1/_4\ H^3/H^2$	three-banded female
	$^1/_4\ H^1/Y$	one-banded male
	$^1/_4\ H^2/Y$	two-banded male

Because both female F_1 progeny obtain allele H^3 from their father, yet only one has a 3-banded pattern, there is obviously a dominance relationship among the alleles. The mother indicates that H^1 is dominant to H^2. The female progeny must be H^1/H^3 and H^3/H^2, and they have a 1-banded and 3-banded pattern, respectively. H^3 must be dominant to H^2 because the daughter with that combination of alleles has to be the 3-banded daughter. In other words, there is a multiple-allelic series with $H^1 > H^3 > H^2$.

b. The cross is

P	$H^3/H^2 \times H^1/Y$	
F_1	$^1/_4\ H^1/H^3$	one-banded female
	$^1/_4\ H^1/H^2$	one-banded female
	$^1/_4\ H^2/Y$	two-banded male
	$^1/_4\ H^3/Y$	three-banded male

49. a. The first two crosses indicate that wild-type is dominant to both platinum and aleutian. The third cross indicates that two genes are involved rather than one gene with multiple alleles because a 9:3:3:1 ratio is observed.

Let platinum be *a*, aleutian be *b*, and wild-type be *A/–* ; *B/–*.

Cross 1:	P	*A/A* ; *B/B* × *a/a* ; *B/B*	wild-type × platinum
	F_1	*A/a* ; *B/B*	all wild-type
	F_2	3 *A/–* ; *B/B*:1 *a/a* ; *B/B*	3 wild-type:1 platinum
Cross 2:	P	*A/A* ; *B/B* × *A/A* ; *b/b*	wild-type × aleutian
	F_1	*A/A* ; *B/b*	all wild-type
	F_2	3 *A/A* ; *B/–*:1 *A/A* ; *b/b*	3 wild-type:1 aleutian
Cross 3:	P	*a/a* ; *B/B* × *A/A* ; *b/b*	platinum × aleutian
	F_1	*A/a* ; *B/b*	all wild-type
	F_2	9 *A/–* ; *B/–*	wild-type
		3 *A/–* ; *b/b*	aleutian
		3 *a/a* ; *B/–*	platinum
		1 *a/a* ; *b/b*	sapphire

b.

	sapphire × platinum		sapphire × aleutian	
P	*a/a* ; *b/b* × *a/a* ; *B/B*		*a/a* ; *b/b* × *A/A* ; *b/b*	
F_1	*a/a* ; *B/b*	platinum	*A/a* ; *b/b*	aleutian
F_2	3 *a/a* ; *B/–* 1 *a/a* ; *b/b*	platinum sapphire	3 *A/–* ; *b/b* 1 *a/a* ; *b/b*	aleutian sapphire

50. a. The genotypes are

P	*B/B* ; *i/i* × *b/b* ; *I/I*	
F_1	*B/b* ; *I/i*	hairless
F_2	9 *B/–* ; *I/–* 3 *B/–* ; *i/i* 3 *b/b* ; *I/–* 1 *b/b* ; *i/i*	hairless straight hairless bent

b. In order to solve this problem, first write as much as you can of the progeny genotypes.

4 hairless	*–/–* ; *I/–*
3 straight	*B/–* ; *i/i*
1 bent	*b/b* ; *i/i*

Each parent must have a *b* allele and an *i* allele. The partial genotypes of the parents are

–/*b* ; –/*i* × –/*b* ; –/*i*

At least one parent carries the *B* allele, and at least one parent carries the *I* allele. Assume for a moment that the first parent carries both. The partial genotypes become

B/b ; *I/i* × –/*b* ; –/*i*

Note that $^1/_2$ the progeny are hairless. This must come from a *I/i* × *i/i* cross. Of those progeny with hair, the ratio is 3:1, which must come from a *B/b* × *B/b* cross. The final genotypes are therefore

B/b ; *I/i* × *B/b* ; *i/i*

51. a. The first question is whether there are two genes or one gene with three alleles. Note that a black × eyeless cross produces black and brown progeny in one instance but black and eyeless progeny in the second instance. Further note that a black × black cross produces brown, a brown × eyeless cross produces brown, and a brown × black cross produces eyeless. The results are confusing enough that the best way to proceed is by trial and error.

There are two possibilities: one gene or two genes.

One gene. Assume one gene for a moment. Let the gene be *E* and assume the following designations:

E^1 = black

E^2 = brown

E^3 = eyeless

If you next assume, based on the various crosses, that black > brown > eyeless, genotypes in the pedigree become

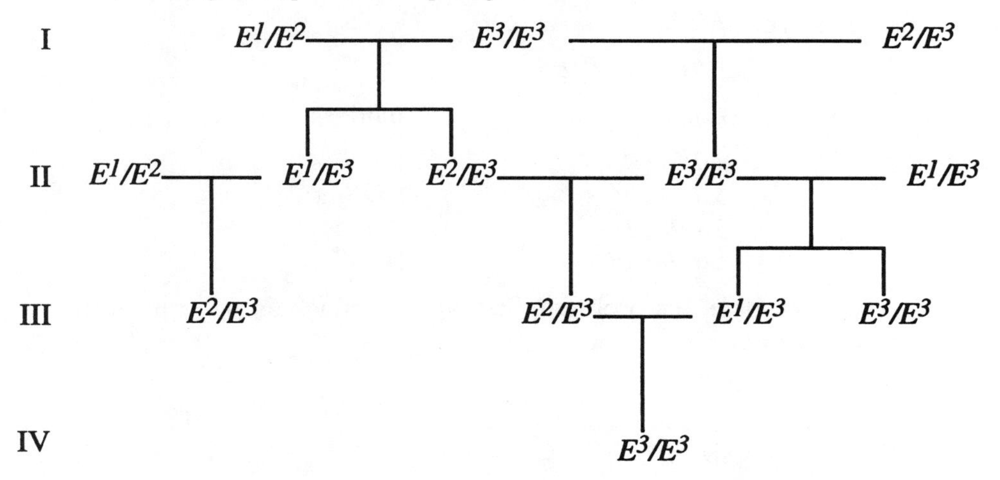

b. The genotype of individual II-3 is E^2/E^3.

Two genes. If two genes are assumed, then questions arise regarding whether both are autosomal or if one might be X-linked. Eye color appears to be autosomal. The presence or absence of eyes could be X-linked. Let

B = black X^E = normal eyes

b = brown X^e = eyeless

The pedigree then is

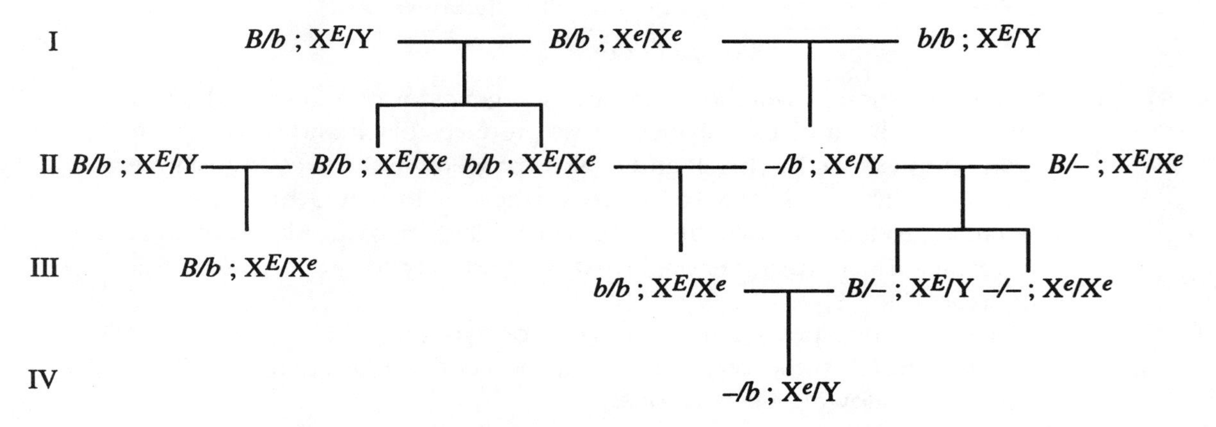

b. With this interpretation of the pedigree, individual II–3 is b/b ; X^E/X^e.

Without further data, it is impossible to choose scientifically between the two possible explanations. A basic "rule" in science is that the more simple answer should be used until such time as data exist that would lead to a rejection of it. In this case, the one-gene explanation is essentially as complex as the two-gene explanation, with one exception. The exception is that eye color is fundamentally different from a lack of eyes. Therefore, the better guess is that two genes are involved.

52. a. and b. Cross 1 indicates that orange is dominant to yellow. Crosses 2–4 indicate that red is dominant to orange, yellow, and white. Crosses 5–7 indicate that there are two genes involved in the production of color. Cross 5 indicates that yellow and white are different genes. Cross 6 indicates that orange and white are different genes.

In other words, epistasis is involved, and the homozygous recessive white genotype seems to block production of color by a second gene.

Begin by explaining the crosses in the simplest manner possible until such time as it becomes necessary to add complexity. Therefore, assume that orange and yellow are alleles of gene *A*, with orange dominant.

Cross 1: P $A/A \times a/a$

F_1 A/a

F_2 $3\ A/\!-:1\ a/a$

Immediately there is trouble in trying to write the genotypes for cross 2, unless red is a third allele of the same gene. Assume the following dominance relationships: red > orange > yellow. Let the alleles be designated as follows:

red A^R
orange A^O
yellow A^Y

Crosses 1–3 now become

P	$A^O/A^O \times A^Y/A^Y$	$A^R/A^R \times A^O/A^O$	$A^R/A^R \times A^Y/A^Y$
F1	A^O/A^Y	A^R/A^O	A^R/A^Y
F2	$3\ A^O/- : 1\ A^Y/A^Y$	$3\ A^R/- : 1\ A^O/A^O$	$3\ A^R/- : 1\ A^Y/A^Y$

Cross 4: To do this cross you must add a second gene. You must also rewrite the above crosses to include the second gene. Let *B* allow color and *b* block color expression, producing white. The first three crosses become

P	$A^O/A^O\ ;\ B/B \times A^Y/A^Y\ ;\ B/B$	$A^R/A^R\ ;\ B/B \times A^O/A^O\ ;\ B/B$	$A^R/A^R\ ;\ B/B \times A^Y/A^Y\ ;\ B/B$
F_1	$A^O/A^Y\ ;\ B/B$	$A^R/A^O\ ;\ B/B$	$A^R/A^Y\ ;\ B/B$
F_2	$3\ A^O/-\ ;\ B/B : 1\ A^Y/A^Y\ ;\ B/B$	$3\ A^R/-\ ;\ B/B : 1\ A^O/A^O\ ;\ B/B$	$3\ A^R/- : 1\ A^Y/A^Y\ ;\ B/B$

The fourth cross is

P $A^R/A^R\ ;\ B/B \times A^R/A^R\ ;\ b/b$

F_1 $A^R/A^R\ ;\ B/b$

F_2 $3\ A^R/A^R\ ;\ B/- : 1\ A^R/A^R\ ;\ b/b$

Cross 5: To do this cross, note that there is no orange appearing. Therefore, the two parents must carry the alleles for red and yellow, and the expression of red must be blocked.

P A^Y/A^Y ; B/B × A^R/A^R ; b/b

F_1 A^R/A^Y ; B/b

F_2 9 $A^R/–$; $B/–$ red
3 $A^R/–$; b/b white
3 A^Y/A^Y; $B/–$ yellow
1 A^Y/A^Y ; b/b white

Cross 6: This cross is identical with cross 5 except that orange replaces yellow.

P A^O/A^O ; B/B × A^R/A^R ; b/b

F_1 A^R/A^O ; B/b

F_2 9 $A^R/–$; $B/–$ red
3 $A^R/–$; b/b white
3 A^O/A^O ; $B/–$ orange
1 A^O/A^O ; b/b white

Cross 7: In this cross, yellow is suppressed by b/b.

P A^R/A^R ; B/B × A^Y/A^Y ; b/b

F_1 A^R/A^Y ; B/b

F_2 9 $A^R/–$; $B/–$ red
3 $A^R/–$; b/b white
3 A^Y/A^Y ; $B/–$ yellow
1 A^Y/A^Y ; b/b white

53. The ratios observed in the F_2 of these crosses are all variations of 9:3:3:1, indicating that two genes are assorting and epistasis is occurring.

(1) F_1 A/a ; B/b × A/a ; B/b

F_2 9 $A/–$; $B/–$ cream
3 $A/–$; b/b cream
3 a/a ; $B/–$ black
1 a/a ; b/b gray

A testcross of the F_1 would be A/a ; B/b × a/a ; b/b and the offspring would be

$^1/_4$ A/a ; B/b cream
$^1/_4$ a/a ; B/b black
$^1/_4$ A/a ; b/b cream
$^1/_4$ a/a ; b/b gray

(2) F_1 *A/a* ; *B/b* × *A/a* ; *B/b*

F_2

9 *A/–* ; *B/–*	orange
3 *A/–* ; *b/b*	yellow
3 *a/a* ; *B/–*	yellow
1 *a/a* ; *b/b*	yellow

A testcross of the F_1 would be *A/a* ; *B/b* × *a/a* ; *b/b* and the offspring would be

$^1/_4$ *A/a* ; *B/b*	orange
$^1/_4$ *a/a* ; *B/b*	yellow
$^1/_4$ *A/a* ; *b/b*	yellow
$^1/_4$ *a/a* ; *b/b*	yellow

(3) F_1 *A/a* ; *B/b* × *A/a* ; *B/b*

F_2

9 *A/–* ; *B/–*	black
3 *A/–* ; *b/b*	white
3 *a/a* ; *B/–*	black
1 *a/a* ; *b/b*	black

A testcross of the F_1 would be *A/a* ; *B/b* × *a/a* ; *b/b* and the offspring would be

$^1/_4$ *A/a* ; *B/b*	black
$^1/_4$ *a/a* ; *B/b*	black
$^1/_4$ *A/a* ; *b/b*	white
$^1/_4$ *a/a* ; *b/b*	black

(4) F_1 *A/a* ; *B/b* × *A/a* ; *B/b*

F_2

9 *A/–* ; *B/–*	solid red
3 *A/–* ; *b/b*	mottled red
3 *a/a* ; *B/–*	small red dots
1 *a/a* ; *b/b*	small red dots

A testcross of the F_1 would be *A/a* ; *B/b* × *a/a* ; *b/b* and the offspring would be

$^1/_4$ *A/a* ; *B/b*	solid red
$^1/_4$ *a/a* ; *B/b*	small red dots
$^1/_4$ *A/a* ; *b/b*	mottled red
$^1/_4$ *a/a* ; *b/b*	small red dots

54. a. Intercrossing mutant strains that all share a common recessive phenotype is the basis of the complementation test. This test is designed to identify the number of different genes that can mutate to a particular phenotype. In this problem, if the progeny of a given cross still express the wiggle phenotype, the mutations fail to complement and are considered alleles of the same gene; if the progeny are wild type, the mutations complement and the two strains carry mutant alleles of separate genes.

b. From the data,

1 and 5	fail to complement	gene *A*
2, 6, 8, and 10	fail to complement	gene *B*
3 and 4	fail to complement	gene *C*
7, 11, and 12	fail to complement	gene *D*
9	complements all others	gene *E*

There are five complementation groups (genes) identified by these data.

c. mutant 1: $a^1/a^1 \cdot b^+/b^+ \cdot c^+/c^+ \cdot d^+/d^+ \cdot e^+/e^+$

mutant 2: $a^+/a^+ \cdot b^2/b^2 \cdot c^+/c^+ \cdot d^+/d^+ \cdot e^+/e^+$

mutant 5: $a^5/a^5 \cdot b/^+b^+ \cdot c^+/c^+ \cdot d^+/d^+ \cdot e^+/e^+$

1/2 hybrid: $a^+/a^1 \cdot b^+/b^2 \cdot c^+/c^+ \cdot d^+/d^+ \cdot e^+/e^+$

1/5 hybrid: $a^1/a^5 \cdot b^+/b^+ \cdot c^+/c^+ \cdot d^+/d^+ \cdot e^+/e^+$ phenotype: wiggles

1 and 5 are both mutant for gene *A*

2 and 5 are mutant for different genes

55. a. If the nonsense mutation is homozygous, then complete translation will occur only with T^S. The traditional definition of a recessive allele is that its phenotype is expressed when it is the only type of allele present. By this definition, T^+ is recessive to the dominant T^S.

b. In a trihybrid cross, the white phenotype will occur only when one or both of the enzymes are homozygous for a nonsense mutation and the wild-type form of *T* is also homozygous. The frequencies and phenotypes seen are

27	$A/- \; ; \; B/- \; ; \; T^S/-$	purple
9	$a^n/a^n \; ; \; B/- \; ; \; T^S/-$	purple
9	$A/- \; ; \; b^n/b^n \; ; \; T^S/-$	purple
9	$A/- \; ; \; B/- \; ; \; T^+/T^+$	purple
3	$a^n/a^n \; ; \; b^n/b^n \; ; \; T^S/-$	purple
3	$a^n/a^n \; ; \; B/- \; ; \; T^+/T^+$	white
3	$A/- \; ; \; b^n/b^n \; ; \; T^+/T^+$	white

1	a^n/a^n ; b^n/b^n ; T^+/T^+	white

or 57 purple to 7 white.

Questions 56–58 all ask you to compare expected results to actual data to determine if a certain hypothesis is correct. In Chapter 5, you will learn how to use a specific analytical test (χ^2) to determine if actual observations deviate from expectations purely on the basis of chance. For now, let's see how to start this process.

56. For the cross *B/b* × *B/b*, you expect $^3/_4$ *B/–* and $^1/_4$ *b/b* or for 400 progeny, 300 and 100, respectively. This should be compared with the actual data of 280 and 120. Alternatively, if the *B/B* genotype was actually lethal, then you would expect $^2/_3$ *B/b* and $^1/_3$ *b/b* or for 400 progeny, 267 and 133, respectively. This compares more favorably with the actual data.

57. There are a total of 159 progeny that should be distributed in 9:3:3:1 ratio if the two genes are assorting independently. You can see that

Observed	Expected
88 *P/–* ; *Q/–*	90
32 *P/–* ; *q/q*	30
25 *p/p–* ; *Q/–*	30
14 *p/p* ; *q/q*	10

58. For 160 progeny, there should be 90:70, 130:30, or 120:40 of phenotype 1:phenotype 2 progeny for the various expected ratios. Only the 9:7 ratio seems feasible, and that would indicate that two genes are assorting independently where

A/– ; *B/–*	phenotype 1
A/– ; *b/b*	phenotype 2
a/a ; *B/–*	phenotype 3
a/a ; *b/b*	phenotype 4

5 BASIC EUKARYOTIC CHROMOSOME MAPPING

1. You perform the following cross and are told that the two genes are 10 m.u. apart.

$$A\ B/a\ b \times a\ b/a\ b$$

Among their progeny, 10% should be recombinant (*A b/a b* and *a B/a b*) and 90% should be parental (*A B/a b* and *a b/a b*). Therefore, *A B/a b* should represent $^1/_2$ the parentals, or 45%.

2.

P	*A d/A d* × *a D/a D*	
F_1	*A d/a D*	
F_2	1 *A d/A d*	phenotype: A d
	2 *A d/a D*	phenotype: A D
	1 *a D/a D*	phenotype: a D

3.

P		*R S/r s* × *R S/r s*	
	gametes	$^1/_2\ (1 - 0.35)$	*R S*
		$^1/_2\ (1 - 0.35)$	*r s*
		$^1/_2\ (0.35)$	*R s*
		$^1/_2\ (0.35)$	*r S*

F_1 genotypes

0.1056 *R S/R S*	0.1138 *r s/r S*
0.1056 *r s/r s*	0.1138 *r s/R s*
0.2113 *R S/r s*	0.0306 *R s/R s*
0.1138 *R S/r S*	0.0306 *r S/r S*
0.1138 *R S/R s*	0.0613 *R s/r S*

F_1 phenotypes

0.6058 R S
0.1056 r s
0.1444 R s
0.1444 r S

4. The cross is *E/e · F/f* × *e/e · f/f*. If independent assortment exists, the progeny should be in a 1:1:1:1 ratio, which is not observed. Therefore, there is linkage. *E f* and *e F* are recombinants equaling one-third of the progeny. The two genes are 33.3 map units (m.u.) apart.

5. Because only parental types are recovered, the two genes must be tightly linked and recombination must be very rare. Knowing how many progeny were looked at would give an indication of how close the genes are.

6. The problem states that a female that is *A/a · B/b* is testcrossed. If the genes are unlinked, they should assort independently and the four progeny classes should be present in roughly equal proportions. This is clearly not the case. The *A/a · B/b* and *a/a · b/b* classes (the parentals) are much more common than the *A/a · b/b* and *a/a · B/b* classes (the recombinants). The two genes are on the same chromosome and are 10 map units apart.

$$RF = 100\% \times (46 + 54)/1000 = 10\%$$

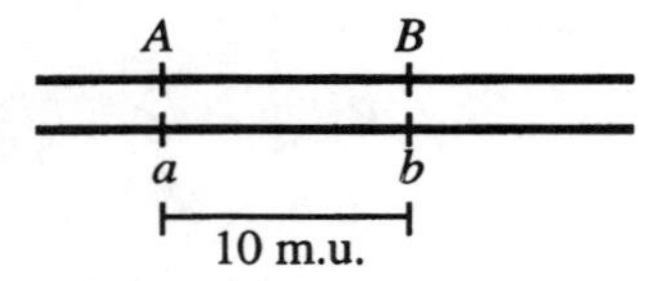

7. The cross is *A/A · b/b* × *a/a · B/B*. The F_1 would be *A/a · B/b*.

a. If the genes are unlinked, all four progeny classes from the testcross (including *a/a* ; *b/b*) would equal 25%.

b. With completely linked genes, the F_1 would produce only *A b* and *a B* gametes. Thus, there would be a 0% chance of having *a b/a b* progeny from a testcross of this F_1.

c. If the two genes are linked and 10 map units apart, 10% of the testcross progeny should be recombinants. Since the F_1 is *A b/a B*, *a b* is one of the recombinant classes (*A B* being the other) and should equal $^1/_2$ the total recombinants, or 5%.

d. 12% (see part c)

8. Meiosis is occurring in an organism that is *C d/c D*, ultimately producing haploid spores. The parental genotypes are *C d* and *c D*, in equal frequency. The recombinant types are *C D* and *c d*, in equal frequency. Eight map units means 8% recombinants. Thus, *C D* and *c d* will each be present at a frequency of 4%, and *C d* and *c D* will each be present at a frequency of (100% – 8%)/2 = 46%.

a. 4%; b. 4%; c. 46%; d. 8%

9. To answer this question, you must realize that

(1) One chiasma involves two of the four chromatids of the homologous pair, so if 16% of the meioses have one chiasma, it will lead to 8% recombinants observed in the progeny (half of the chromosomes of such a meiosis are still parental), and

(2) Half the recombinants will be *B r*, so the correct answer is 4%.

10. ***Unpacking the Problem***

1. There is no correct drawing; any will do. Pollen from the tassels is placed on the silks of the females. The seeds are the F_1 corn kernels.
2. The +'s all look the same because they signify wild type for each gene. The information is given in a specific order, which prevents confusion, at least initially. However, as you work the problem, which may require you to reorder the genes, errors can creep into your work if you do not make sure that you reorder the genes for each genotype in exactly the same way. You may find it easier to write the complete genotype, p^+ instead of +, to avoid confusion.
3. The phenotype is purple leaves and brown midriff to seeds. In other words, the two colors refer to different parts of the organism.
4. There is no significance in the original sequence of the data.
5. A tester is a homozygous recessive for all genes being studied. It is used so that the meiotic products in the organism being tested can be seen directly in the phenotype of the progeny.
6. The progeny phenotypes allow you to infer the genotypes of the plants. For example, *gre* stands for "green," the phenotype of $p^+/–$; *sen* stands for "virus-sensitive," the phenotype of $v^+/–$; and *pla* stands for "plain seed," the phenotype of $b^+/–$. In this testcross, all progeny have at least one recessive allele so the "gre sen pla" progeny are actually $p^+/p \cdot v^+/v \cdot b^+/b$.
7. *Gametes* refers to the gametes of the two purebreeding parents. *F_1 gametes* refers to the gametes produced by the completely heterozygous F_1 progeny. They indicate whether crossing-over and independent assortment have occurred. In this case, because there is either independent assortment or crossing-over, or both, the data indicate that the three genes are not so tightly linked that zero recombination occurred.
8. The main focus is meiosis occurring in the F_1 parent.

9. The gametes from the tester are not shown because they contribute nothing to the phenotypic differences seen in the progeny.

10. Eight phenotypic classes are expected for three autosomal genes, whether or not they are linked, when all three genes have simple dominant-recessive relationships among their alleles. The general formula for the number of expected phenotypes is 2^n, where n is the number of genes being studied.

11. If the three genes were on separate chromosomes, the expectation is a 1:1:1:1:1:1:1:1 ratio.

12. The four classes of data correspond to the parentals (largest), two groups of single crossovers (intermediate), and double crossovers (smallest).

13. By comparing the parentals with the double crossovers, gene order can be determined. The gene in the middle "flips" with respect to the two flanking genes in the double-crossover progeny. In this case, one parental is + + + and one double crossover is *p* + +. This indicates that the gene for leaf color (*p*) is in the middle.

14. If only two of the three genes are linked, the data can still be grouped, but the grouping will differ from that mentioned in (12) above. In this situation, the unlinked gene will show independent assortment with the two linked genes. There will be one class composed of four phenotypes in approximately equal frequency, which combined will total more than half the progeny. A second class will be composed of four phenotypes in approximately equal frequency, and the combined total will be less than half the progeny. For example, if the cross were *a b*/+ + ; *c*/+ × *a b*/*a b* ; *c*/*c*, the parental class (more frequent class) would have four components: *a b c*, *a b* +, + + *c*, and + + +. The recombinant class would be *a* + *c*, *a* + +, + *b c*, and + *b* +.

15. *Point* refers to locus. The usage does not imply linkage, but rather a testing for possible linkage. A fourpoint testcross would look like the following: *a*/+ · *b*/+ · *c*/+ · *d*/+ × *a*/*a* · *b*/*b* · *c*/*c* · *d*/*d*.

16. A *recombinant* refers to an individual who has alleles inherited from two different grandparents, both of whom were the parents of the individual's heterozygous parent. Another way to think about this term is that in the recombinant individual's heterozygous parent, recombination took place among the genes that were inherited from his or her parents. In this case, the recombination took place in the F_1 and the recombinants are among the F_2 progeny.

17. The "recombinant for" columns refer to specific gene pairs and progeny that exhibit recombination between those gene pairs.

18. There are three "recombinant for" columns because three genes can be grouped in three different gene pairs.

19. *R* refers to recombinant progeny, and they are determined by reference back to the parents of their heterozygous parent.

20. Column totals indicate the number of progeny that experience crossing-over between the specific gene pairs. They are used to calculate map units between the two genes.

21. The diagnostic test for linkage is a recombination frequency of less than 50%.

22. A map unit represents 1% crossing-over and is the same as a centimorgan.

23. In the tester, recombination cannot be detected in the gamete contribution to the progeny because the tester is homozygous. The F_1 individuals have genotypes fixed by their parents' homozygous state and, again, recombination cannot be detected in them, simply because their parents were homozygous.

24. Interference I = 1 – coefficient of coincidence = 1 – (observed double crossovers/expected double crossovers). The expected double crossovers are equal to *p*(frequency of crossing-over in the first region, in this case between *v* and *p*) × *p*(frequency of crossing-over in the second region, between *p* and *b*) × number of progeny. The probability of crossing-over is equal to map units converted back to percentage.

25. If the three genes are not all linked, then interference cannot be calculated.

26. A great deal of work is required to obtain 10,000 progeny in corn because each seed on a cob represents one progeny. Each cob may contain as many as 200 seeds. While seed characteristics can be assessed at the cob stage, for other characteristics, each seed must separately be planted and assessed after germination and growth. The bookkeeping task is also enormous.

Solution to the Problem

a. The three genes are linked.

b. Comparing the parentals (most frequent) with the double crossovers (least frequent), the gene order is *v p b*. There were 2200 recombinants between *v* and *p*, and 1500 between *p* and *b*. The general formula for map units is

m.u. = 100%(number of recombinants)/total number of progeny

Therefore, the map units between *v* and *p* = 100%(2200)/10,000 = 22 m.u., and the map units between *p* and *b* = 100%(1500)/10,000 = 15 m.u.

The map is

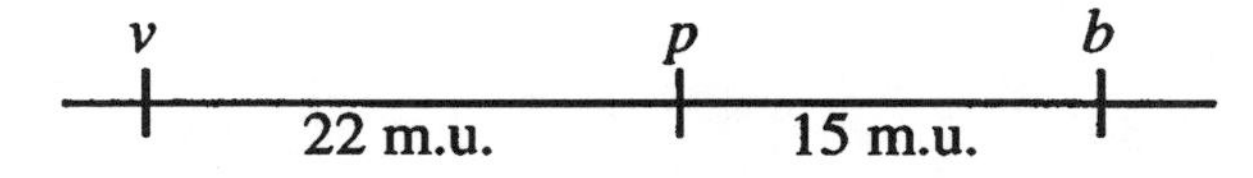

c. I = 1 – observed double crossovers/expected double crossovers
= 1 – 132/(0.22)(0.15)(10,000)
= 1 – 0.4 = 0.6

11. a. All these genes are linked. To determine this, each gene pair is examined separately. For example, are *A* and *B* linked?

$A\ B = 140 + 305 = 445$
$a\ b = 145 + 310 = 455$
$a\ B = 42 + 6 = 48$
$A\ b = 43 + 9 = 52$

Conclusion: the two genes are linked and 10 m.u. apart.

Are *A* and *D* linked?

$A\ D = 0$
$a\ d = 0$
$A\ d = 43 + 140 + 9 + 305 = 497$
$a\ D = 42 + 145 + 6 + 310 = 503$

Conclusion: the two genes show no recombination and at this resolution, are 0 m.u. apart.

Are *B* and *C* linked?

$B\ C = 42 + 140 = 182$
$b\ c = 43 + 145 = 188$
$B\ c = 6 + 305 = 311$
$b\ C = 9 + 310 = 319$

Conclusion: the two genes are linked and 37 m.u. apart.

Are *C* and *D* linked?

$C\ D = 42 + 310 = 350$
$c\ d = 43 + 305 = 348$
$C\ d = 140 + 9 = 149$
$c\ D = 145 + 6 = 151$

Conclusion: the two genes are linked and 30 m.u. apart.

Therefore, all four genes are linked.

b. and c. Because *A* and *D* show no recombination, first rewrite the progeny omitting *D* and *d* (or omitting *A* and *a*).

a B C	42
A b c	43
A B C	140
a b c	145
a B c	6
A b C	9
A B c	305
a b C	310
	1000

Note that the progeny now look like those of a typical three point testcross, with *A B c* and *a b C* the parental types (most frequent) and *a B c* and *A b C* the double recombinants (least frequent). The gene order is *B A C*. This is determined either by the map distances or by comparing double recombinants with the parentals; the gene that "switches" in reference with the other two is the gene in the center ($B\ A\ c \rightarrow B\ a\ c$, $b\ a\ C \rightarrow b\ A\ C$).

Next, rewrite the progeny, this time putting the genes in the proper order, and classify the progeny.

B a C	42	CO *A–B*
b A c	43	CO *A–B*
B A C	140	CO *A–C*
b a c	145	CO *A–C*
B a c	6	DCO
b A C	9	DCO
B A c	305	parental
b a C	310	parental

To construct the map of these genes, use the following formula:

$$\text{distance between two genes} = \frac{(100\%)(\text{number of single CO} + \text{number of DCO})}{\text{total number of progeny}}$$

For the *A* to *B* distance:

$$= \frac{(100\%)(42 + 43 + 6 + 9)}{1000} = 10 \text{ m.u.}$$

For the *A* to *C* distance:

$$= \frac{(100\%)(140 + 145 + 6 + 9)}{1000} = 30 \text{ m.u.}$$

The map is:

B —— 10 m.u. —— *A, D* —— 30 m.u. —— *C*

The parental chromosomes actually were *B (A,d) c/b (a,D) C*, where the parentheses indicate that the order of the genes within is unknown.

d. Interference = 1 – (observed DCO/expected DCO)
= 1 – (6 + 9)/[(0.10)(0.30)(1000)]
= 1 – 15/30 = 0.5

12. **a.** Males must be heterozygous for both genes, and the two must be closely linked: *M F/ m f*.

b. *m f/m f*

c. Sex is determined by the male contribution. The two parental gametes are *M F*, determining maleness (*M F/m f*), and *m f*, determining femaleness (*m f/m f*). Occasional recombination would yield *M f*, determining a hermaphrodite (*M f/m f*), and *m F*, determining total sterility (*m F/m f*).

d. recombination in the male yielding *M f*

e. Hermaphrodites are rare because the genes must be tightly linked.

13. The verbal description indicates the following cross and result:

P $\quad N/- \cdot A/- \times n/n \cdot O/O$

F_1 $\quad N/n \cdot A/O \times N/n \cdot A/O$

The results indicate linkage, so the cross and results can be rewritten:

P $\quad N\,A/-\,- \times n\,O/n\,O$

F_1 $\quad N\,A/n\,O \times N\,A/n\,O$

F_2 $\quad$ 66% $\quad N\,A/-\,-$ or $N\,-/-\,A$
16% $\quad n\,O/n\,O$
9% $\quad n\,A/n\,-$
9% $\quad N\,O/-\,O$

Only one genotype is fully known: 16% $n\,O/n\,O$, a combination of two parental gametes. The frequency of two parental gametes coming together is the frequency of the first times the frequency of the second. Therefore, the frequency of each $n\,O$ gamete is the square root of 0.16, or 0.4. Within an organism the two parental gametes occur in equal frequency. Therefore, the frequency of $N\,A$ is also 0.4. The parental total is 0.8, leaving 0.2 for all recombinants. Therefore, $N\,O$ and $n\,A$ occur at a frequency of 0.1 each. The two genes are 20 m.u. apart.

14. The original cross was

P $\quad P\,L/P\,L \times p\,l/p\,l$

F_1 $\quad P\,L/p\,l \times P\,L/p\,l$

Before proceeding, recognize that crossing-over can occur in both parents and that some crossovers cannot be detected by phenotype. The gametes from each plant are as follows:

parental types: $\quad P\,L$, $p\,l$
recombinants: $\quad P\,l$, $p\,L$

The F_2 is as follows:

$P\,L/P\,L$ purple, long	$p\,l/p\,L$ red, long
$P\,L/p\,l$ purple, long	$p\,l/p\,l$ red, round
$P\,L/P\,l$ purple, long	$P\,l/p\,L$ purple, long
$P\,L/p\,L$ purple, long	$P\,l/P\,l$ purple, round
$P\,l/p\,l$ purple, round	$p\,L/p\,L$ red, long

Of these 10 different genotypes, only red, round can be identified unambiguously. It consists of two parental-type gametes and occurs at a frequency of 19.2 percent. The probability of such a genotype can be calculated by multiplying the probability of a gamete from the first parent times the probability of the same gamete from the second parent. Thus, the square root of 19.2 percent will yield the frequency of this one type of parental gamete, or roughly 44 percent. Because parental types occur with equal frequency, the parentals are 88 percent and the recombinants are 12 percent. Therefore, there are approximately 12 m.u. between the two genes.

15. P $a\ b\ c/a\ b\ c \times a^+\ b^+\ c^+/a^+\ b^+\ c^+$

F_1 $a^+\ b^+\ c^+/a\ b\ c \times a^+\ b^+\ c^+/a\ b\ c$

F_2		
	1364	$a^+\ b^+\ c^+$
	365	$a\ b\ c$
	87	$a\ b\ c^+$
	84	$a^+\ b^+\ c$
	47	$a\ b^+\ c^+$
	44	$a^+\ b\ c$
	5	$a\ b^+\ c$
	4	$a^+\ b\ c^+$

This problem is somewhat simplified by the fact that recombination does not occur in male *Drosophila*. Also, only progeny that received the *a b c* chromosome from the male will be distinguishable among the F_2 progeny.

a. Because you cannot distinguish between $a^+\ b^+\ c^+/a^+\ b^+\ c^+$ and $a^+\ b^+\ c^+/a\ b\ c$, use the frequency of $a\ b\ c/a\ b\ c$ to estimate the frequency of $a^+\ b^+\ c^+$ (parental) gametes from the female.

parentals	730	(2 × 365)
CO *a–b*:	91	($a\ b^+\ c^+$, $a^+\ b\ c$ = 47 + 44)
CO *b–c*:	171	($a\ b\ c^+$, $a^+\ b^+\ c$ = 87 + 84)
DCO:	9	($a^+\ b\ c^+$, $a\ b^+\ c$ = 4 + 5)
	1001	

a–b: 100%(91 + 9)/1001 = 10 m.u.

b–c: 100%(171 + 9)/1001 = 18 m.u.

b. Coefficient of coincidence = (observed DCO)/(expected DCO)

= 9/[(0.1)(0.18)(1001)] = 0.5

16. a. By comparing the two most frequent classes (parentals: $an\ br^+\ f^+$, $an^+\ br\ f$) to the least frequent classes (DCO: $an^+\ br\ f^+$, $an\ br^+\ f$), the gene order can be determined. The gene in the middle switches with respect to the other two (the order is *an f br*). Now the crosses can be written fully.

P $an\ f^+\ br^+/an\ f^+\ br^+ \times an^+\ f\ br/an^+\ f\ br$

F_1 $an^+\ f\ br/an\ f^+\ br^+ \times an\ f\ br/an\ f\ br$

F_2			
	355	$an\ f^+\ br^+/an\ f\ br$	parental
	339	$an^+\ f\ br/an\ f\ br$	parental
	88	$an^+\ f^+\ br^+/an\ f\ br$	CO *an–f*
	55	$an\ f\ br/an\ f\ br$	CO *an–f*
	21	$an^+\ f\ br^+/an\ f\ br$	CO *f–br*
	17	$an\ f^+\ br/an\ f\ br$	CO *f–br*
	2	$an^+\ f^+\ br/an\ f\ br$	DCO
	2	$an\ f\ br^+/an\ f\ br$	DCO
	879		

b. *an–f:* 100% (88 + 55 + 2 + 2)/879 = 16.72 m.u.

f–br: 100% (21 + 17 + 2 + 2)/879 = 4.78 m.u.

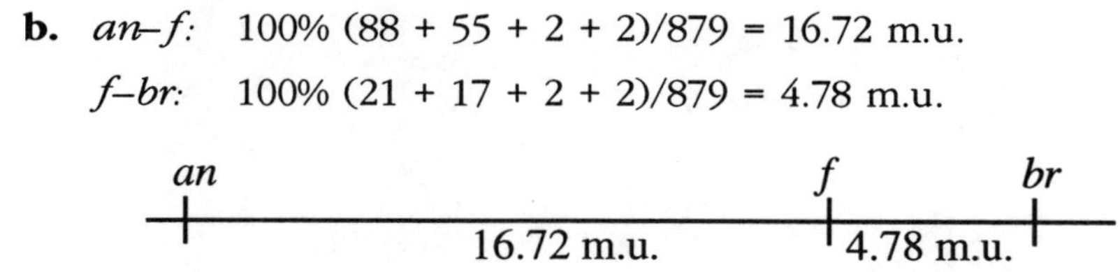

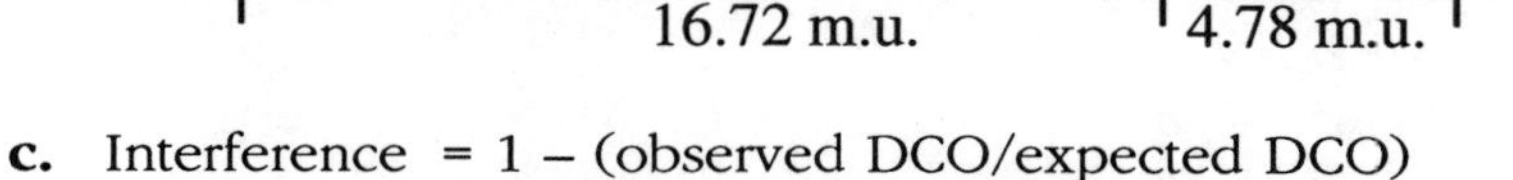

c. Interference = 1 – (observed DCO/expected DCO)
= 1 – 4/(0.1672)(0.0478)(879) = 0.431

17. By comparing the most frequent classes (parental: + *v lg*, *b* + +) with the least frequent classes (DCO: + + +, *b v lg*) the gene order can be determined. The gene in the middle switches with respect to the other two, yielding the following sequence: *v b lg*. Now the cross can be written:

P $v\ b^+\ lg/v^+\ b\ lg^+ \times v\ b\ lg/v\ b\ lg$

F_1	305	$v\ b^+\ lg/v\ b\ lg$	parental
	275	$v^+\ b\ lg^+/v\ b\ lg$	parental
	128	$v^+\ b\ lg/v\ b\ lg$	CO *b–lg*
	112	$v\ b^+\ lg^+/v\ b\ lg$	CO *b–lg*
	74	$v^+\ b^+\ lg/v\ b\ lg$	CO *v–b*
	66	$v\ b\ lg+/v\ b\ lg$	CO *v–b*
	22	$v+\ b+\ lg+/v\ b\ lg$	DCO
	18	$v\ b\ lg/v\ b\ lg$	DCO

v–b: 100%(74 + 66 + 22 + 18)/1000 = 18.0 m.u.

b–lg: 100%(128 + 112 + 22 + 18)/1000 = 28.0 m.u.

c.c. = observed DCO/expected DCO = (22 + 18)/(0.28)(0.18)(1000) = 0.79

18. Let *F* = fat, *L* = long tail, and *Fl* = flagella. The gene sequence is *F L Fl* (compare most frequent with least frequent). The cross is

P *F L Fl/f l fl* × *f l fl/f l fl*

F1	398	*F L Fl/f l fl*	parental
	370	*f l fl/f l fl*	parental
	72	*F L fl/f l fl*	CO *L–Fl*
	67	*f l Fl/f l fl*	CO *L–Fl*
	44	*f L FL/f l fl*	CO *F–L*
	35	*F l fl/f l fl*	CO *F–L*
	9	*f L fl/f l fl*	DCO
	5	*F l Fl/f l fl*	DCO

L–Fl: 100%(72 + 67 + 9 + 5)/1000 = 15.3 m.u.

F–L: 100%(44 + 35 + 9 + 5)/1000 = 9.3 m.u.

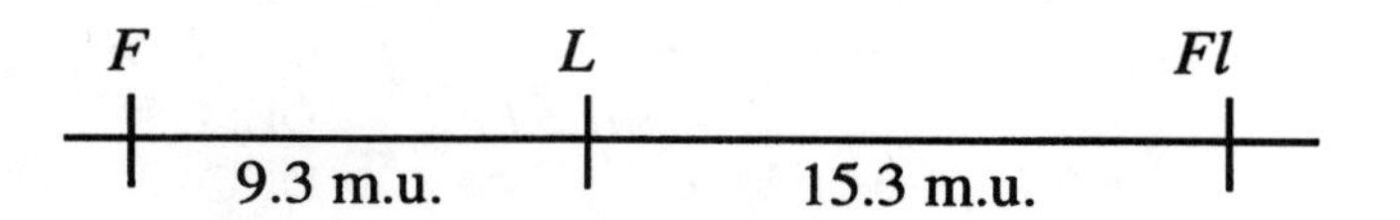

19. **a. and b.** The data indicate that the progeny males have a different phenotype than the females. Therefore, all the genes are on the X chromosome.

The two most frequent phenotypes in the males indicate the genotypes of the X chromosomes in the female, and the two least frequent phenotypes in the males indicate the gene order. Data from only the males are used to determine map distances. The cross is

P $\quad x\ z\ y^+/x^+\ z^+\ y \times x^+\ z^+\ y^+/Y$

F_1 males

430	$x\ z\ y^+/Y$	parental
441	$x^+\ z^+\ y/Y$	parental
39	$x\ z\ y/Y$	CO z–y
30	$x^+\ z^+\ y^+/Y$	CO z–y
32	$x^+\ z\ y^+/Y$	CO x–z
27	$x\ z^+\ y/Y$	CO x–z
1	$x^+\ z\ y/Y$	DCO
0	$x\ z^+\ y^+/Y$	DCO

c. z–y: 100%(39 + 30 + 1)/1000 = 7.0 m.u.

x–z: 100%(32 + 27 + 1)/1000 = 6.0 m.u.

c.c. = observed DCO/expected DCO

= 1/[(0.06)(0.07)(1000)] = 0.238

20. The data given for each of the three-point testcrosses can be used to determine the gene order by realizing that the rarest recombinant classes are the result of double cross-over events. By comparing these chromosomes with the "parental" types, the alleles that have switched represent the gene in the middle.

For example, in (1), the most common phenotypes (+ + + and *a b c*) represent the parental allele combinations. Comparing these to the rarest phenotypes of this data set (+ *b c* and *a* + +) indicates that the *a* gene is recombinant and must be in the middle. The gene order is *b a c*.

For (2), + *b c* and *a* + + (the parentals) should be compared with + + + and *a b c* (the rarest recombinants) to indicate that the *a* gene is in the middle. The gene order is *b a c*.

For (3), compare + *b* + and *a* + *c* with *a b* + and + + *c*, which gives the gene order *b a c*.

For (4), compare + + *c* and *a b* + with + + + and *a b c*, which gives the gene order *a c b*.

For (5), compare + + + and *a b c* with + + *c* and *a b* +, which gives the gene order *a c b*.

21. The gene order is *a c b d*.

Recombination between *a* and *c* occurred at a frequency of

100%(139 + 3 + 121 + 2)/(669 + 139 + 3 + 121 + 2 + 2,280 + 653 + 2,215)

=100%(265/6,082) = 4.36%.

Recombination between *b* and *c* in cross 1 occurred at a frequency of

100%(669 + 3 + 2 + 653)/(669 + 139 + 3 + 121 + 2 + 2,280 + 653 + 2,215)

=100%(1,327/6,082) = 21.82%.

Recombination between b and c in cross 2 occurred at a frequency of

100%(8 + 14 + 153 + 141)/(8 + 441 + 90 + 376 + 14 + 153 + 64 + 141)

=100%(316/1,287) = 24.55%.

The difference between the two calculated distances between b and c is not surprising because each set of data would not be expected to yield exactly identical results. Also, many more offspring were analyzed in cross 1. Combined, the distance would be

100%[(316 + 1,327)/(1,287 + 6,082)] = 22.3%

Recombination between b and d occurred at a frequency of

100%(8 + 90 + 14 + 64)/(8 + 441 + 90 + 376 + 14 + 153 + 64 + 141)

=100%(176/1,287) = 13.68%.

The general map is

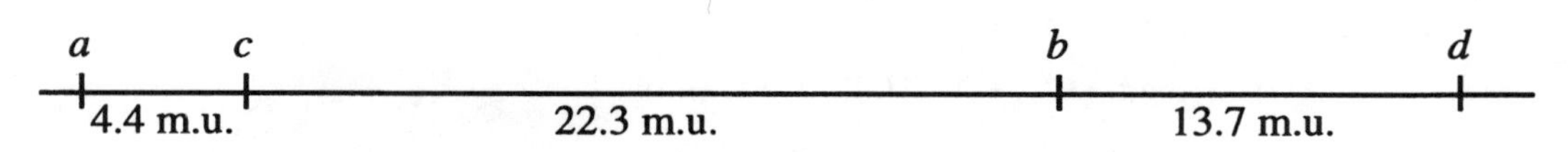

22. **a.** The hypothesis is that the genes are not linked. Therefore, a 1:1:1:1 ratio is expected.

b. $\chi^2 = (54-50)^2/50 + (47-50)^2/50 + (52-50)^2/50 + (47-50)^2/50$

$= 0.32 + 0.18 + 0.08 + 0.18 = 0.76$

c. With 3 degrees of freedom, the p value is between 0.50 and 0.90.

d. Between 50% and 90% of the time values this extreme from the prediction would be obtained by chance alone.

e. Accept the initial hypothesis.

f. Because the χ^2 value was insignificant, the two genes are assorting independently. The genotypes of all individuals are

P	dp^+/dp^+ ; $e/e \times dp/dp$; e^+/e^+	
F_1	dp^+/dp ; e^+/e	
tester	dp/dp ; e/e	
progeny	long, ebony	dp^+/dp ; e/e
	long, gray	dp^+/dp ; e^+/e
	short, gray	dp/dp ; e^+/e
	short, ebony	dp/dp ; e/e

23. The cross was $asp \cdot gal \cdot rad^+ \cdot aro^+ \times asp^+ \cdot gal^+ \cdot rad \cdot aro$.

a. The first task is to decide if there is any linkage between the four genes. Arrange the results according to frequencies:

0.136	0.064	0.034	0.016
asp gal + +	*asp gal* + *rad*	*asp gal aro* +	*asp gal aro rad*
+ + *aro rad*	*asp* + + +	*asp* + *aro rad*	*asp* + *aro* +
asp + + *rad*	+ *gal aro rad*	+ *gal* + +	+ *gal* + *rad*
+ *gal aro* +	+ + *aro* +	+ + + *rad*	+ + + +

If all four genes were independently assorting, then all the classes would occur at an equal frequency. This is not observed. Therefore, there is some linkage between these genes.

The two parental classes are present in the highest frequency, as are two other classes: *asp* + + *rad* and + *gal aro* +. Note that *asp* and *gal* are assorting independently of each other, which means that they are unlinked. Also note that *aro* and *rad* are assorting independently of each other and are, therefore, unlinked. Finally, note that *gal* and *rad* are not assorting independently and *asp* and *aro* are not assorting independently. There are two linkage groups, and the original cross was

$$gal\ rad^{+}\ ;\ asp\ aro^{+} \times gal^{+}\ rad\ ;\ asp^{+}\ aro.$$

Reclassify the data with regard to the *gal–rad* linkage.

Parentals		Recombinants	
0.136	*asp gal* + +	0.064	*asp gal* + *rad*
0.136	+ + *aro rad*	0.064	*asp* + + +
0.136	*asp* + + *rad*	0.064	+ *gal aro rad*
0.136	+ *gal aro* +	0.064	+ + *aro* +
0.034	*asp gal aro* +	0.016	*asp gal aro rad*
0.034	*asp* + *aro rad*	0.016	*asp* + *aro* +
0.034	+ *gal* + +	0.016	+ *gal* + *rad*
0.034	+ + + *rad*	0.016	+ + + +
0.680		0.320	

The frequency of recombination between *gal* and *rad* is 32%. Reclassify the data with regard to the *aro–asp* linkage.

Parentals		Recombinants	
0.136	*asp gal* + +	0.034	*asp gal aro* +
0.136	+ + *aro rad*	0.034	*asp* + *aro rad*
0.136	*asp* + + *rad*	0.034	+ *gal* + +
0.136	+ *gal aro* +	0.034	+ + + *rad*
0.064	*asp gal* + *rad*	0.016	*asp gal aro rad*
0.064	*asp* + + +	0.016	*asp* + *aro* +
0.064	+ *gal aro rad*	0.016	+ *gal* + *rad*
0.064	+ + *aro* +	0.016	+ + + +
0.80		0.200	

The frequency of recombination between *aro* and *asp* is 20%.

b. The map of the four genes is

24. a. and b. The wild type is dominant for both traits. First, notice that the F_2 offspring differ with regard to sex. At least one of the genes is X-linked. Second, notice that the two genes are not assorting independently. This means that both genes are on the X chromosome and are 10 m.u. apart. The cross was

P $\quad$ *G A/G A* × *g a*/Y

F_1 $\quad$ *G A/g a* females and *G A*/Y males

F_2 $\quad$ females $\quad$ males

females		males	
45% *G A/G A*	45%	*G A*/Y	parental
45% *g a/G A*	45%	*g a*/Y	parental
5% *g A/G A*	5%	*g A*/Y	recombinant
5% *G a/G A*	5%	*G a*/Y	recombinant

25. a.

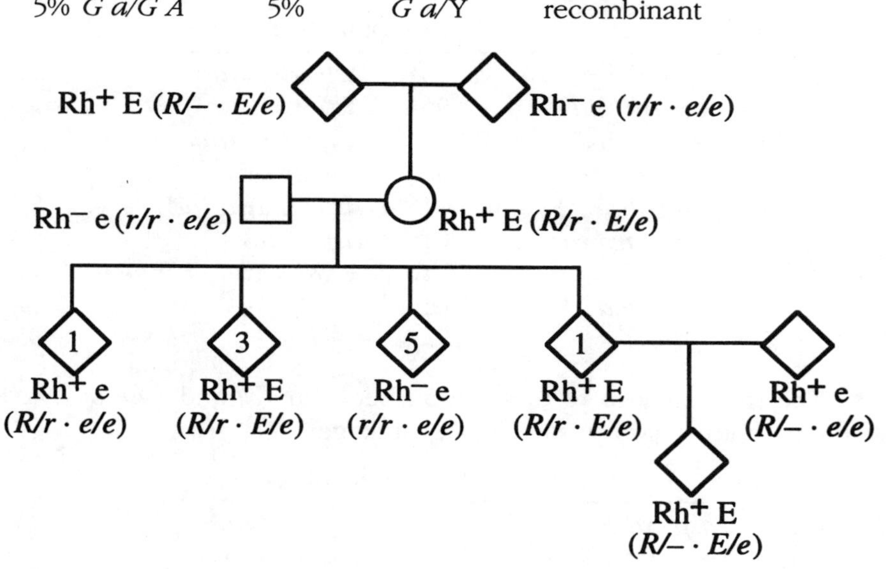

b. Yes.

c. Dominant.

d. As drawn, the pedigree hints at linkage. If unlinked, expect that the phenotypes of the 10 children should be in a 1:1:1:1 ratio of Rh^+ E, Rh^+ e, Rh^- E, and Rh^- e. There are actually 5 Rh^- e, 4 Rh^+ E, and 1 Rh^+ e. If linked, this last phenotype would represent a recombinant and the distance between the two genes would be 100%(1/10) = 10 m.u. However, there is just not enough data to strongly support that conclusion.

26. Part (a) of this problem is solved two ways, once in the standard way, once in a way that emphasizes a more mathematical approach.

The cross is

P $\quad P\ A\ R/P\ A\ R \times p\ a\ r/p\ a\ r$

F_1 $\quad P\ A\ R/p\ a\ r \times p\ a\ r/p\ a\ r$, a three-point testcross

a. In order to find what proportion will have the Vulcan phenotype for all three characteristics, we must determine the frequency of parentals. Crossing-over occurs 15% of the time between P and A, which means it does not occur 85% of the time. Crossing-over occurs 20% of the time between A and R, which means that it does not occur 80% of the time.

p(no crossover between either gene)

= *p*(no crossover between *P* and *A*) × *p*(no crossover between A and *R*)

= (0.85)(0.80) = 0.68

Half the parentals are Vulcan, so the proportion that are completely Vulcan is $^1/_2$(0.68) = 0.34.

Mathematical method

Number of parentals

= 1 – (single CO individuals – DCO individuals)
= 1 – {[0.15 + 0.20 – 2(0.15)(0.20)] – (0.15) (0.20)} = 0.68

Because half the parentals are Earth alleles and half are Vulcan, the frequency of children with all three Vulcan characteristics is $^1/_2$(0.68) = 0.34.

b. Same as above, 0.34.

c. To yield Vulcan ears and hearts and Earth adrenals, a crossover must occur in both regions, producing double crossovers. The frequency of Vulcan ears and hearts and Earth adrenals will be half the DCOs, or $^1/_2$(0.15)(0.20) = 0.015.

d. To yield Vulcan ears and an Earth heart and adrenals, a single crossover must occur between *P* and *A*, and no crossover can occur between *A* and *R*. The frequency will be

$$p(\text{CO } P\text{–}A) \times p(\text{no CO } A\text{–}R) = (0.15)(0.80) = 0.12$$

Of these, $^1/_2$ are *P a r* and 1/2 are *p A R*. Therefore, the proportion with Vulcan cars and an Earth heart and adrenals is 0.06.

27. a. For a plant to be *a b c/a b c* from selfing of *A b c/a B C*, both gametes must be derived from a crossover between *A* and *B*. The frequency of the *a b c* gamete is

$^1/_2\ p(\text{CO } A\text{–}B) \times p(\text{no CO } B\text{–}C) = {}^1/_2(0.20)(0.70) = 0.07$

Therefore, the frequency of the homozygous plant will be $(0.07)^2 = 0.0049$.

b. The cross is $A\ b\ c/a\ B\ C \times a\ b\ c/a\ b\ c$.

To calculate the progeny frequencies, note that the parentals are equal to all those that did not experience a crossover. Mathematically this can be stated as

$$\text{parentals} = p(\text{no CO } A\text{–}B) \times p(\text{no CO } B\text{–}C) = (0.80)(0.70) = 0.56$$

Since each parental should be represented equally,

$$A\ b\ c = {}^{1}/_{2}(0.56) = 0.28$$

$$a\ B\ C = {}^{1}/_{2}(0.56) = 0.28$$

As calculated above, the frequency of the $a\ b\ c$ gamete is

$${}^{1}/_{2}\ p(\text{CO } A\text{–}B) \times p(\text{no CO } B\text{–}C) = {}^{1}/_{2}(0.20)(0.70) = 0.07$$

as is the frequency of $A\ B\ C$.

The frequency of the $A\ b\ C$ gamete is

$${}^{1}/_{2}\ p(\text{CO } B\text{–}C) \times p(\text{no CO } A\text{–}B) = {}^{1}/_{2}(0.30)(0.80) = 0.12$$

as is the frequency of $a\ B\ c$.

Finally, the frequency of the $A\ B\ c$ gamete is

$${}^{1}/_{2}\ p(\text{CO } A\text{–}B) \times p(\text{CO } B\text{–}C) = {}^{1}/_{2}(0.20)(0.30) = 0.03$$

as is the frequency of $a\ b\ C$.

So for 1,000 progeny, the expected results are

$A\ b\ c$	280
$a\ B\ C$	280
$A\ B\ C$	70
$a\ b\ c$	70
$A\ b\ C$	120
$a\ B\ c$	120
$A\ B\ c$	30
$a\ b\ C$	30

c. Interference = 1 – observed DCO/expected DCO

$$0.2 = 1 - \text{observed DCO}/(0.20)(0.30)$$

$$\text{observed DCO} = (0.20)(0.30) - (0.20)(0.20)(0.30) = 0.048$$

The A–B distance = 20% = 100% [p(CO A–B) + p(DCO)]. Therefore, p(CO A–B) = 0.20 – 0.048 = 0.152

Similarly, the B–C distance = 30% = 100% [p(CO B–C) + p(DCO)]. Therefore, p(CO B–C) = 0.30 – 0.048 = 0.252

The p(parental) = 1 – p(CO A–B) – p(CO B–C) – p(observed DCO) = 1 – 0.152 – 0.252 – 0.048 = 0.548.

So for 1,000 progeny, the expected results are

A b c	274
a B C	274
A B C	76
a b c	76
A b C	126
a B c	126
A B c	24
a b C	24

28. Assume there is no linkage. (This is your hypothesis. If it can be rejected, the genes are linked.) The expected values would be that genotypes occur with equal frequency. There are four genotypes in each case ($n = 4$) so there are 3 degrees of freedom.

$$\chi^2 = \Sigma\ (\text{observed-expected})^2/\text{expected}$$

Cross 1: $\chi^2 = [(310–300)^2 + (315–300)^2 + (287–300)^2 + (288–300)^2]/300$
$= 2.1266$; $p > 0.50$, nonsignificant; hypothesis cannot be rejected

Cross 2: $\chi^2 = [(36–30)^2 + (38–30)^2 + (23–30)^2 + (23–30)^2]/300$
$= 6.6$; $p > 0.10$, nonsignificant; hypothesis cannot be rejected

Cross 3: $\chi^2 = [(360–300)^2 + (380–300)^2 + (230–300)^2 + (230–300)^2]/300$
$= 66.0$; $p < 0.005$, significant; hypothesis must be rejected

Cross 4: $\chi^2 = [(74–60)^2 + (72–60)^2 + (50–60)^2 + (44–60)^2]/300$
$= 11.60$; $p < 0.01$, significant; hypothesis must be rejected

29. The data approximate a 9:3:3:1 ratio, which suggests two unlinked genes. Let A = resistance to rust 24, a = susceptibility to rust 24, B resistance to rust 22, b = susceptibility to rust 22.

a. P *A/A ; b/b* (770B) × *a/a ; B/B* (Bombay)

F_1 *A/a ; B/b* × *A/a ; B/b*

F_2

184	*A/– ; B/–*
63	*A/– ; b/b*
58	*a/a ; B/–*
15	*a/a ; b/b*
320	

b. Expect:

$(320)(^9/_{16}) = 180$	*A/– ; B/–*
$(320)(^9/_{16}) = 60$	*A/– ; b/b*
$(320)(^9/_{16}) = 60$	*a/a ; B/–*
$(320)(^9/_{16}) = 20$	*a/a ; b/b*

$$\chi^2 = (184 - 180)^2/180 + (63 - 60)^2/60 + (58 - 60)^2/60 + (15 - 50)^2/20 = 1.55$$

P(3 df) > 0.58, nonsignificant. A P value is the probability that the result could be observed by chance alone. Therefore, the hypothesis of two independently assorting genes cannot be rejected.

30. a. Both disorders must be recessive to yield the patterns of inheritance that are observed. Notice that only males are affected, strongly suggesting X linkage for both disorders. In the first pedigree there is a 100% correlation between the presence or absence of both disorders, indicating close linkage. In the second pedigree, the presence and absence of both disorders are inversely correlated, again indicating linkage. In the first pedigree, the two defective alleles must be cis within the heterozygous females to show 100% linkage in the affected males, while in the second pedigree the two defective alleles must be trans within the heterozygous females.

b. and c. Let *a* stand for the allele giving rise to steroid sulfatase deficiency (vertical bar) and *b* stand for the allele giving rise to ornithine transcarbamylase deficiency (horizontal bar). Crossing-over cannot be detected without attaching genotypes to the pedigrees. When this is done, it can be seen that crossing-over need not occur in either of the pedigrees to give rise to the observations.

First pedigree:

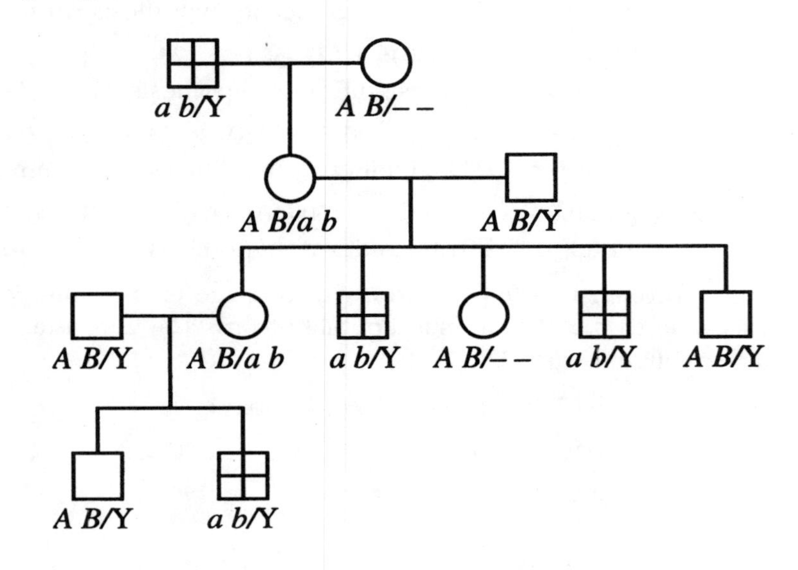

Second pedigree:

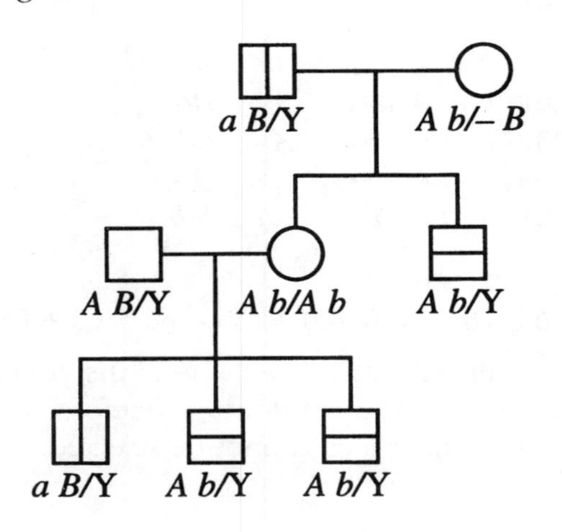

31. a. Blue sclerotic (*B*) appears to be an autosomal dominant disorder. Hemophilia (*h*) appears to be an X-linked recessive disorder.

b. If the individuals in the pedigree are numbered as generations I through IV and the individuals in each generation are numbered clockwise, starting from the top-right-hand portion of the pedigree, their genotypes are

I: *b/b* ; *H/h*, *B/b* ; *H/*Y

II: *B/b* ; *H/*Y, *B/b* ; *H/*Y, *b/b* ; *H/*Y, *B/b* ; *H/h*, *b/b* ; *H/*Y, *B/b* ; *H/h*, *B/b* ; *H/h*, *B/b* ; *H/–*, *b/b* ; *H/–*

III: *b/b* ; *H/–*, *B/b* ; *H/–*, *b/b* ; *h/*Y, *b/b* ; *H/*Y, *B/b* ; *H/*Y, *B/b* ; *H/–*, *B/b* ; *H/*Y, *B/b* ; *h/*Y, *B/b* ; *H/–*, *b/b* ; *H/*Y, *B/b* ; *H/–*, *b/b* ; *H/*Y, *B/b* ; *H/–*, *B/b* ; *H/*Y, *B/b* ; *h/*Y, *b/b* ; *H/*Y, *b/b* ; *H/*Y, *b/b* ; *H/–*, *b/b* ; *H/*Y, *b/b* ; *H/*Y, *B/b* ; *H/–*, *B/b* ; *H/*Y, *B/b* ; *h/*Y

IV: *b/b* ; *H/–*, *B/b* ; *H/–*, *B/b* ; *H/–*, *b/b* ; *H/h*, *b/b* ; *H/h*, *b/b* ; *H/*Y, *b/b* ; *H/H*, *b/b* ; *H/*Y, *b/b* ; *H/h*, *b/b* ; *H/H*, *b/b* ; *H/H*, *b/b* ; *H/*Y, *b/b* ; *H/*Y, *b/b* ; *H/H*, *b/b* ; *H/*Y, *b/b* ; *H/*Y, *B/b* ; *H/*Y, *b/b* ; *H/*Y, *b/b* ; *H/–*, *b/b* ; *H/*Y, *b/b* ; *H/*Y, *b/b* ; *H/–*, *b/b* ; *H/H*, *b/b* ; *H/–*, *b/b* ; *H/–*, *b/b* ; *H/*Y, *b/b* ; *H/*Y, *b/b* ; *H/*Y, *b/b* ; *H/h*, *B/b* ; *H/h*, *B/b* ; *H/*Y, *b/b* ; *H/*Y, *B/b* ; *H/*Y, *b/b* ; *H/h*

c. There is no evidence of linkage between these two disorders. Because of the modes of inheritance for these two genes, no linkage would be expected.

d. The two genes exhibit independent assortment.

e. No individual could be considered intrachromosomally recombinant. However, a number show interchromosomal recombination, for example, all individuals in generation III that have both disorders.

32. a. Note that only males are affected by both disorders. This suggests that both are X-linked recessive disorders. Using *p* for protan and *P* for non-protan and *d* for deutan and *D* for non-deutan, the inferred genotypes are listed on the pedigree below. The Y chromosome is shown, but the X is represented by the alleles carried.

b. Individual II-2 must have inherited both disorders in the trans configuration (on separate chromosomes). Therefore, individual III-2 inherited both traits as the result of recombination (crossing-over) between his mother's X chromosomes.

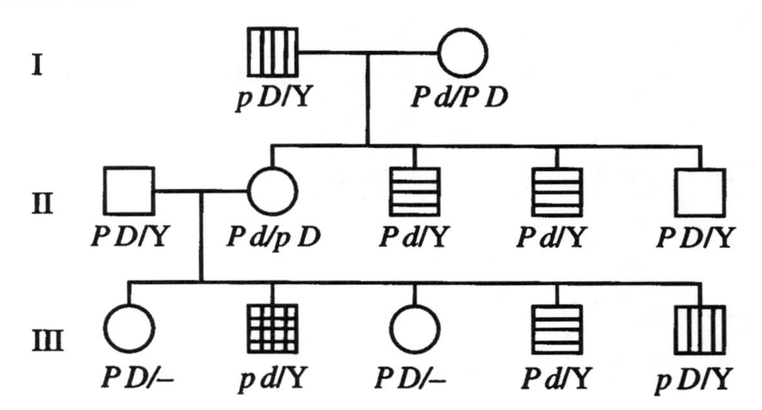

c. Because both genes are X-linked, this represents crossing-over. The progeny size is too small to give a reliable estimate of recombination.

33. If *h* = hemophilia and *b* = colorblindness, the genotypes for individuals in the pedigree can be written as

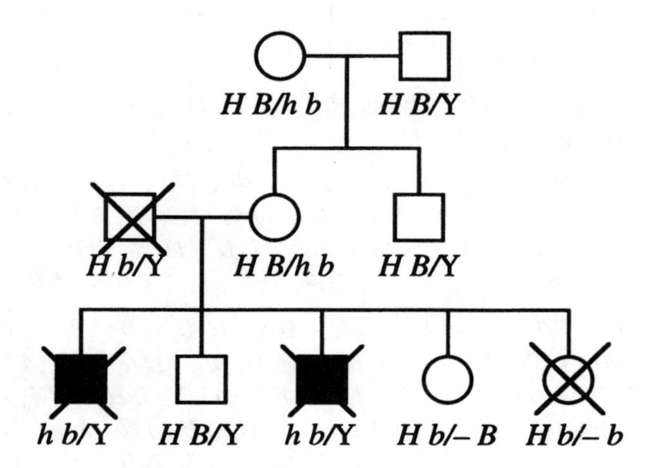

The mother of the two women in question would produce the following gametes:

0.45	*H B*
0.45	*h b*
0.05	*h B*
0.05	*H b*

Woman III-4 can be either *H b/H B* (0.45 chance) or *H b/h B* (0.05 chance), because she received *B* from her mother. If she is *H b/h B* [0.05/(0.45 + 0.05) = 0.10 chance], she will produce the parental and recombinant gametes with the same probabilities as her mother. Thus, her child has a 45 percent chance of receiving *h B*, a 5 percent chance of receiving *h b*, and a 50 percent chance of receiving a Y from his father. The probability that her child will be a hemophiliac son is (0.1)(0.5)(0.5) = 0.025 = 2.5 percent.

Woman III-5 can be either *H b/H b* (0.05 chance) or *H b/h b* (0.45 chance), because she received *b* from her mother. If she is *H b/h b* [0.45/(0.45 + 0.05) = 0.90 chance], she has a 50 percent chance of passing h to her child, and there is a 50 percent chance that the child will be male. The probability that she will have a son with hemophilia is (0.9)(0.5)(0.5) = 0.225 = 22.5 percent.

34. a. Cross 1 reduces to

P	$A/A \cdot B/B \cdot D/D \times a/a \cdot b/b \cdot d/d$
F_1	$A/a \cdot B/b \cdot D/d \times a/a \cdot b/b \cdot d/d$

The testcross progeny indicate these three genes are linked.

testcross	*A B D*	316	parental
progeny	*a b d*	314	parental
	A B d	31	CO *B–D*
	a b D	39	CO *B–D*
	A b d	130	CO *A–B*
	a B D	140	CO *A–B*
	A b D	17	DCO
	a B d	13	DCO

A–B: 100%(130 + 140 + 17 + 13)/1000 = 30 m.u.

B–D: 100%(31 + 39 + 17 + 13)/1000 = 10 m.u.

Cross 2 reduces to

P $A/A \cdot C/C \cdot E/E \times a/a \cdot c/c \cdot e/e$

F_1 $A/a \cdot C/c \cdot E/e \times a/a \cdot c/c \cdot e/e$

The testcross progeny indicate these three genes are linked.

Testcross	*A C E*	243	parental
progeny	*a c e*	237	parental
	A c e	62	CO *A–C*
	a C E	58	CO *A–C*
	A C e	155	CO *C–E*
	a c E	165	CO *C–E*
	a C e	46	DCO
	A c E	34	DCO

A–C: 100% (62 + 58 + 46 + 34)/1000 = 20 m.u.

C–E: 100% (155 + 165 + 46 + 34)/1000 = 40 m.u.

The map that accommodates all the data is

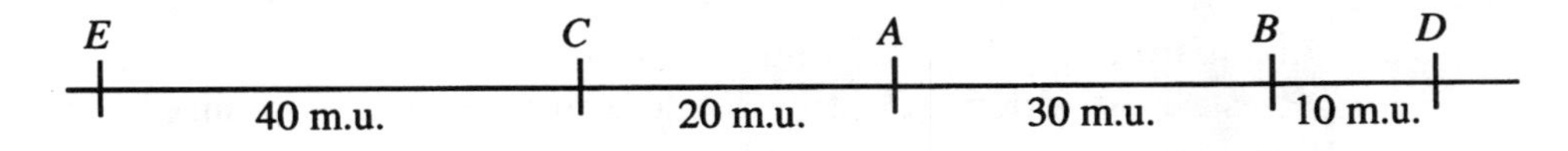

b. Interference (I) = 1 – [(observed DCO)/(expected DCO)]

For cross 1: I = 1 – {30/[(0.30)(0.10)(1000)]} = 1 – 1 = 0, no interference.

For cross 2: I = 1 – {80/[(0.20)(0.40)(1000)]} = 1 – 1 = 0, no interference.

35. a–c. The cross is $G/G \cdot y/y \times g/g \cdot Y/Y$. If the two are unlinked, a 9:3:3:1 ratio should be observed in the progeny. The proper way to decide on linkage is to conduct a chi-square analysis of the data. The chi-square value indicates that it is highly unlikely that the two genes are unlinked.

In the absence of chi-square analysis, the best way to decide whether the data indicate linkage is to realize that a 9:3:3:1 ratio would predict 562.5:187.5:187.5:62.5 and that the data are far from the predicted ratio. Because the observations do not match that expected for independent assortment, linkage is a viable alternative.

Assuming linkage, the cross becomes

P $\quad$ *G y/G y* × *g Y/g Y*

F_1 $\quad$ *G y/g Y* × *G y/g Y*

F_2 data as in problem

In this situation, it is necessary to realize that the phenotype yellow, glabrous represents the double homozygous recessive. To obtain a double homozygous recessive, both the male and female gametic contribution must be recombinant. In other words, the frequency of

$$(1/2 \text{ recombinants})^2 \times 1000 = 15$$

Let $1/2$ recombinants = x, and divide each side of the equation by 1000. Then the equation becomes

$$x^2 = 0.015$$

$$x = 0.122$$

If $1/2$ the recombinants is 0.122, then the recombinants total 0.244, or 24.4% of the progeny. The map distance between the two genes is 24.4 m.u.

36. a. The first F_1 is *L H/l h* and the second is *l H/L h*. For progeny that are *l h/l h*, they have received a "parental" chromosome from the first F_1 and a "recombinant" chromosome from the second F_1. The genes are 16% apart, so the chance of a parental chromosome is $1/2(100 - 16\%) = 42\%$ and the chance of a recombinant chromosome is $1/2(16\%) = 8\%$.

The chance of both events = 42% × 8% = 3.36%

b. To obtain *Lh/l h* progeny, either a parental chromosome from each parent was inherited *or* a recombinant chromosome from each parent was inherited. The total probability will therefore be

(42% × 42%) + (8% × 8%) = (17.6% + 0.6%) = 18.2%

37. a. and b. Again, the best way to determine whether there is linkage is through chi-square analysis, which indicates that it is highly unlikely that the three genes assort independently. To determine linkage by simple inspection, look at gene pairs. Because this is a testcross, independent assortment predicts a 1:1:1:1 ratio.

Comparing shrunken and white, the frequencies are

+ +	(113 + 4)/total
s wh	(116 + 2)/total
+ *wh*	(2708 + 626)/total
s +	(2538 + 601)/total

There is not independent assortment between shrunken and white, which means that there is linkage.

Comparing shrunken and waxy, the frequencies are

+ +	(626 + 4)/total
s wa	(601 + 2)/total
+ *wa*	(2708 + 113)/total
s +	(2538 + 1160/total

There is not independent assortment between shrunken and waxy, which means that there is linkage.

Comparing white and waxy, the frequencies are

+	+	(2538 + 4)/total
wh	*wa*	(2708 + 2)/total
wh	+	(626 + 116)/total
+	*wa*	(601 + 113)/total

There is not independent assortment between waxy and white, which means that there is linkage.

Because all three genes are linked, the strains must be + *s* +/*wh* + *wa* and *wh s wa*/*wh s wa* (compare most frequent, parentals, to least frequent, double crossovers, to obtain the gene order). The cross can be written as

P + *s* +/*wh* + *wa* × *wh s wa*/*wh s wa*

F_1 as in problem

Crossovers between white and shrunken and shrunken and waxy are

113	601
116	626
4	4
2	2
235	1233

Dividing by the total number of progeny and multiplying by 100% yields the following map:

white — 3.5 m.u. — **shrunken** — 18.4 m.u. — **waxy**

c. Interference = 1 − (observed double crossovers/expected double crossovers)

= 1 − 6/(0.035)(0.184)(6,708) = 0.86

38. The cross is *A b*/*a B* × *A b*/*a B*. The *A* and *B* genes are 10 m.u. apart. Therefore, the gametes should occur in the following frequencies:

0.45	*A b*
0.45	*a B*
0.05	*a b*
0.05	*A B*

Construct a Punnett square, and sum the values for each progeny phenotype. These values are

0.5025	*A*/– · *B*/–
0.0025	*a b*/*a b*
0.2475	*A b*/– *b*
0.2475	*a B*/*a* –

The z value is calculated using the following formula:

$$z = (A/- \cdot B/-)(a\ b/a\ b)/(A\ b/-\ b)(a\ B/a\ -)$$

Substituting the values from the Punnett square, this becomes

$$z = (0.5025)(0.0025)/(0.2475)\ (0.2475) = 0.020$$

Using the table of z values, 0.020 is equivalent to 9.9 m.u.

39. a. The results of this cross indicate independent assortment of the two genes. This might be diagrammed as

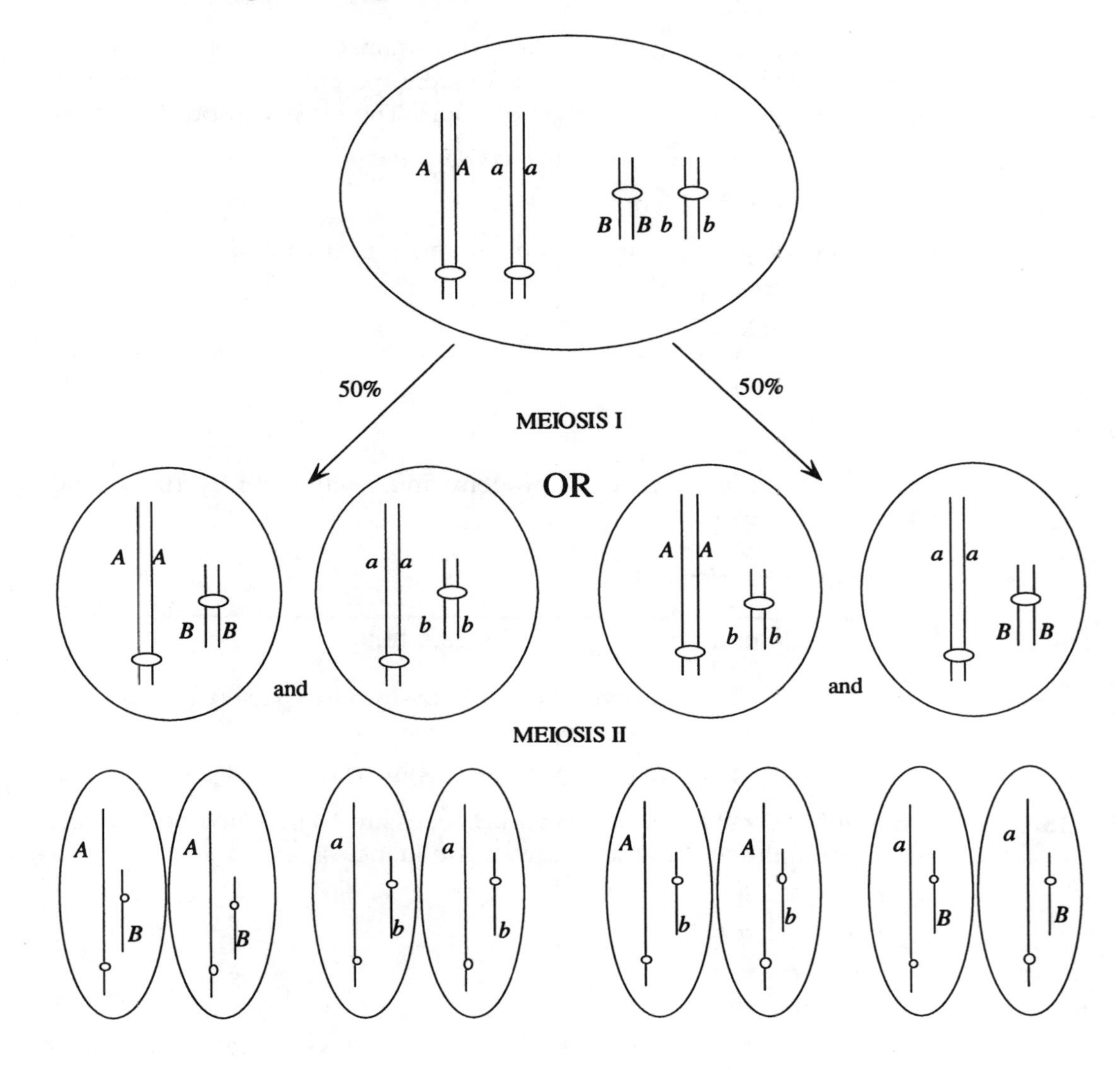

b. The results of this cross indicate that the two genes are linked and 10 m.u. apart. Further, the recessive alleles are in repulsion in the dihybrid (*C d/c D* × *c d/c d*). This might be diagrammed as

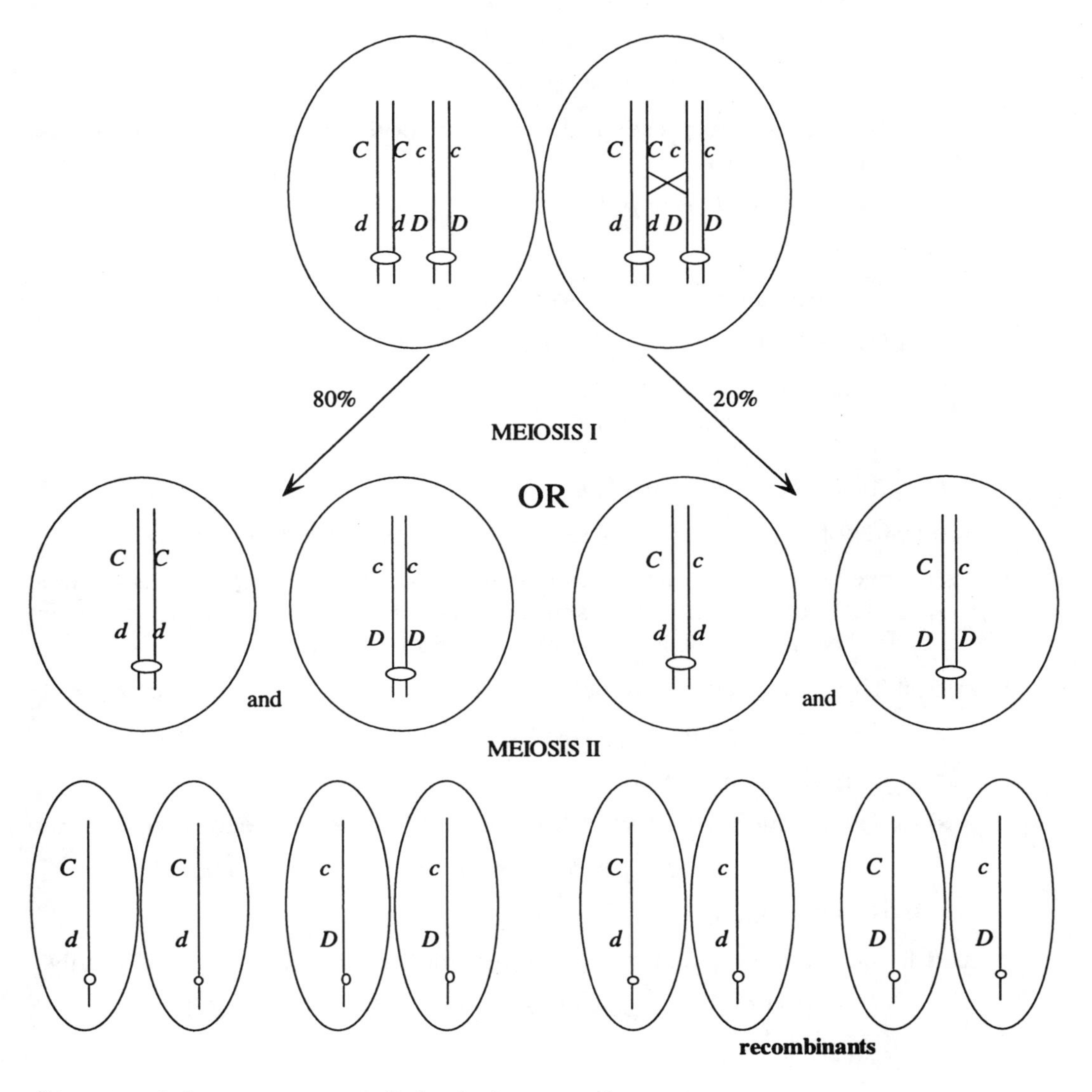

40. a. If the genes are unlinked, the cross becomes

P $\quad$ *hyg/hyg* ; *her/her* × hyg^+/hyg^+ ; her^+/her^+

F_1 $\quad$ hyg^+/hyg ; her^+/her × hyg^+/hyg ; her^+/her

F_2 $\quad$ $^9/_{16}$ $\quad$ $hyg^+/–$; $her^+/–$

$^3/_{16}$ $\quad$ $hyg^+/–$; *her/her*

$^3/_{16}$ $\quad$ *hyg/hyg* ; $her^+/–$

$^1/_{16}$ $\quad$ *hyg/hyg* ; *her/her*

So only $^1/_{16}$ (or 6.25%) of the seeds would be expected to germinate.

b. and c. No. More than twice the expected seeds germinated, so assume the genes are linked. The cross then becomes

P $\quad$ *hyg her/hyg her* × *hyg*$^+$ *her*$^+$/*hyg*$^+$ *her*$^+$

F_1 $\quad$ *hyg*$^+$ *her*$^+$/*hyg her* × *hyg*$^+$ *her*$^+$/*hyg her*

F_2 $\quad$ 13% *hyg her/hyg her*

Because this class represents the combination of two parental chromosomes, it is equal to

$$p(hyg\ her) \times p(hyg\ her) = (^1/_2 \text{ parentals})^2 = 0.13$$

and

$$\text{parentals} = 0.72 \qquad \text{so recombinants} = 1 - 0.72 = 0.28$$

Therefore, a testcross of *hyg*$^+$ *her*$^+$/*hyg her* should give

36% *hyg*$^+$ *her*$^+$/*hyg her*
36% *hyg her/hyg her*
14% *hyg*$^+$ *her/hyg her*
14% *hyg her*$^+$/*hyg her*

and 36% of the progeny should grow (the *hyg her/hyg her* class).

41. Crossing-over occurs 8% of the time between *w* and *s*, which means it does not occur 92% of the time. Crossing-over occurs 14% of the time between *s* and *e*, which means that it does not occur 86% of the time.

a. and b. The frequency of parentals = p(no crossover between either gene)

$$= p(\text{no CO } w\text{–}s) \times p(\text{no CO } s\text{–}e) = (0.92)(0.86) = 0.791,$$

or $^1/_2(0.791) = 0.396$ each

c. and d. The frequency that will show recombination between *w* and *s* only

$$= p(\text{CO } w\text{–}s) \times p(\text{no CO } s\text{–}e) = (0.08)(0.86) = 0.069,$$

or $^1/_2(0.069) = 0.035$ each

e. and f. The frequency that will show recombination between *s* and *e* only

$$= p(\text{CO } s\text{–}e) \times p(\text{no CO } w\text{–}s) = (0.14)(0.92) = 0.128,$$

or $^1/_2(0.128) = 0.064$ each

g. and h. The frequency that will show recombination between *w* and *s* and *s* and *e*

$$= p(\text{CO } w\text{–}s) \times p(\text{CO } s\text{–}e) = (0.08)(0.14) = 0.011,$$

or $^1/_2(0.011) = 0.006$ each

42. The cross is

	B L/b l	×	*B l/b L*
gametes	0.40 *B L*		0.40 *B l*
	0.40 *b l*		0.40 *b L*
	0.10 *B l*		0.10 *B L*
	0.10 *b L*		0.10 *b l*

By filling in the appropriate Punnett square, it can be determined that among the progeny

Frequency	Phenotype
0.54	B L
0.21	B l
0.21	b L
0.04	b l

43. (1), (2), and (3). Because all eight allelic combinations are equally likely in the gametes, it can be inferred that the three genes are on separate chromosomes. The following figures use chromosome size and centromere placement to distinguish the three nonhomologous chromosomes:

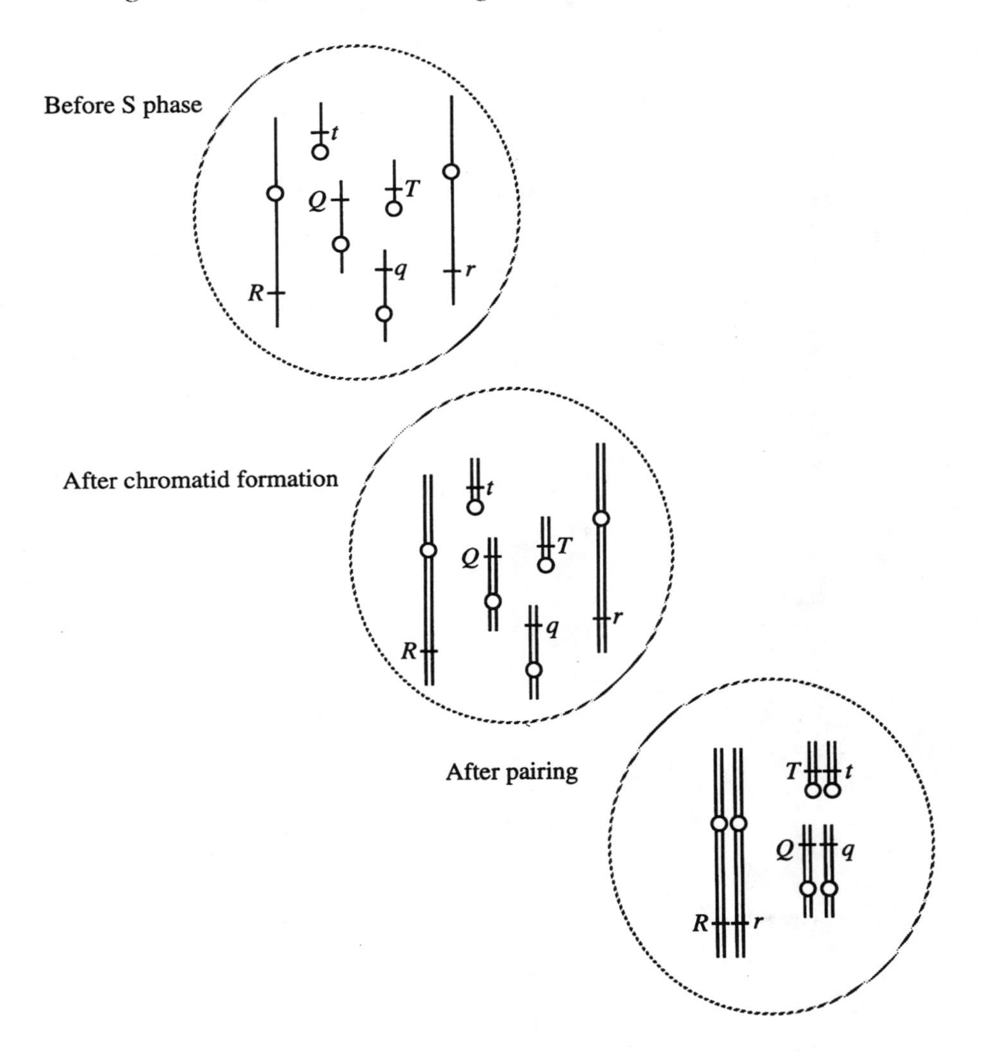

43. (4) and (5). Independent assortment of nonhomologous chromosomes will give 8 possible allelic combinations. This is graphically represented by showing the various alignments the separate chromosomes may take during meiosis. The segregation of homologous chromosomes during anaphase 1 and then the splitting of sister chromatids during anaphase II is schematically indicated. Although it is very likely that crossing-over will take place during prophase I, it will not affect the genotypic ratios of the gametes and thus is ignored.

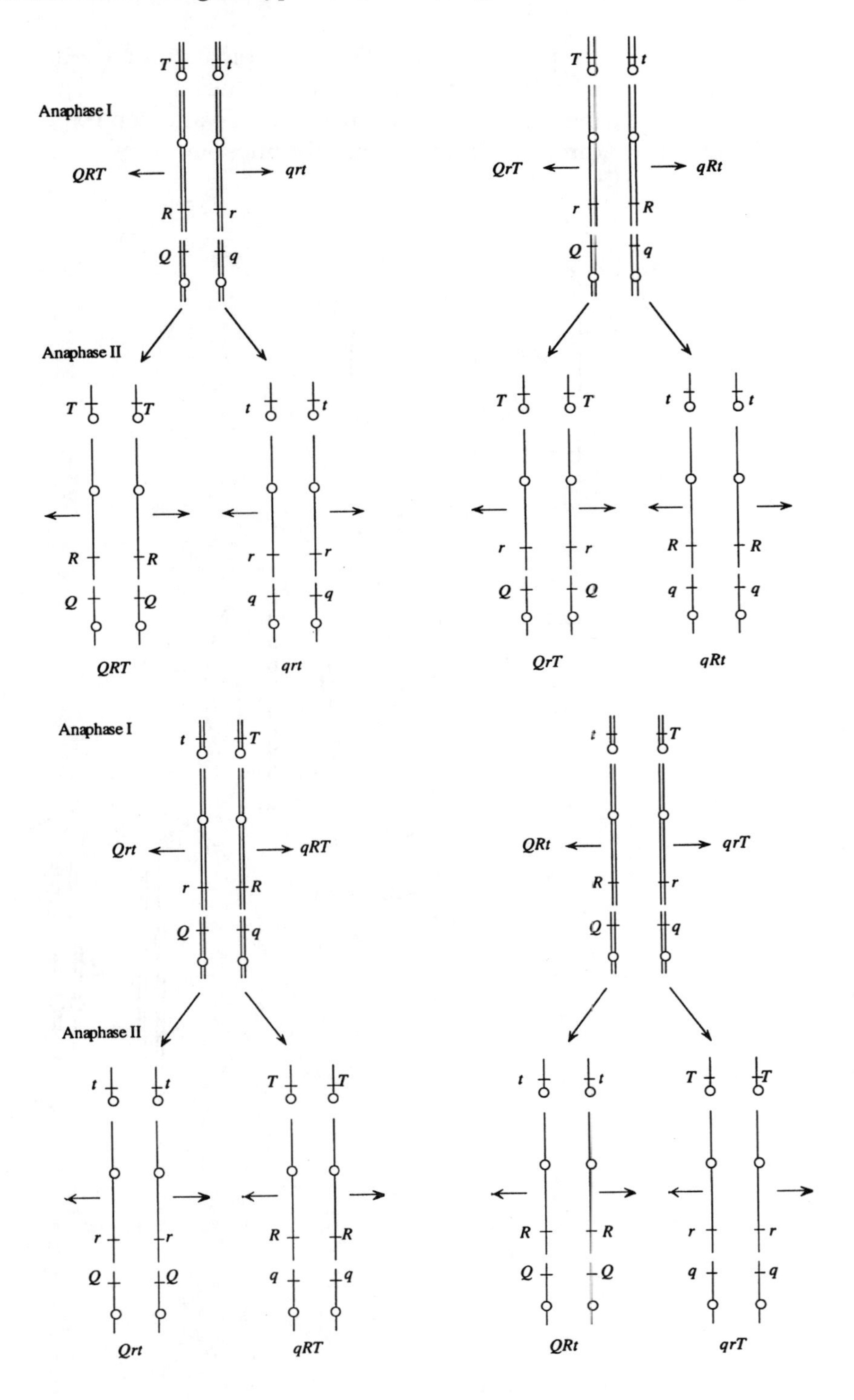

44. a. From the data, it can be concluded that the genes for flower color and plant height are linked (on the same chromosome). The position of the gene for leaf width cannot be determined from this data, since all progeny express the dominant trait. In the tall, red parent, the alleles for tall and white are in the cis configuration (on the same chromosome) as are the alleles for short and red. Thus, the progeny that are tall, white or short, red represent the parental chromosomes and the tall, red or short, white are recombinants. The two genes are 100%(21+19)/total = 4 map units apart.

tall, red, wide × short, white, narrow

$$\frac{T \quad r}{t \quad R} \cdot \frac{W}{W} \times \frac{t \quad r}{t \quad r} \cdot \frac{w}{w}$$

b. The chance of obtaining short, white, wide progeny = $p(t\ r)^2 = (^{4\%}/_{2})^2 =$ 0.04%.

45. Comparing independent assortment (RF = 0.5) to a recombination frequency of 34% (RF = 0.34)

RF	0.5	0.34
P	0.25	0.33
P	0.25	0.33
R	0.25	0.17
R	0.25	0.17

the probability of obtaining these results assuming independent assortment will be equal to

$0.25 \times 0.25 \times 0.25 \times 0.25 \times 0.25 \times 0.25 \times B = 0.00024B$ (where B = number of possible birth orders for four parental and two recombinant progeny)

For an RF = 0.34, the probability of this result will be equal to

$$0.33 \times 0.17 \times 0.33 \times 0.33 \times 0.17 \times 0.33 \times B = 0.00034B$$

The ratio of the two = 0.00034B/0.00024B = 1.42 and the Lod = 0.15.

46. a. meiosis I (crossing-over has occurred between all genes and their centromeres)

b. impossible

c. meiosis I (crossing-over has occurred between gene B and its centromere)

d. meiosis I

e. meiosis II

f. meiosis II (crossing-over has occurred between genes A and B and their centromeres)

g. meiosis II

h. impossible

i. mitosis

j. impossible

6 Specialized Eukaryotic Chromosome Mapping Techniques

1. **a.**

al-2$^+$	*al-2*
al-2$^+$	*al-2*
al-2	*al-2*$^+$
al-2	*al-2*$^+$
al-2$^+$	*al-2*
al-2$^+$	*al-2*
al-2	*al-2*$^+$
al-2	*al-2*$^+$

b. The 8 percent value can be used to calculate the distance between the gene and the centromere. That distance is $^1/_2$ the percentage of second-division segregation, or 4 percent.

2. **a.** *arg-6* · *al-2*, *arg-6* · *al-2*$^+$, *arg-6*$^+$ · *al-2*$^+$, *arg-6*$^+$ · *al-2*$^+$

b. *arg-6*$^+$ · *al-2*$^+$, *arg-6*$^+$ · *al-2*, *arg-6* · *al-2*$^+$, *arg-6* · *al-2*

c. *arg-6*$^+$ · *al-2*, *arg-6*$^+$ · *al-2*, *arg-6* · *al-2*$^+$, *arg-6* · *al-2*$^+$

3. The formula for this problem is $f(i) = e^{-m}m^i/i!$, where $m = 2$ and $i = 0, 1$, or 2.

a. $f(0) = e^{-2}2^0/0! = e^{-2} = 0.135$ or 13.5%

b. $f(1) = e^{-2}2^1/1! = e^{-2}(2) = 0.27$ or 27%

c. $f(2) = e^{-2}2^2/2! = e^{-2}(2) = 0.27$ or 27%

4. **a.** A region 1 crossover will yield the following results:

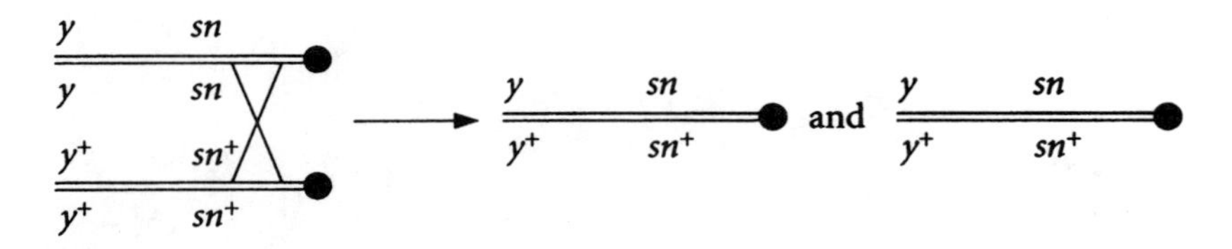

It is now possible that through random alignment and separation during mitosis that both *y sn* chromatids end up in the same daughter cell. As this *y sn/y sn* cell continues to divide, it will give rise to a yellow, singed spot in an otherwise wild-type fly.

b. A region 2 crossover will yield the following chromosomes:

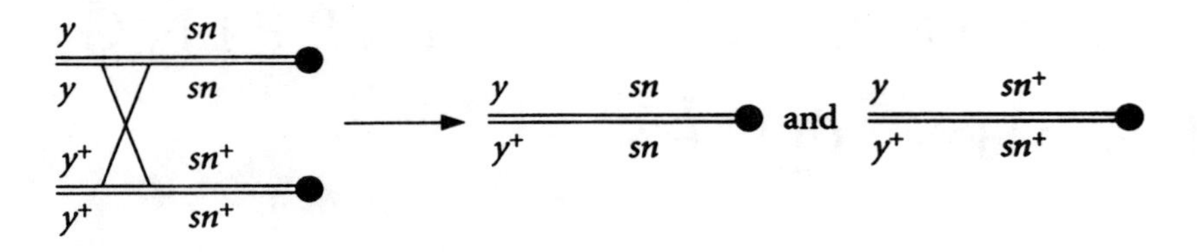

Mitosis can yield the following combinations:

$y\ sn/y\ sn^+$ and $y^+\ sn/y^+\ sn^+$

or

$y\ sn/y^+\ sn^+$ and $y^+\ sn/y\ sn^+$

Of these, the $y\ sn/y\ sn^+$ cell will give rise to a yellow spot in an otherwise wild-type fly.

5. This problem is analogous to meiosis in organisms that form linear tetrads. Let red = *R* and blue = *r*. This can now be compared to meiosis in an organism that is *R/r*. The patterns, their frequencies, and the division of segregation are given below. Notice that the probabilities change as each ball/allele is selected. This occurs when there is sampling without replacement.

$^1/_2\ R$
- $\times\ ^1/_3\ R \times\ ^1/_1\ r \times\ ^1/_1\ r =\ ^1/_6\ RRrr$ — first division
- $\times\ ^2/_3\ r$
  - $\times\ ^1/_2\ R \times\ ^1/_1\ r =\ ^1/_6\ RrRr$ — second division
  - $\times\ ^1/_2\ r \times\ ^1/_1\ R =\ ^1/_6\ RrrR$ — second division

$^1/_2\ r$
- $\times\ ^1/_3\ r \times\ ^1/_1\ R \times\ ^1/_1\ R =\ ^1/_6\ rrRR$ — first division
- $\times\ ^2/_3\ R$
  - $\times\ ^1/_2\ R \times\ ^1/_1\ r =\ ^1/_6\ rRRr$ — second division
  - $\times\ ^1/_2\ r \times\ ^1/_1\ R =\ ^1/_6\ RrRr$ — second division

These results indicate one-third first-division segregation and two-thirds second-division segregation.

6. **a.** The formula is RF = $^1/_2(1 - e^{-m})$. Since RF = 0.20, $e^{-m} = 1 - 0.4 = 0.6$, and $m = 0.51$. Because an m value of 1.0 = 50 map units (m.u.), 0.51 × 50 m.u. = 25.5 m.u.

b. The problem is the interpretation of 45 m.u. That could represent two loci approximately 45 m.u. apart, or it could represent two unlinked loci. A χ^2 test should be used to determine if the data indicate linkage. Assume the hypothesis of no linkage, which gives the expectation of a 1:1:1:1 ratio.

$$\chi^2 = \frac{(58 - 50)^2 + (52 - 50)^2 + (47 - 50)^2 + (43 - 50)^2}{50}$$

$$= 2.52$$

With 3 degrees of freedom, the probability is greater than 10% that the genes are not linked. Therefore, the hypothesis of no linkage can be accepted.

7. Rewrite the column headings to note what is missing from the media, and then count the different types of patterns. Growth will occur if the wild-type gene is present or if the medium supplies whatever gene product is missing:

–Leu	–Nic	–Ad	–Arg	Total
+	+	–	–	6
–	–	+	+	4
–	+	–	+	5
+	–	+	–	4
+	–	–	–	1
+	–	–	+	0
–	+	+	–	0

The first four categories are approximately in a 1:1:1:1 ratio, indicating independent assortment of two chromosomes. Furthermore, the pattern of growth on two of the media types and no growth on two of the media types indicates that two genes are located on each of the two chromosomes, rather than one gene on one chromosome and three genes on the second.

The two missing categories indicate which genes are linked. Remember that the cross is

$$arg^- \cdot ad^- \cdot nic^+ \cdot leu^+ \times arg^+ \cdot ad^+ \cdot nic^- \cdot leu^-$$

Growth is not seen in the (–Arg and –Leu) or (–Nic and –Arg) media simultaneously, which, based on the genotypes of the parents, indicates that *nic* is linked to *ad* and *leu* is linked to *arg*. Both *nic* and *ad* assort independently from *leu* and *arg*.

a. The parents were

$$ad^- \ nic^+ \ ; \ leu^+ \ arg^- \times ad^+ \ nic^- \ ; \ leu^- \ arg^+$$

b. Culture 16 resulted from a crossover between *ad* and *nic*. The reciprocal did not show up in this small sample.

8. Recall that larger RF values are less accurate measures of map distance than smaller RF values. Also recall that multiple crossovers reduce the calculated map distance. Therefore, an RF of 36% is an underestimate of the true map distance between waxy and shrunken.

The formula that is needed to calculate the proportion of meiocytes with different numbers of crossovers is $f(i) = e^{-m}m^i/i!$, where $i = 0$, 1, or 2, e is the base of natural logarithms, and m is the mean number of crossovers.

To calculate m, use the formula RF = $(1 - e^{-m})/2$.

$$0.36 = (1 - e^{-m})/2$$
$$e^{-m} = 0.28$$
$$m = 1.27$$

a. The probability of no crossover is

$f(0) = e^{-1.27}(1.27)^0/0! = 0.28$, or 28%

b. The probability of one crossover is

$f(1) = e^{-1.27}(1.27)^1/1! = 0.36$, or 36%

c. The probability of two crossovers is

$f(2) = e^{-1.27}(1.27)^2/2! = 0.23$, or 23%

d. The probability of at least one crossover is

p(at least 1 CO) = 1 – p(0 CO) = 1 – 0.28 = 0.72, or 72%

9. To work this problem, it is helpful to draw the chromosomes and explore the consequences of crossing-over in different regions. Here, the genes are assumed to be coupled.

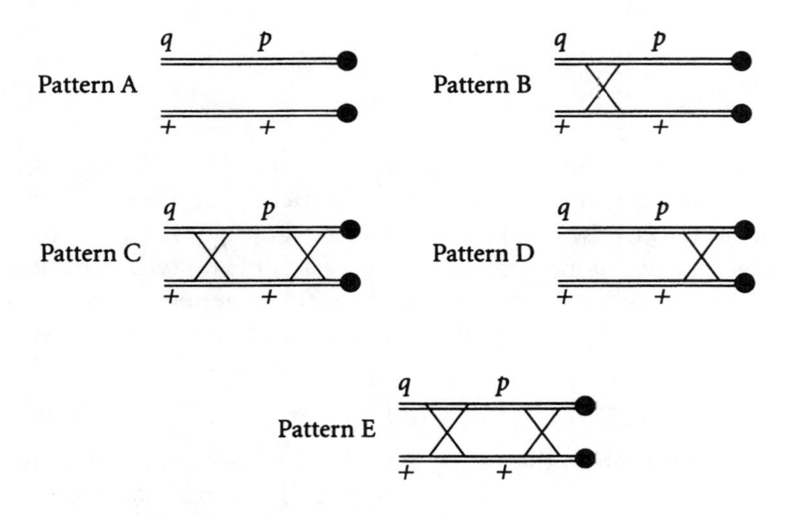

With no crossovers, the pattern is $p\ q$, $p\ q$, + +, + +, a parental ditype. Both genes show an M_I pattern.

With one CO between q and p, the pattern is $p\ q$, p +, + q, + +. This is a T, with gene p showing M_I and gene q showing M_{II} segregation.

With one CO between p and the centromere, the pattern is $p\ q$, + +, $p\ q$, + +. This is a PD, with both genes showing M_{II} segregation.

With one CO between p and the centromere and one CO between p and q, for a two-strand double, the pattern is $p\ q$, $+\ q$, $p\ +$, $+\ +$. This is a T, with M_{II} for p and M_I for q. (Note: a four-strand double would give similar results.) For a three-strand double, the pattern is $p\ +$, $+\ q$, $p\ q$, $+\ +$. This is a T, with both genes showing M_{II} segregation.

a. M_IM_I, PD is pattern A, no crossovers. The probability is

p(no CO from centromere to p) × p(no CO from p to q)

= (0.88)(0.80) = 0.704

b. M_IM_I, NPD requires two crossovers between q and p, which does not occur according to the rules of the problem. The probability is 0.

c. M_IM_{II}, T is pattern B. The probability is

p(no CO from p to centromere) × p(CO between q and p)

= (0.88)(0.2) = 0.176

d. $M_{II}M_I$, T is pattern C. Because there are two out of four outcomes that achieve this result (the two- and four-strand double), in the following calculation there is an adjustment by a factor of 1/2. The probability is

1/2p(CO between p and centromere) × p(CO between q and p) = (0.5)(0.12)(0.2) = 0.012

e. $M_{II}M_{II}$, PD is pattern D. The probability is

p(CO between p and centromere) × p(no CO from q to p)

= (0.12)(0.8) = 0.096

f. $M_{II}M_{II}$, NPD requires one crossover between p and the centromere and two crossovers between q and p, which does not occur according to the rules of the problem. The probability is 0.

g. $M_{II}M_{II}$, T is pattern E. Because there are two out of four outcomes that achieve this result (the two three-strand doubles), in the following calculation there is an adjustment by a factor of 1/2. The probability is

1/2p(CO between p and centromere) × p(CO between q and p) = (0.5)(0.12)(0.2) = 0.012

10. a.

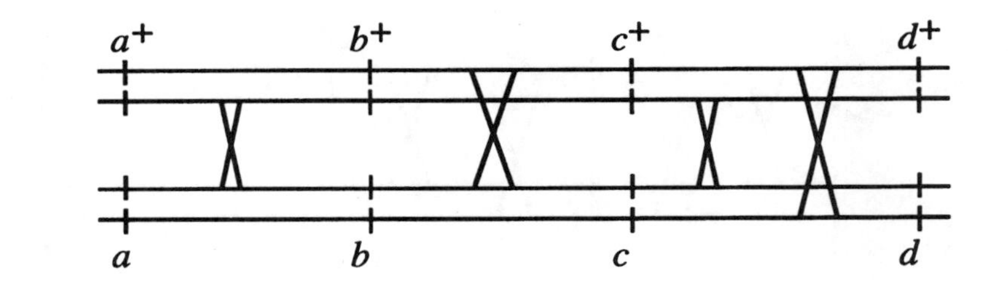

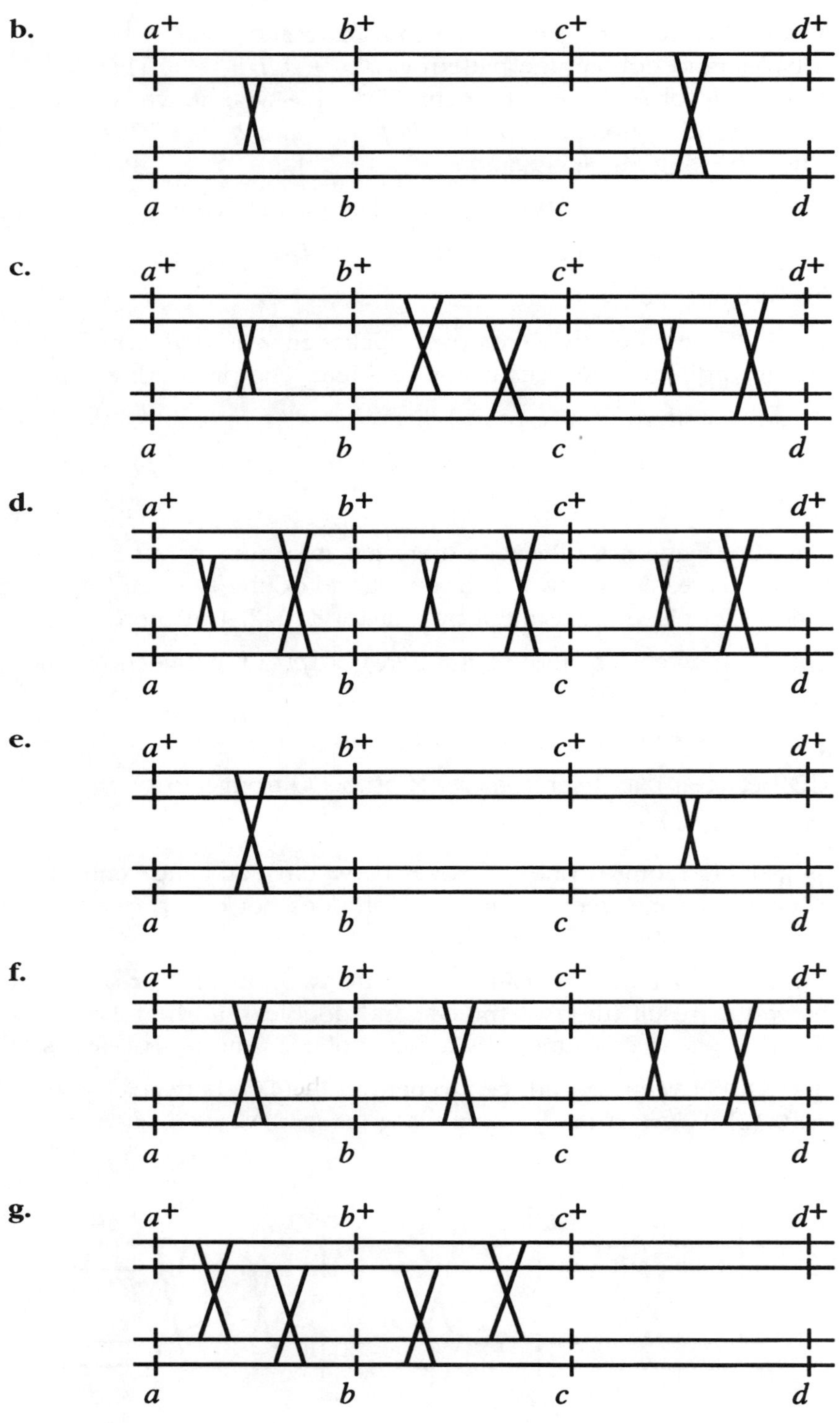
b.
a+ b+ c+ d+
a b c d
c.
a+ b+ c+ d+
a b c d
d.
a+ b+ c+ d+
a b c d
e.
a+ b+ c+ d+
a b c d
f.
a+ b+ c+ d+
a b c d
g.
a+ b+ c+ d+
a b c d

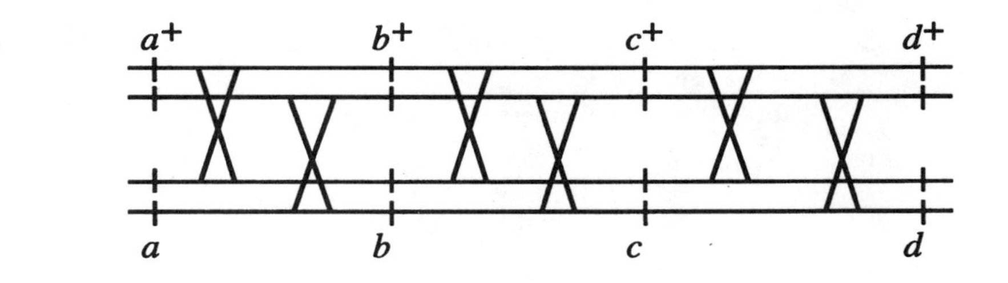

11. ***Unpacking the Problem***

1. Fungi are generally haploid.
2. There are four pairs, or eight ascospores, in each ascus. One member of each pair is presented in the data.
3. A mating type in fungi is analogous to sex in humans, in that the mating types of two organisms must differ in order to have a mating that produces progeny. Mating type is determined experimentally simply by seeing if progeny result from specific crosses.
4. The mating types *A* and *a* do not indicate dominance and recessiveness. They simply symbolize the mating-type difference.
5. *arg1* indicates that the organism requires arginine for growth. Testing for the genotype involves isolating nutritional mutants and then seeing if arginine supplementation will allow for growth.
6. $arg1^+$ indicates that the organism is wild-type and does not require supplemental arginine for growth.
7. *Wild type* refers to the common form of an organism in its natural population.
8. *Mutant* means that, for the trait being studied, an organism differs from the wild type.
9. The actual function of the alleles in this problem does not matter in solving the problem.
10. Linear tetrad analysis refers to the fact that the ascospores in each ascus are in a linear arrangement that reflects the order in which the two meiotic divisions occurred to produce them. By tracking traits and correlating them with position, it is possible to detect crossing-over events that occurred at the tetrad (four-strand, homologous pairing) stage prior to the two meiotic divisions.
11. Linear tetrad analysis allows for the mapping of centromeres in relation to genes, which cannot be done with unordered tetrad analysis.
12. A cross is made in *Neurospora* by placing the two organisms in the same test tube or petri dish and allowing them to grow. Gametes develop and fertilization, followed by meiosis, mitosis, and ascus formation, occurs. The asci are isolated, and the ascospores are dissected out of them with the aid of a microscope. The ascus has an octad, or eight spores, within it, and the spores are arranged in four (tetrad) pairs.

13. Meiosis occurs immediately following fertilization in *Neurospora.*

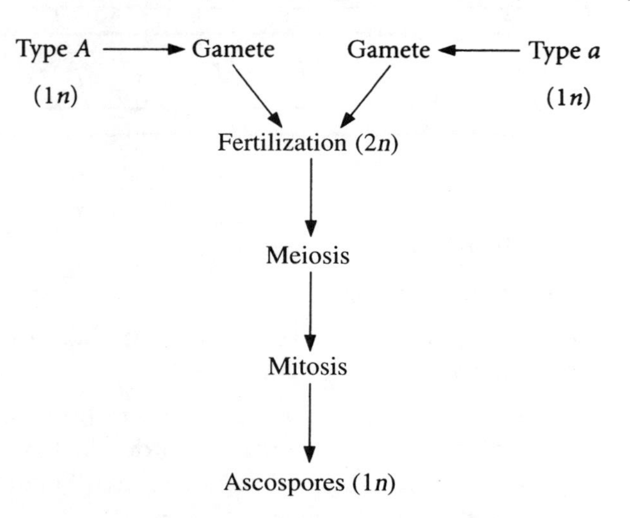

14. Meiosis produced the ascospores that were analyzed.
15. The cross is $A \cdot arg\text{-}1 \times a \cdot arg\text{-}1^+$.
16. Although there are eight ascospores, they occur in pairs. Each pair represents one chromatid of the originally paired chromosomes. By convention, both members of a pair are represented by a single genotype.
17. The seven classes represent the seven types of outcomes. The specific outcomes can be classified as follows:

Class	1	2	3	4	5	6	7
Outcome	PD	NPD	T	T	PD	NPD	T
A/a	I	I	I	II	II	II	II
$arg\text{-}1^+/arg\text{-}1$	I	I	II	I	II	II	II

where PD = parental ditype, NPD = nonparental ditype, T = tetratype, I = first-division segregation, and II = second-division segregation.

Other classes can be detected, but they indicate the same underlying process. For example, the following three asci are equivalent.

1	2	3
A arg	$A\ arg^+$	$A\ arg^+$
$A\ arg^+$	*A arg*	*A arg*
a arg	*a arg*	$a\ arg^+$
$a\ arg^+$	$a\ arg^+$	*a arg*

In the first ascus, a crossover occurred between chromatids 2 and 3, while in the second ascus it occurred between chromatids 1 and 3, and in the third ascus the crossover was between chromatids 1 and 4. A fourth equivalent ascus would contain a crossover between chromatids 2 and 3. All four indicate a crossover between the second gene and its centromere and all are tetratypes.

18. This is exemplified in the answer to (17) above.

19. The class is identical with class 1 in the problem, but inverted.
20. *Linkage arrangement* refers to the relative positions of the two genes and the centromere along the length of the chromosome.
21. A *genetic interval* refers to the region between two loci, whose size is measured in map units.
22. It is not known whether the two loci are on separate chromosomes or are on the same chromosome. The general formula for calculating the distance of a locus to its centromere is to measure the percentage of tetrads that show second-division segregation patterns for that locus and divide by two.
23. Recall that there are eight ascospores per ascus. By inspection, the frequency of recombinant *A arg-1*$^+$ ascospores is 4(125) + 2(100) + 2(36) + 4(4) + 2(6) = 800. There is also the reciprocal recombinant genotype *a arg-1*.
24. Class 1 is parental; class 2 is nonparental ditype. Because they occur at equal frequencies, the two genes are not linked.

Solution to the Problem

a. The cross is $A \cdot arg\text{-}1 \times a \cdot arg\text{-}1^+$. Use the classification of asci in part (17) above. First decide if the two genes are linked by using the formula PD>>NPD, when the genes are linked, while PD = NPD when they are not linked. PD = 127 + 2 = 129 and NPD = 125 + 4 = 129, which means that the two genes are not linked. Alternatively,

$$\text{RF} = 100\%(^1/_2\ \text{T} + \text{NPD})/\text{total asci}$$

$$= 100\%[(^1/_2)(100 + 36 + 6) + (125 + 4)]/400 = 50\%.$$

Next calculate the distance between each gene and its centromere using the formula RF = 100%($^1/_2$ number of tetrads exhibiting M_{II} segregation)/(total number of asci).

$$A\text{–centromere} = 100\%(^1/_2)(36 + 2 + 4 + 6)/400$$

$$= 100\%(24/400) = 6\ \text{m.u.}$$

$$arg^+\text{–centromere} = 100\%(^1/_2)(100 + 2 + 4 + 6)/400$$

$$= 100\%(56/400) = 14\ \text{m.u.}$$

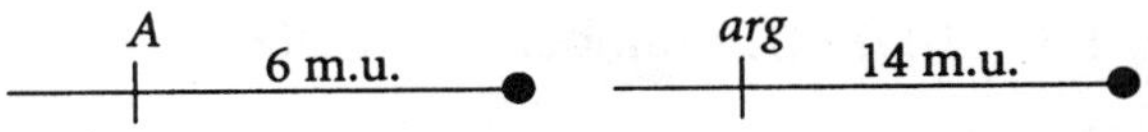

b. Class 6 can be obtained if a single crossover occurred between chromatids 2 and 3 between each gene and its centromere.

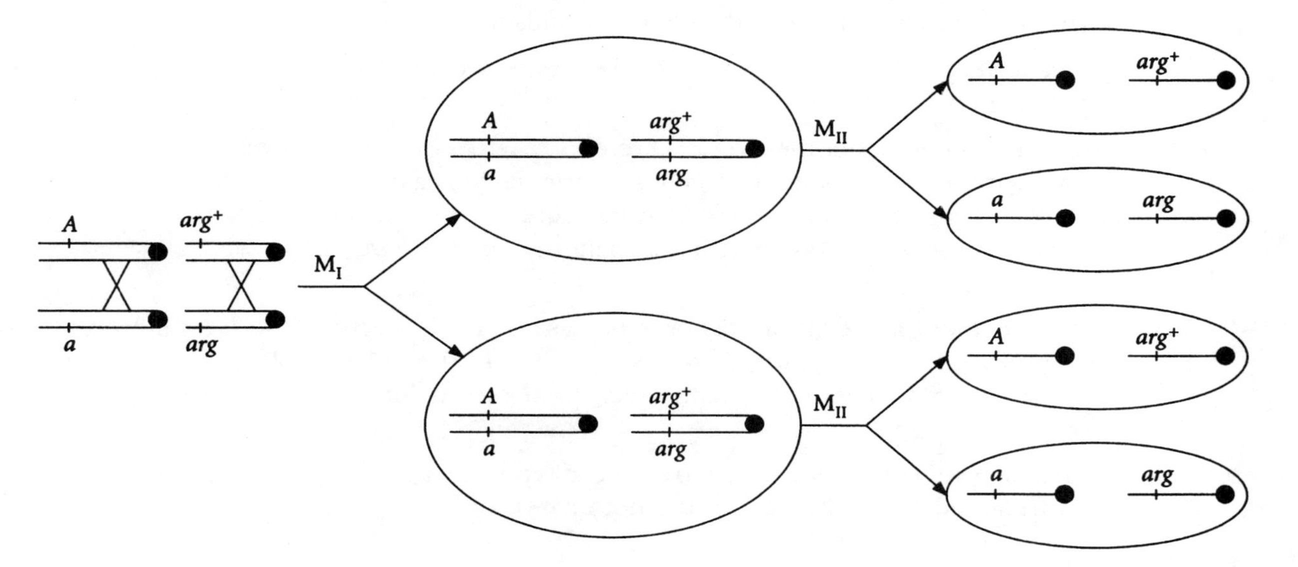

12. a. The cross is $t \cdot m^+ \times t^+ \cdot m$. The six asci types can be classified as follows:

Class	PD	T	PD	T	NPD	T
t	I	I	II	II	I	II
m	I	II	II	I	I	II
Number	260	76	4	54	1	5

where PD = parental ditype, NPD = nonparental ditype, T = tetratype, I = first-division segregation, and II = second-division segregation.

First decide whether the two genes are linked. If PD = NPD, genes are not linked and if PD >> NPD, linkage is indicated. In this case, 264 PD >> 1 NPD so genes are linked.

Next, calculate the distance between the two genes. The formula to use is

$$\text{RF} = 100\%(^1/_2\text{T} + \text{NPD})/\text{total asci}$$
$$= 100\%\ [^1/_2(76 + 54 + 5) + 1]400 = 17.1\%.$$

The distance between each gene and the centromere is calculated using the following formula:

$$\text{RF} = 100\%(1/2 \text{ number of } \text{M}_{\text{II}} \text{ asci})/\text{total number of asci}.$$

In this case,

$$t\text{–centromere} = 100\%\ [^1/_2(4 + 54 + 5)]/400 = 7.9 \text{ m.u.}$$

$$m\text{–centromere} = 100\%\ [^1/_2(76 + 4 + 5)]/400 = 10.6 \text{ m.u.}$$

The map is

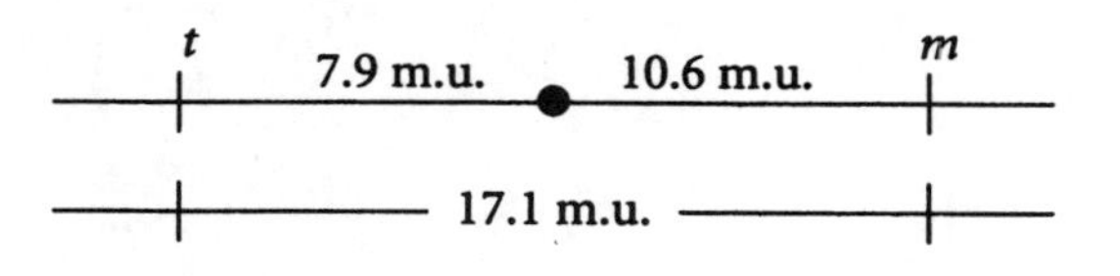

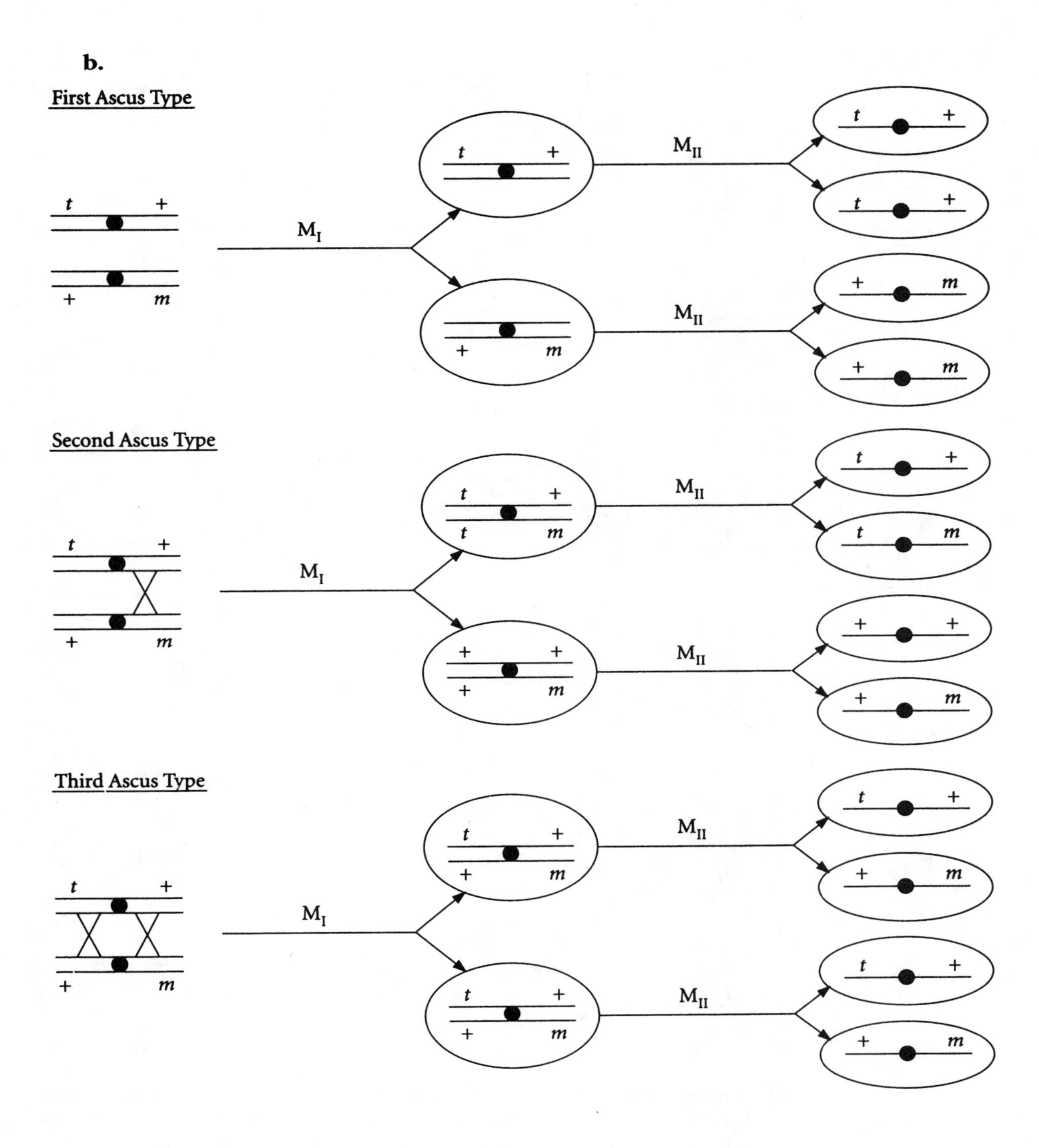
b.
First Ascus Type
t
+
+
m
M_I
M_{II}
Second Ascus Type
Third Ascus Type

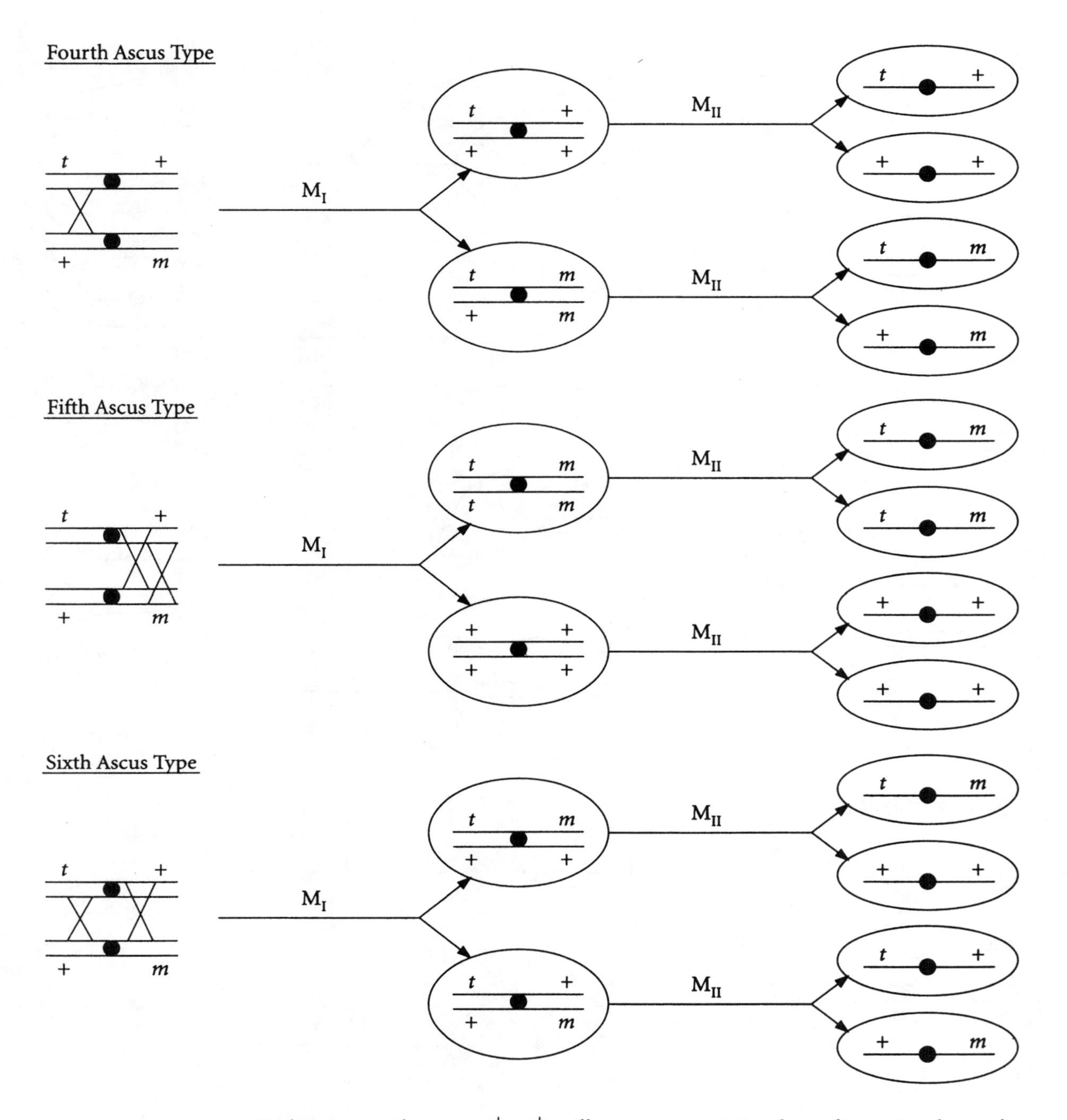

c. Only spores that are $t^+ m^+$ will grow on minimal medium. Look at class 2. One-fourth of the spores will grow in minimal medium. The 76 asci represent 19% of the total number of asci. Therefore, of 1000 randomly chosen spores, 19% × 1000 = 190 will be class 2 and $^1/_4$ ($^1/_4$ × 190 = 48) of these will be able to grow in minimal medium. In a similar manner, the contribution of the other three classes containing $t^+ m^+$ asci can be determined, as follows:

$$100\%(1000)[^1/_4(76/400) + ^1/_4(54/400) + ^1/_2(1/400) + ^1/_4(5/400)] = 86$$

13. Before beginning this problem, classify all asci as PD, NPD, or T and determine whether there is M_I or M_{II} segregation for each gene:

	ASCI TYPE						
	1	2	3	4	5	6	7
Type	PD	NPD	T	T	PD	NPD	T
gene a	I	I	I	II	II	II	II
gene b	I	I	II	I	II	II	II

If PD >> NPD, linkage is indicated. The distance between a gene and its centromere = 100% $(^1/_2)(M_{II})$/total. The distance between two genes = 100% $(^1/_2T + NPD)$/total.

Cross 1: PD = NPD and RF = 50%; the genes are not linked.

a–centromere: 100% $(^1/_2)(0)/100$ = 0 m.u. Gene *a* is close to the centromere.

b–centromere: 100% $(^1/_2)(32)/100$ = 16 m.u.

Cross 2: PD >> NPD; the genes are linked.

a–b: 100% $[^1/_2(15) + 1]/100$ = 8.5 m.u.

a–centromere: 100% $(^1/_2)(0)/100$ = 0 m.u. Gene *a* is close to the centromere.

b–centromere: 100% $(^1/_2)(15)/100$ = 7.5 m.u.

Cross 3: PD >> NPD; the genes are linked.

a–b: 100% $[^1/_2(40) + 3]/100$ = 23 m.u.

a–centromere: 100% $(^1/_2)(2)/100$ = 1 m.u.

b–centromere: 100% $(^1/_2)(40 + 2)/100$ = 21 m.u.

Cross 4: PD > > NPD; the genes are linked.

a–b: 100% $[^1/_2(20) + 1]/100$ = 11 m.u.

a–centromere: 100% $(^1/_2)(10)/100$ = 5 m.u.

b–centromere: 100% $(^1/_2)(18 + 8 + 1)/100$ = 13.5 m.u.

Cross 5: PD = NPD (and RF = 49%); the genes are not linked.

a–centromere: 100% $(^1/_2)(22 + 8 + 10 + 20)/99$ = 30.3 m.u.

b–centromere: 100% $(^1/_2)(24 + 8 + 10 + 20)/99$ = 31.3 m.u.

These values are approaching the 67% theoretical limit of loci exhibiting M_{II} patterns of segregation and should be considered cautiously.

Cross 6: PD >> NPD; the genes are linked.

a–b: 100% $[^1/_2(1 + 3 + 4) + 0]/100$ = 4 m.u.

a–centromere: 100% $(^1/_2)(3 + 61 + 4)/100$ = 34 m.u.

b–centromere: 100% $(^1/_2)(1 + 61 + 4)/100$ = 33 m.u.

These values are at the 67% theoretical limit of loci exhibiting M_{II} patterns of segregation and therefore both loci can be considered unlinked to the centromere.

Cross 7: PD >> NPD; the genes are linked.

a–b: 100% $[^1/_2(3 + 2) + 0]/100 = 2.5$ m.u.

a–centromere: 100% $(^1/_2)(2)/100 = 1$ m.u.

b–centromere: 100% $(^1/_2)(3)/100 = 1.5$ m.u.

Cross 8: PD = NPD; the genes are not linked.

a–centromere: 100% $(^1/_2)(22 + 12 + 11 + 22)/100 = 33.5$ m.u.

b–centromere: 100% $(^1/_2)(20 + 12 + 11 + 22)/100 = 32.5$ m.u.

Same as cross 5.

Cross 9: PD >> NPD; the genes are linked.

a–b: 100% $[^1/_2(10 + 18 + 2) + 1]/100 = 16$ m.u.

a–centromere: 100% $(^1/_2)(18 + 1 + 2)/100.= 10.5$ m.u.

b–centromere: 100% $(^1/_2)(10 + 1 + 2)/100 = 6.5$ m.u.

Cross 10: PD = NPD; the genes are not linked.

a–centromere: 100% $(^1/_2)(60 + 1 + 2 + 5)/100 = 34$ m.u.

b–centromere: 100% $(^1/_2)(2 + 1 + 2 + 5)/100 = 5$ m.u.

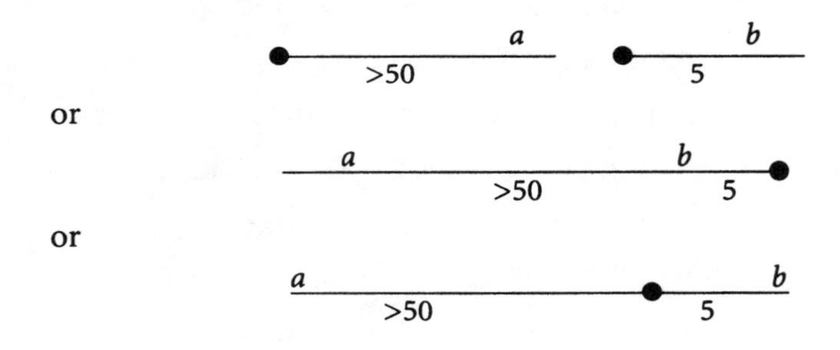

Cross 11: PD = NPD; the genes are not linked.

a–centromere: 100% $(^1/_2)(0)/100 = 0$ m.u.

b–centromere: 100% $(^1/_2)(0)/100 = 0$ m.u.

14. The cross is $a\,;\,b^+ \times a^+\,;\,b$. Each gene has a given probability of a crossover between it and the centromere. Remember that map units = 100% $[^1/_2(M_{II})]$/total asci. Therefore, if there are 5 m.u. between a gene and its centromere, there will be 10 percent second-division asci and 90 percent noncrossover, or first-division asci (M_I).

When there is no crossing-over, two equally likely patterns are possible: $a\ b$, $a\ b$, $a^+\ b^+$, $a^+\ b^+$ (NPD) and $a\ b^+$, $a\ b^+$, $a^+\ b$, $a^+\ b$ (PD).

If one gene (assume gene *a*) experiences a crossover, the pattern will be: a, a^+, a, a^+ or a, a^+, a^+, a. Combined with M_I segregation for gene *b*, this would result in four equally likely outcomes: $a\ b$, $a^+\ b$, $a\ b^+$, $a^+\ b^+$ (T); or *a b*, a^+ *b*, a^+ b^+, *a* b^+ (T); or a^+ *b*, *a b*, a^+ b^+, *a* b^+ (T); or *a b*, a^+ *b*, a^+ b^+, *a* b^+ (T). Note that patterns 2 and 4 are identical.

If both genes experience crossovers, the pattern is $^1/_4$ PD: $^1/_2$ T: $^1/_4$ NPD (you should check to see if you can produce this ratio).

The following table presents the probabilities for both genes:

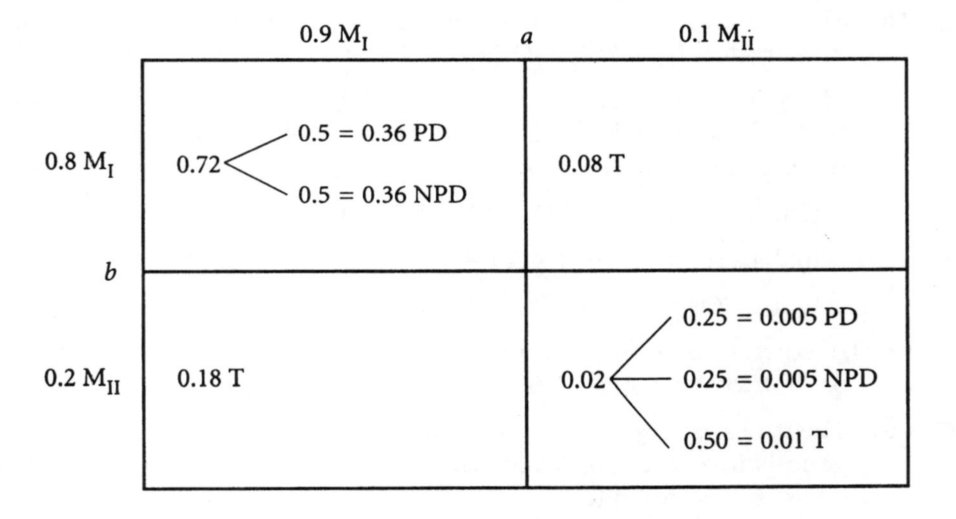

To understand the table, look at box 1. This represents the frequency that no crossovers occur between either gene and its centromere ($p = 0.8 \times 0.9 = 0.72$). The pattern for *a* is thus *a*, *a*, a^+, a^+ and the pattern for *b* is *b*, *b*, b^+, b^+. The *b* gene can line up in either orientation with the *a* gene: *a b*, *a b*, a^+ b^+, a^+ b^+ or *a* b^+, *a* b^+, a^+ *b*, a^+ *b*. Therefore, there is a 1:1 chance of *a* b^+, *a* b^+, a^+ *b*, a^+ *b* (PD) and *a b*, *a b*, a^+ b^+, a^+ b^+ (NPD). The other boxes can be interpreted in a similar manner.

a. 36.5% PD: the total of PD from the table

b. 36.5% NPD: the total of NPD from the table

c. 27% T: the total of T from the table

d. 50% recombinants: $^1/_2$T + NPD

e. 25% wild type

15. The number of recombinants is equal to NPD + $^1/_2$T. The uncorrected map distance is based on RF = (NPD + $^1/_2$T)/total. The corrected map distance = 50(T + 6NPD)/total.

Cross 1:
recombinant frequency = 4% + $^1/_2$(45%) = 26.5%
uncorrected map distance = [4% + $^1/_2$(45%)]/100% = 26.5 m.u.
corrected map distance = 50[45% + 6(4%)]/100% = 34.5 m.u.

Cross 2:
recombinant frequency = 2% + $^1/_2$(34%) = 19%
uncorrected map distance = [2% + $^1/_2$(34%)]/100% = 19 m.u.
corrected map distance = 50[34% + 6(2%)]/100% = 29 m.u.

Cross 3:
recombinant frequency = 5% + $^1/_2$(50%) = 30%
uncorrected map distance = [5% + $^1/_2$(50%)]/100% = 30 m.u.
corrected map distance = 50[50% + 6(5%)]/100% = 40 m.u.

16. First, classify the asci: 138 T, 12 NPD, and 150 PD, for a total of 300 asci.

a. The number of recombinants is equal to NPD + $^1/_2$T, which leads to an uncorrected RF of 100% (12+ 69)/300 = 27 m.u.

b. The general formula is DCOs = 4(NPD). Therefore, DCOs = 4(12) = 48. Half of these will result in tetratypes, and one-quarter will result in parental ditypes. Therefore

actual 0 crossovers = PD – 12 = 150 – 12 = 138, or 46%

actual 1 crossovers = T – 24 = 138 – 24 = 114, or 38%

actual DCOs = 48, or 16%

c. To correct for double crossovers, the corrected map distance = 50(T + 6NPD)/total = 50[138 + 6(12)]/300 = 35 m.u.

17. a. The cross is arg^- × arg^-. Because ARG^+ progeny result, more than one gene is involved, and each mutation is recessive (complementation). The cross can be rewritten

$$arg\text{-}1^+ \cdot arg\text{-}2^- \times arg\text{-}1^- \cdot arg\text{-}2^+$$

A 4:0 (4 ARG^-:0 ARG^+) ascus is a PD ascus, because all spores require arginine. The 3:1 ascus must represent a T ascus. The 2:2 ascus is an NPD ascus.

b. The data support independent assortment of the two genes: PD = NPD = 40%.

18. Because PD = NPD, the *his-?* is not linked to *ad-3*. The T-type ascospores require one crossover between the gene and the centromere. Because only 10 tetrads were analyzed, *his-2* would be expected to produce one-tenth of a T ascus [10(0.01) = 0.1], *his-3* would be expected to produce one T ascus [10(0.1) = 1], and *his-4* would be expected to produce four T asci [10(0.4) = 4]. Only *his-4* is located far enough from its centromere to result in 60 percent T asci. Therefore, *his-?* is most likely *his-4*.

19. Because the two mutants, when crossed, result in some black spores, two separate genes are involved. Let mutant 1 = $w \cdot t^+$, mutant 2 = $w^+ \cdot t$, and wild type = $w^+ \cdot t^+$. The cross involving the two mutants is

P $w \cdot t^+$ (white) × $w^+ \cdot t$ (tan)

Asci types are

4 black : 4 white = 4 $w^+ \cdot t^+$: 4 $w \cdot t^+$ (NPD)

4 tan : 4 white = 4 $w^+ \cdot t$: 4 $w \cdot t^+$ (PD)

4 white : 2 black : 2 tan = 2 $w \cdot t^+$ (white) : 2 $w\ t$ (white) : 2 $w^+ \cdot t^+$ (black) : 2 $w^+ \cdot t$ (tan)

Notice that there are two types of white, indicating that there is an epistatic relationship between w and t. The w allele blocks expression of the t^+ allele.

20. At meiosis I, normal segregation is between homologous chromosomes. The two attached-X chromosomes segregate together because they structurally are the equivalent of a single chromosome. This produces one daughter cell with two copies of X (disomic) and one daughter cell with no copies of X (nullisomic):

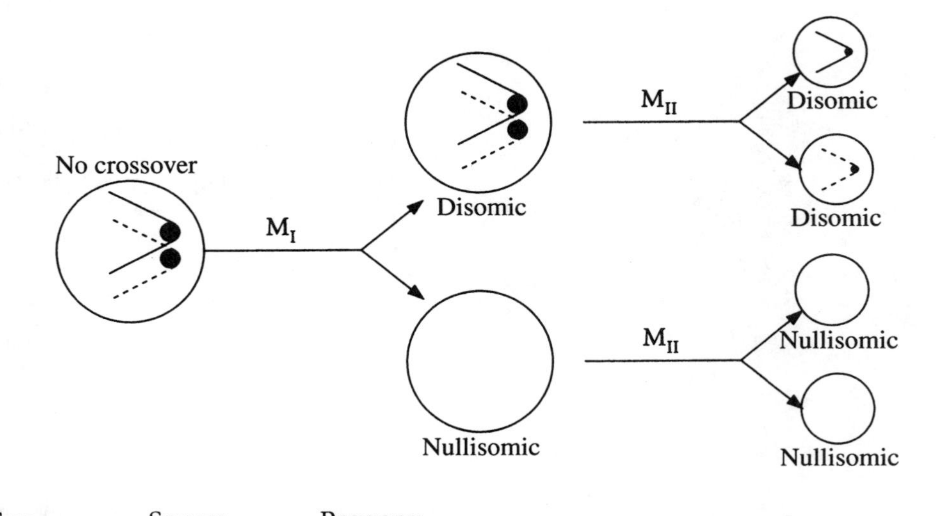

Egg	Sperm	Progeny
XX	X	XXX, nonviable
XX	Y	XXY, viable female
O	X	XO, sterile male
O	Y	OY, nonviable

a. A single crossover between a and b, involving strands 1 and 2 or involving strands 3 and 4, does not result in recombination. However, a 1-3 (illustrated below), 1-4, 2-3, or 2-4 crossover between a and b results in recombination.

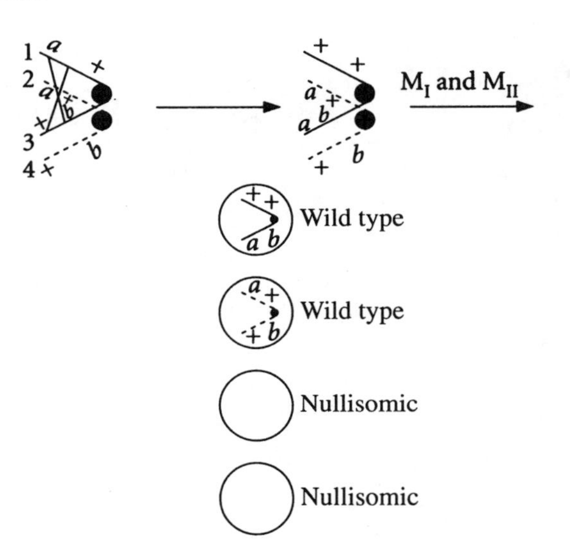

A single crossover between *b* and the centromere, involving strands 1 and 2 or involving strands 3 and 4, does not result in recombination. A 1-3 (illustrated below) or 2-4 crossover between *b* and the centromere also does not result in recombination.

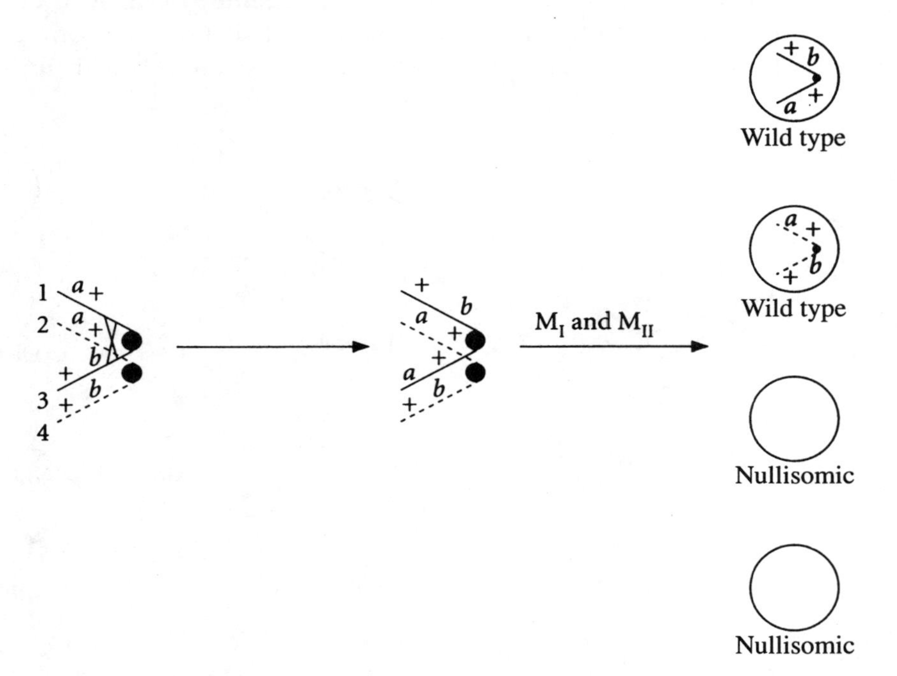

However, a 2-3 (illustrated below) or a 1-4 single crossover between *b* and the centromere does result in recombination.

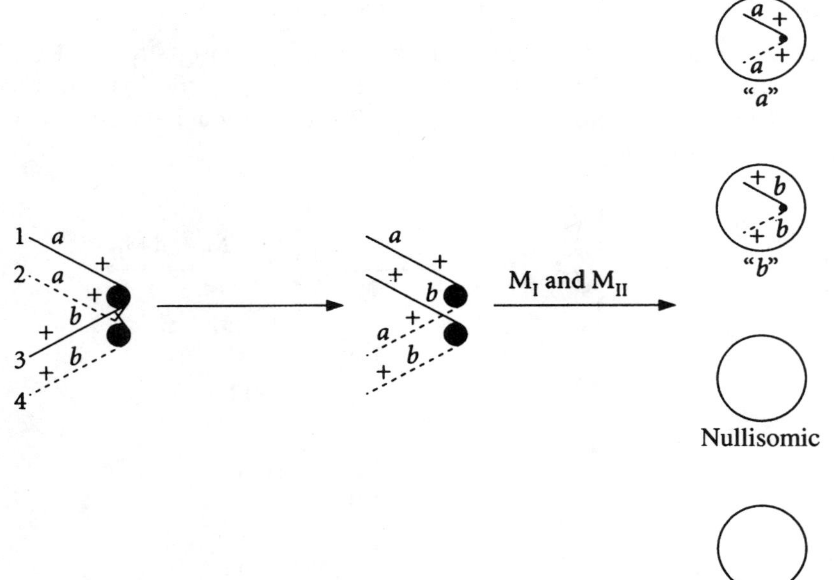

b. With double crossovers, the following are produced:

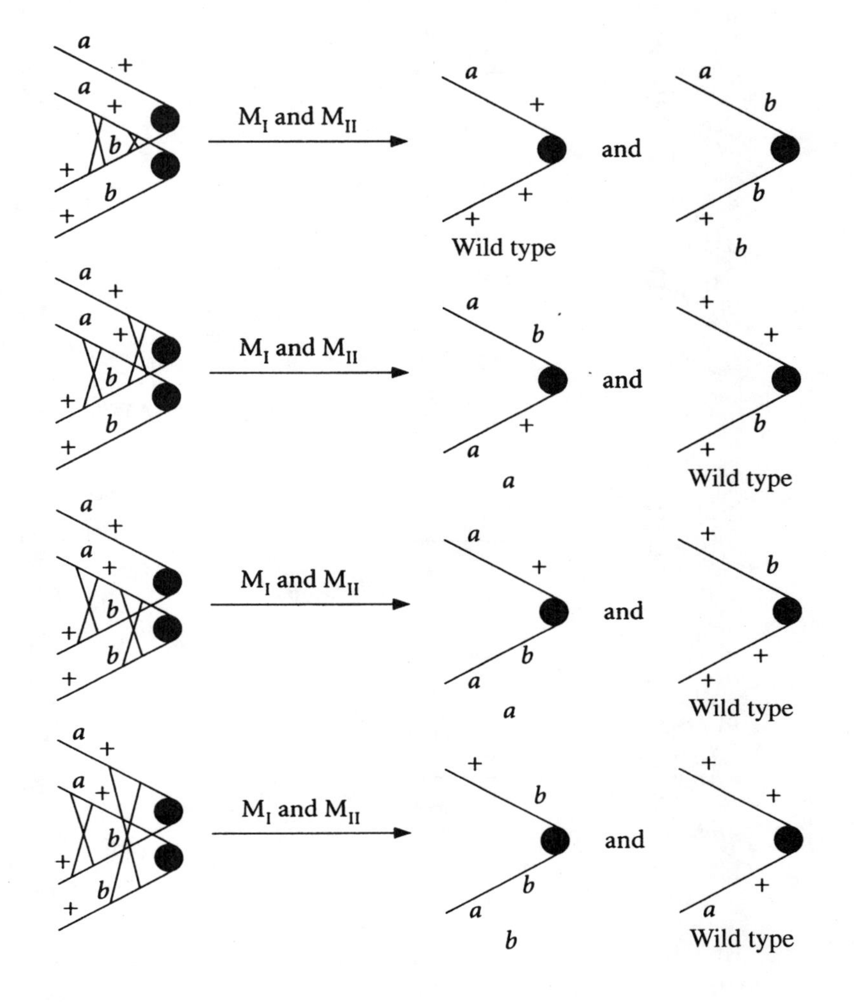

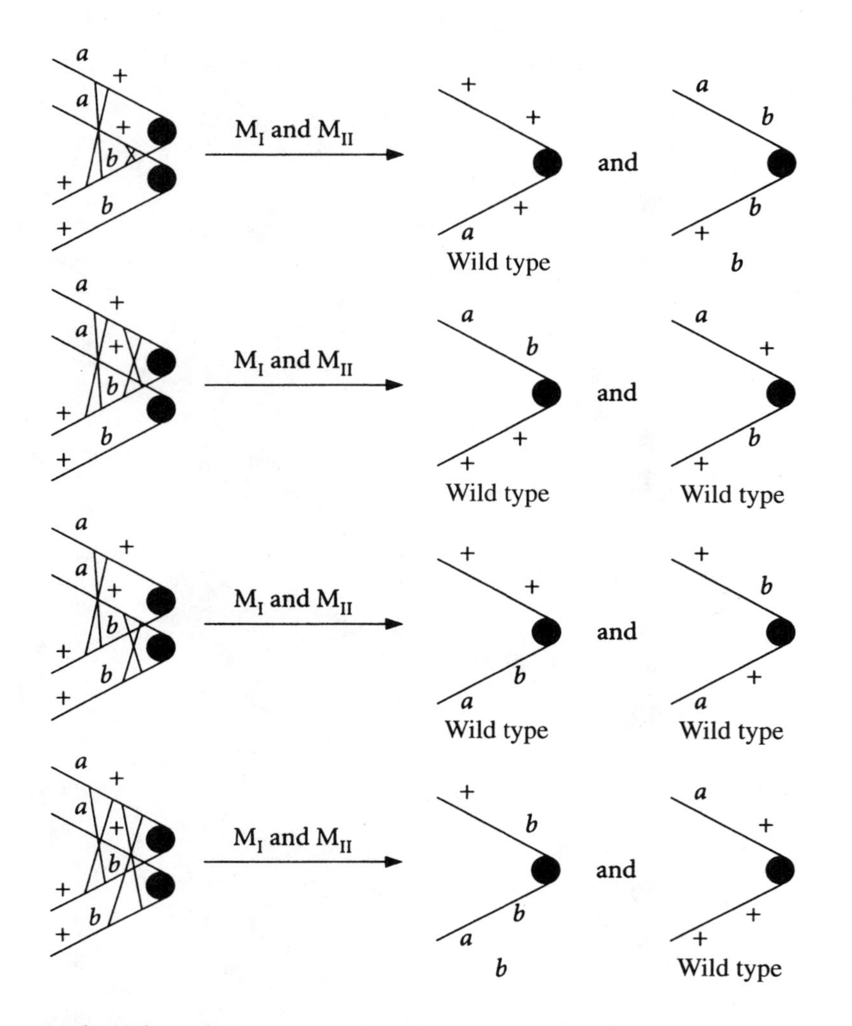

Note that the phenotype in each female (XXY) shows half the meiotic products. Thus, each female can be regarded as a half-tetrad.

21. First classify the asci types. Number 1 is a tetratype (T), number 2 is a parental ditype (PD), and number 3 is a nonparental ditype (NPD).

Recall that tetratypes arise from single and double crossovers. A SCO yields 100% T asci, and a DCO yields 50% T asci. Parental ditypes arise from no crossovers (100% PD) and double crossovers (25% PD). Nonparental ditypes arise from 25% of the double recombination events.

Also recall the formula RF = $100\%(^1/_2\text{T} + \text{NPD})/\text{total asci}$. In other words, the recombination frequency is a function of $^1/_2$ the tetratype asci. If a gene is 5 map units from the centromere or another gene, that means that the crossover frequency is actually 10%. In order to understand this, draw two homologous chromosomes with a single crossover. Only half the chromatids are involved.

Consider the following map derived from the problem:

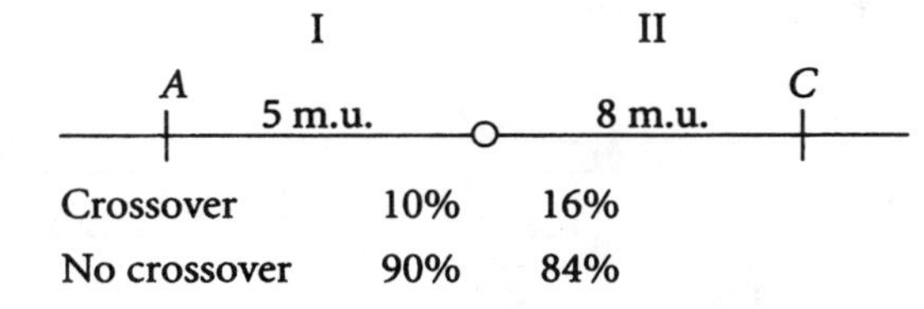

Now consider the following combination of events and their frequencies:

(CO in 1)(no CO in II) = 0.10 × 0.84 = 0.084, all T

(no CO in I)(CO in II) = 0. 90 × 0.16 = 0.144, all T

(CO in I)(CO in II) = 0.10 × 0.16 = 0.016, with $^1/_4$ PD : $^1/_2$ T : $^1/_4$ NPD

(no CO in I)(no CO in II) = 0.90 × 0.84 = 0.756, all PD

Total the frequency of each asci type:

PD = 0.756 + 0.004 = 0.760

T = 0.084 + 0.144 + 0.008 = 0.236

NPD = 0.004

Number 1 is a T ascus, which occurs at a frequency of 0.236.

Number 2 is a PD ascus, which occurs at a frequency of 0.760.

Number 3 is a NPD ascus, which occurs at a frequency of 0.004.

22. First classify the asci types and determine whether they are M_I or M_{II} asci for each gene:

Type	*un*	*cyh*
1. T	II	I
2. T	I	II
3. PD	I	I
4. PD	II	II
5. NPD	II	II
6. T	II	II

Notice that PD (47 + 2) >> NPD (2) and RF = 26.5%. The two genes are linked.

a. *un – cyh* = 100% ($^1/_2$T+ NPD)/total asci

$$= \frac{100\%\ [1/2\ (15 + 29 + 5) + 2]}{15 + 29 + 47 + 2 + 2 + 5}$$

= 26.5 m.u.

un – centromere = 100% (1/2)(MII asci)/total asci

= 100% (1/2)(15 + 2 + 2 + 5)/100 = 12 m.u.

cyh – centromere = 100% (1/2)(MII asci)/total asci

= 100% (1/2)(29 + 2 + 2 + 5)/100 = 19 m.u.

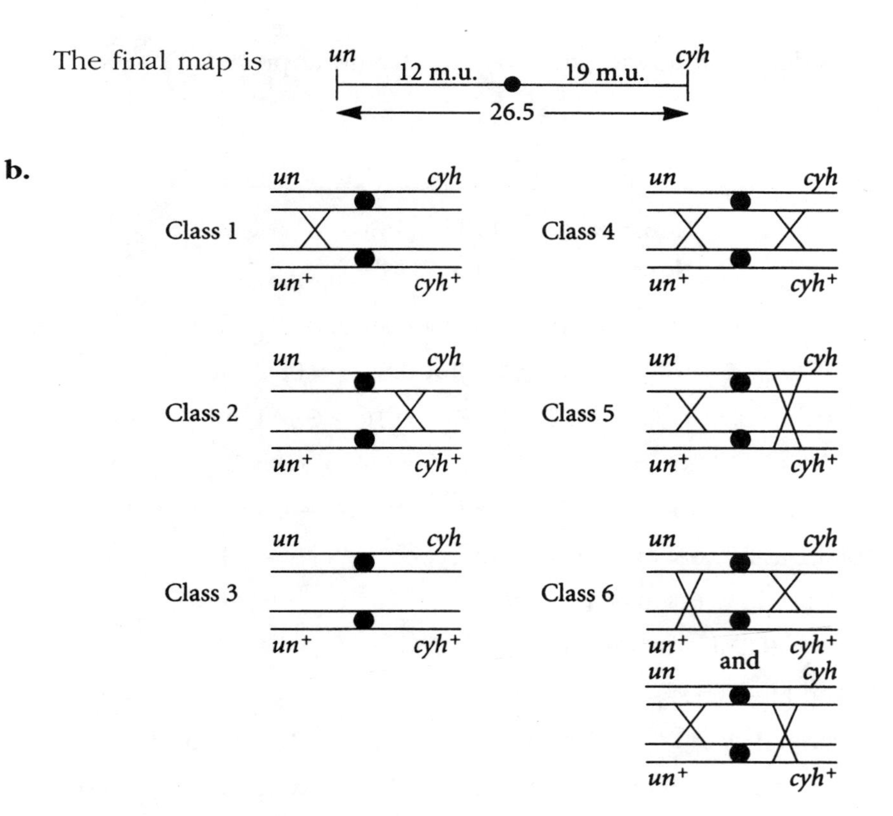

c. Figure 6-14 in the text can be used as a guide to help classify the asci types as to the number of crossovers involved:

Type	un	cyh	CO
1. T	II	I	SCO
2. T	I	II	SCO
3. PD	I	I	NCO
4. PD	II	II	DCO
5. NPD	II	II	DCO
6. T	II	II	DCO

The preceding table indicates that the six basic asci types that result from 0 or 1 crossovers between the centromere and the two genes are present. However, no asci that are the result of 2 crossovers between a gene I and the centromere are represented (e.g., an NPD ascus that exhibits M_I segregation for both genes). The missing asci types probably did not occur because of the relatively low number of asci that were studied.

d. Classes 4, 5, and 6 all involve double crossovers. Class 4 involves a two-strand double crossover, with both crossovers occurring between the centromere and each gene. There is only one way to achieve this result. Class 5 involves a four-strand double crossover, one between the centromere and *un* and one between the centromere and *cyh*. There is only one way of achieving this result. Class 6 involves a three-strand double crossover, with one between *un* and the centromere and one between *cyh* and the centromere. There are two ways of achieving this result. Therefore, class 6 should occur at twice the frequency of either of the other two classes.

23. The chromosome map is

leu3 ——●—— 8 m.u. ——|—— *cys2*

The cross is $leu3^+$ *cys2* × *leu3* $cys2^+$. First classify the asci types:

Type
a. NPD
b. PD
c. T
d. T
e. NPD
f. PD
g. T

Because *leu3* always segregates with the centromere, there are only two possible outcomes:

1. If a crossover occurs between *cys2* and the centromere, only tetratypes occur and they are at a frequency of 16%.
2. If no crossover occurs in that region, only parental ditypes occur and they are at a frequency of 84%.

a. The ascus is a NPD with M_I segregation for both loci. This can happen only when a double crossover occurs. The problem states that double crossovers should not be considered. Therefore, the answer is 0%.

b. The ascus is a PD with M_I segregation for both loci. This occurs only with no crossovers. The frequency is 84%.

c. The ascus is a T, with a crossover between *cys2* and the centromere. The frequency is 16%.

d–g. All four ascus types show M_{II} segregation for *leu3*, which cannot occur. The probability of each type is 0%.

24. The cross is

P	*g C s/g C s* × *G c S/G c S*	
F_1	*g C s/G c S*	wild type

Observation of the phenotypically distinct patches of cells in an otherwise wild-type fly indicate that mitotic crossing-over occurred.

a. The cells of the twin spot must be, respectively, $g/g \cdot C/– \cdot s/s$ and $G/– \cdot c/c \cdot S/–$. A single crossover between the centromere and *S* can yield this result:

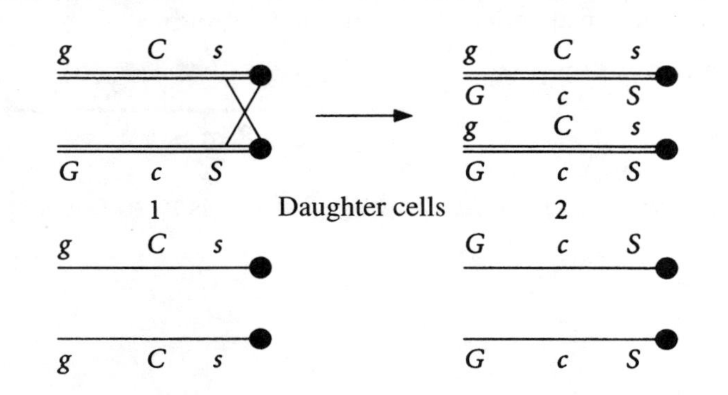

b. The cells of the twin spot must be, respectively, $g/g \cdot C/– \cdot S/–$ and $G/– \cdot c/c \cdot S/–$. A single crossover between *S* and *C* can yield this result:

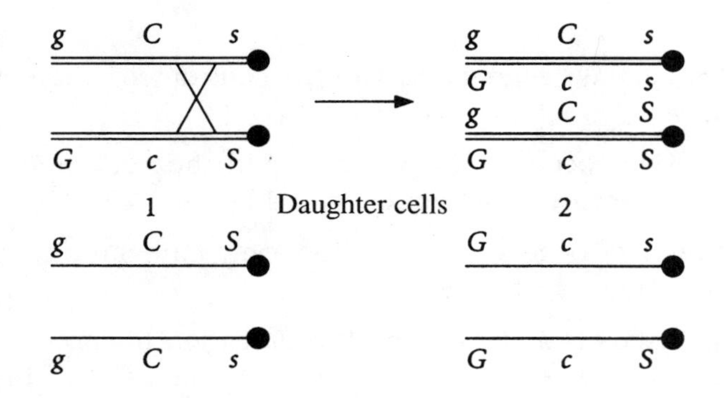

c. The cells of the single spot must be $g/g \cdot C/– \cdot S/–$. A single crossover between *C* and *G* can yield this result:

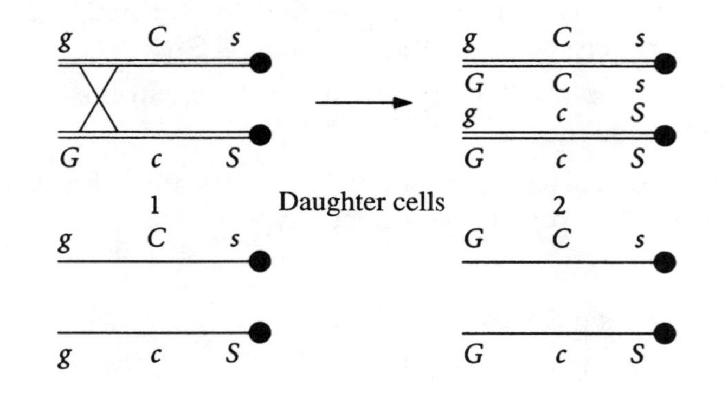

25. Recall that white diploid sectors will have two copies of the chromosome and two copies of each gene. The original chromosomes were

ad col phe pu sm w/ad^+ col^+ phe^+ pu^+ sm^+ w^+ (gene order is alphabetical)

a. Mitotic crossing-over might have occurred, causing *w* to become homozygous in one of the daughter cells. The following diagram illustrates this for alleles w/w^+ only:

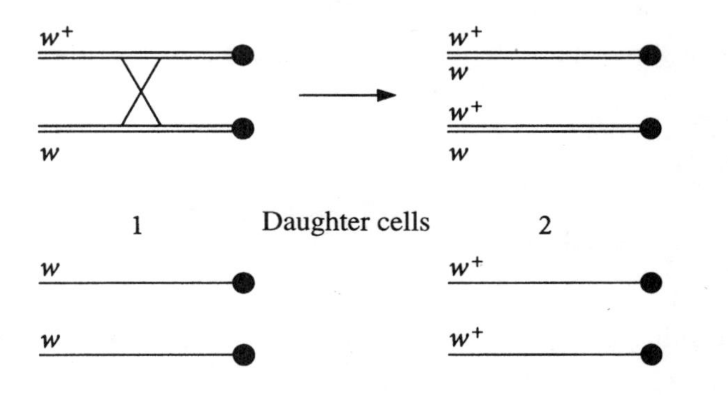

b. Note that any gene distal to the crossover will remain with the *w* allele if no other crossovers occur, while any gene proximal to the crossover (between the centromere and the crossover) will be separated from the *w* allele if no other crossovers occur. Therefore, the closer a proximal gene is to the *w* allele, the less likely it will be separated from *w* by the crossing-over event.

Using this logic, *pu* (with *w* in 100% of the sectors) is distal to *w*. The *col* allele is never observed with *w*, suggesting that it is proximal to it. In other words, *col* lies very close to the centromere. However, as will be discussed below, on which side of the centromere *col* is located cannot be determined. The other genes are between *w* and the centromere: *ad* (with *w* in 95% of the sectors), *sm* (with *w* in 65% of the sectors), and *phe* (with *w* in 41% of the sectors). The final gene order is

pu – *w* – *ad* – *sm* – *phe* – *col* – centromere

c. The relative map distances can be calculated directly from the cosegregation of phenotypes. If two genes are found together X% of the time, then crossing-over occurs between them (100 – X)% of the time.

w – *pu*: 100% – 100% = 0 relative map units. This finding indicates that the relative order determined above may not be exactly correct. It is possible that the gene order is

w – *pu* – *ad* – *sm* – *phe* – *col* – centromere

Without further information, no choice can be made between the two possibilities.

w – *ad*: 100% – 95% = 5 relative map units
w – *sm*: 100% – 65% = 35 relative map units
w– *phe*: 100% – 41% = 59 relative map units

The *w* – centromere distance cannot be calculated.

d. The phenotype associated with homozygous *col* was never observed. This means that *col* was very far from *w*. If *col* were between *w* and the centromere, then some of the sectors should have had a *col* phenotype unless *col* is so close to the centromere that no crossing-over occurs between them. If *col* were on the other side of the centromere from *w*, but close to the centromere so that no mitotic crossing-over occurred, then it would never be observed with *w*. Therefore, *col* is very close to the centromere but could be on either arm.

26. Let *g/g* = green, *G/g* = yellowish, and *G/G* = yellow. Mitotic crossing-over can account for the observations:

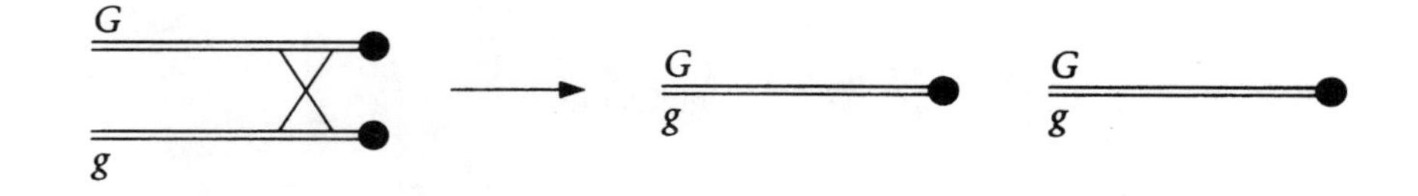

The resulting daughter cells would be *G/G* (yellow) and *g/g* (green).

27. Ignoring the white allele for a moment, the 1:1:1:1 ratio indicates independent assortment of two units, or chromosomes. The *a* and *c* alleles are found in equal proportions with *b* and b^+, indicating that *b* assorts independently of *a* and *c*. If *w* were linked on either chromosome, then the white allele would be found more frequently with one gene combination than another. Because it is not, *w* is not linked to either chromosome.

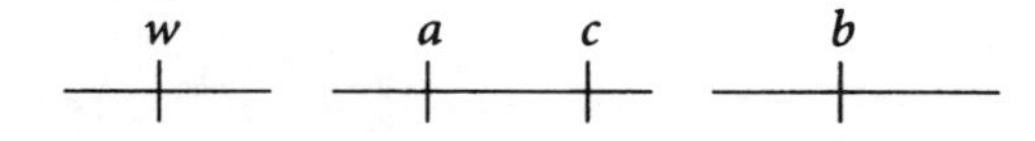

28. Remember that the segregants are yellow and diploid (80 percent $y/y \cdot r^+/-$, 20 percent $y/y \cdot r/r$). To get *y/y* segregants, crossing-over must occur between *y* and the centromere. There are three possible arrangements:

1. yellow — centromere — ribo
2. centromere — yellow — ribo
3. centromere — ribo — yellow

If 1 is correct, after crossing-over the chromosomes would be

y^+ r^+ / y r^+ and y r / y^+ r

Without a crossover between *r* and the centromere, no *r/r* segregants would be observed. Because the two genes are far away from each other, a 1:1 ratio of riboflavin-requiring and non-riboflavin-requiring (wild type for ribofalvin) would be expected.

If 2 is correct, after crossing-over the chromosomes would be

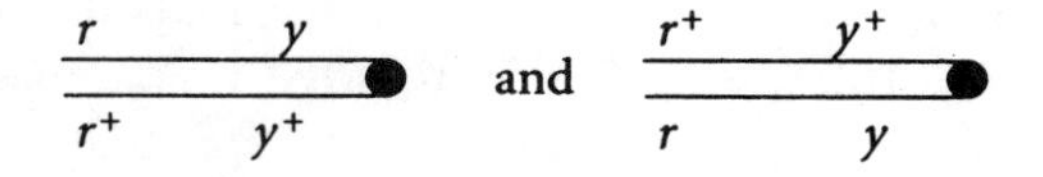

Unless the two genes are more than 50 m.u. apart, most of the *y/y* segregants would be riboflavin-requiring.

If 3 is correct, crossing-over between *r* and the centromere would give riboflavin-requiring segregants, and crossing-over between *y* and the centromere would give wild type for riboflavin. Therefore, both types of segregants would be observed with only one crossover. Furthermore, the 80:20 ratio suggests that crossing-over occurs more frequently between *y* and the centromere.

29. a. All *fpa/fpa* diploid segregants experienced a crossover between *fpa* and the centromere. The data indicate that *pro* and *paba* are linked to *fpa*. The different frequencies are a measure of the distances involved. Because there are no progeny requiring only proline, *pro* is closer to the centromere than is *paba*.

b.

pro – paba: 100%(110)/154 = 71.4 relative m.u.

paba – fpa: 100%(35)/154 = 22.7 relative m.u.

pro – centromere: 100%(9)/154 = 5.8 relative m.u.

c. *pro paba fpa*

30. a. To work this problem, you must first realize that resistance to fluorophenylalanine is recessive and not wild type. Thus, selection was for *fpa* or *fpa/fpa*, depending on whether cells were haploid or diploid.

Note that all the haploid colonies were *fpa leu ribo*, which suggests linkage between these three genes because neither *leu* nor *ribo* was selected. There were approximately equal numbers of colonies that were either *ad w* or ad^+ w^+, which suggests that these two genes are linked and independently assorting from the *fpa*-bearing chromosome.

b. You are given information regarding only one chromosome for the colonies that appeared among the diploids. That chromosome is the *fpa*-bearing chromosome. The phenotypes of the colonies are

24	*fpa* leu^+ $ribo^+$
17	*fpa* leu^+ *ribo*
14	*fpa leu ribo*
55	

Because the colonies are diploid, mitotic crossing-over had to have occurred. All colonies represent a mitotic crossover between *fpa* and the centromere. This is diagrammed below, with the assumption that the gene order is centromere – *leu* – *ribo* – *fpa*. If order existed other than the one assumed, more complicated crossovers would be required to yield one or more of the phenotypes.

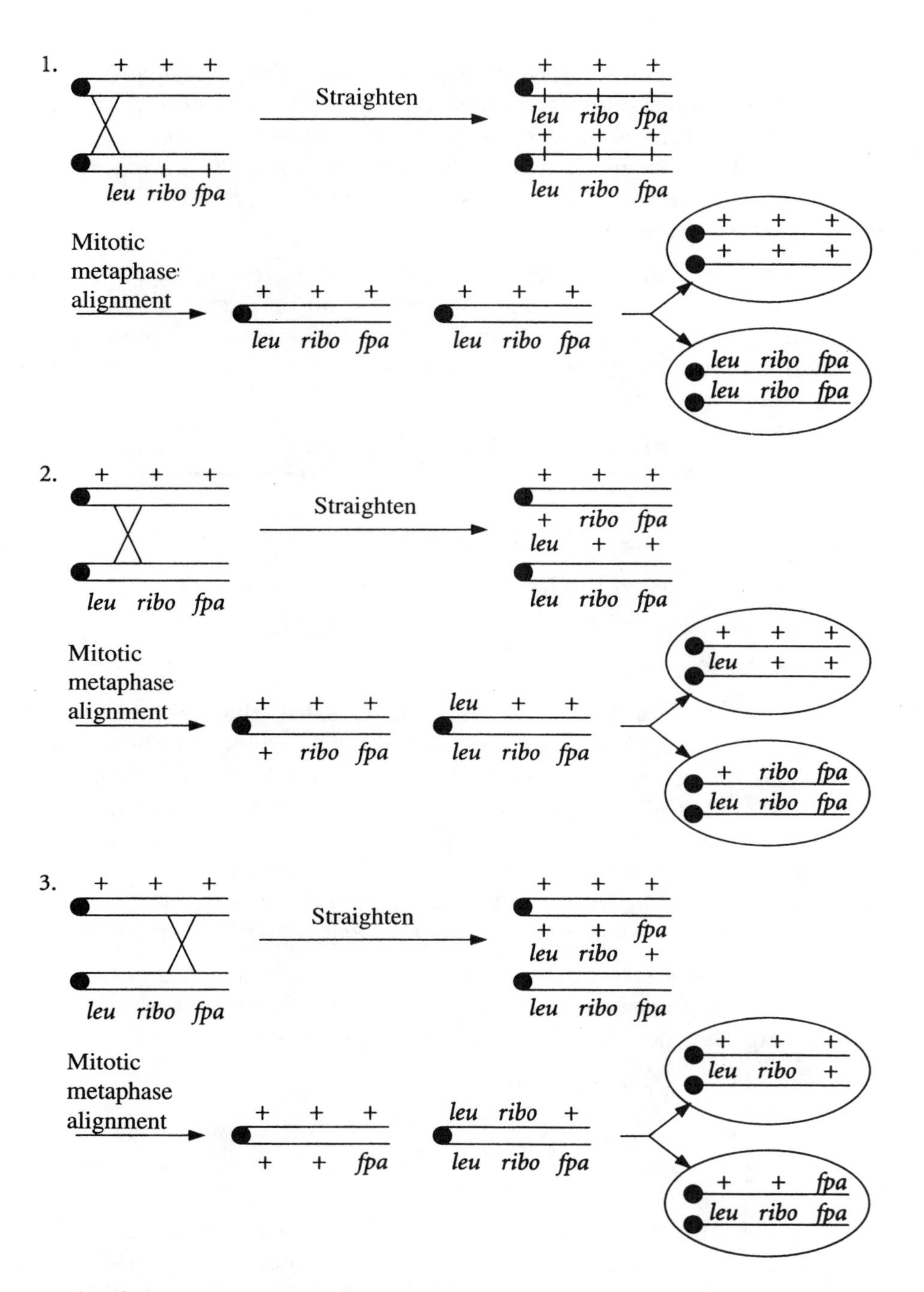

The frequency of each type reflects the relative size for each region.

31. a. and b. There are four patterns that can be observed in the comparisons that can be made between these six markers: + +, – –, + –, and – +. The first two indicate concordance, and the second two indicate a lack of concordance. Ideally, data would show either 100% concordance for the seven hybrids, indicating linkage, or 100% discordance for the seven hybrids, indicating a lack of linkage.

Because radiation hybrids involve chromosome breakage, two genes that are located very close together on the same chromosome may show some discordance despite the close linkage. Two genes that are located on different chromosomes may also show some concordance owing to the chance that two separate fragments may become established within a single hybrid line. Therefore, the problem is how to distinguish between reduced concordance because of chromosome fragmentation and chance concordance because two fragments from different chromosomes are in the same hybrid. Obviously, a statistical solution is needed, but there are not enough data in this problem for a statistical analysis.

Sort the data into three groups: 100% concordance, 100% discordance, and mixed (concordance/discordance). This follows below:

100% concordance	100% discordance	Mixed
EF	None	A-B 2/5
		A-C 2/5
		A-D 6/1
		A-E 2/5
		A-F 2/5
		B-C 5/2
		B-D 1/6
		B-E 3/4
		B-F 3/4
		C-D 3/4
		C-E 3/4
		C-F 3/4
		D-E 3/4
		D-F 3/4

Markers E and F are most likely located on the same chromosome. Markers B and D may be located on different chromosomes.

In the absence of statistical analysis, with so few total hybrids, it is important to pay more attention to the + + patterns than the – – patterns simply because – – can arise either from linkage, with the specific chromosome missing in the hybrid, or from lack of linkage, with the two chromosomes (or fragments) lacking in the hybrid. Therefore, going back to the mixed category and focusing on those marker pairs that had a high degree, but not 100%, of concordance, one sees that the 6/1 pattern of A-D and the 5/2 pattern of B-C stand out. For the A-D pair, 3 of the 6 concordances are + +, while only 2 of the 5 concordances for B-C are + +. It is unclear from the data whether this is a significant difference, and significance cannot be determined in any fashion. Therefore, it would be important to collect more data before drawing further conclusions.

32. α. The only chromosome missing in A and B and present in C is 7.

β. The only chromosome present in all colonies is 1.

γ. The only chromosome missing in A and present in B and C is 5.

δ. The only chromosome present only in B is 6.

ε. Not on chromosomes 1 through 8.

33. For each enzyme, the goal is to determine which chromosome or chromosome arm is found in all positive cell lines and is also absent in all negative cell lines. The data indicate the following gene locations:

Steroid sulfatase: Xp
Phosphoglucomutase-3: 6q
Esterase D: 13q
Phosphofructokinase: 21
Amylase: 1p
Galactokinase: 17q.

34. **a.** Breaks in different regions of 17R result in deletion of all genes from the breakpoint.

b. Because the only human DNA is from 17R, all the human genes expressed must be on 17R. Notice that only gene *c* is expressed by itself. This means that gene *c* is closest to the mouse DNA. Next notice that if *c* and one other gene are expressed, that other gene is always *b*. This means *b* is closer to the mouse DNA than *a* is. The gene order is mouse—*c*—*b*—*a*.

The probability of a break between two genes is a function of the distance between them. Of the 200 lines tested, 48 expressed no human activity. Thus, the breakpoint was between the *c* gene and the mouse centromere. Similarly, a break between *c* and *b* (cells express c only) occurred in 12 lines, and a break between *b* and *a* (cells express c and b) occurred in 80 lines. Finally, 60 lines expressed all three genes, placing the breakpoint between *a* and the end of the chromosome.

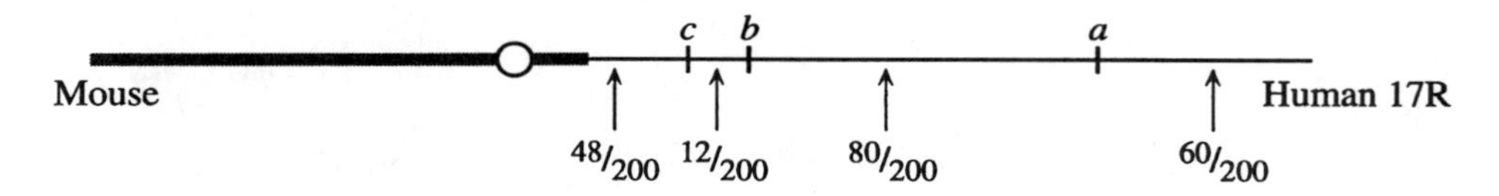

c. The banding patterns of 17R could be used to verify the suggested map as well as provide a more accurate position for the three genes. For example, the 60 lines that express all three genes could be analyzed to see how much of 17R can be removed without deleting gene *a*. The line with the most removed will have the closest breakpoint to the gene from the end. By analyzing the 80 lines that express *c* and *b* but not *a*, the line with the least deleted will define the breakpoint that is closest to *a* from the centromere. By comparing both, the region in between defines the location of gene *a*.

35. Since the problem stipulates that crossing-over does not occur between the two genes, asci that show T or NPD patterns are not possible. Single crossovers between the genes and the centromere will generate PD asci that show $M_{II}M_{II}$ segregation. Three-strand double crossovers will generate the same while two- and four-strand double crossovers will generate PD asci that show M_IM_I segregation.

The formula that is needed to calculate the proportion of meiocytes with different numbers of crossovers is $f(i) = e^{-m}m^i/i!$, where $i = 0$, 1, or 2, e is the base of natural logarithms, and m is the mean number of crossovers.

To calculate m, use the formula RF = $(1 - e^{-m})/2$.

$$0.08 = (1 - e^{-m})/2$$
$$e^{-m} = 0.84$$
$$m = 0.17$$

The probability of no crossover is

$$f(0) = e^{-0.17}(0.17)^0/0! = 0.84, \text{ or } 84\%$$

The probability of one crossover is

$$f(1) = e^{-0.17}(0.17)^1/1! = 0.143, \text{ or } 14.3\%$$

The probability of two crossovers is

$$f(2) = e^{-0.17}(0.17)^2/2! = 0.012, \text{ or } 1.2\%$$

Higher numbers of crossovers (totaling 0.5%) will be ignored.

a. no crossover + $^1/_2$ the double crossovers = 84% + 0.6% = 84.6%

b. 0%

c. 0%

d. 0%

e. single crossover + $^1/_2$ the double crossovers = 14.3% + 0.6% = 14.9%

f. 0%

g. 0%

Again, because there is no crossing-over between the genes, 50% of random spores will be wild type and grow on minimal medium.

36. The cross is $al \cdot cp^+ \times al^+ \cdot cp$,and the two unordered tetrads can be classified as NPD and T.

a. A single crossover between the genes will give rise to a T ascus.

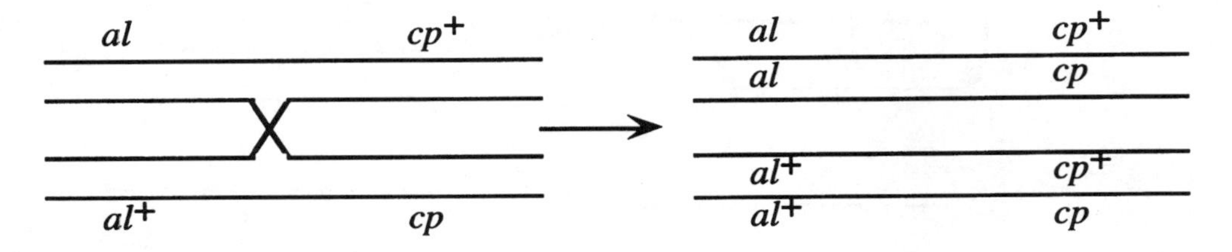

A four-strand double crossover between the genes is necessary for an NPD ascus.

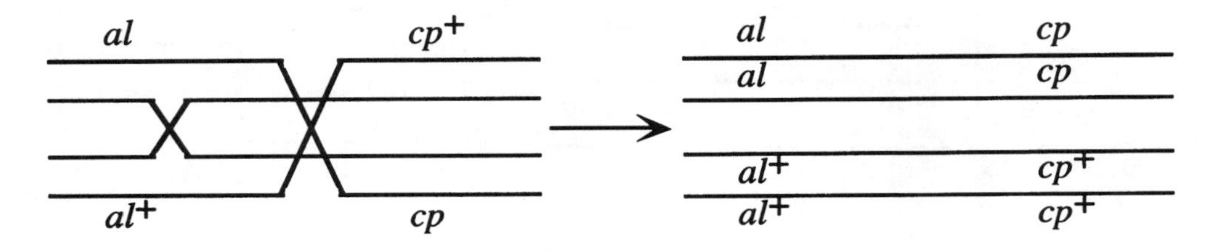

b. Independent assortment of unlinked genes can generate either a T or NPD locus.

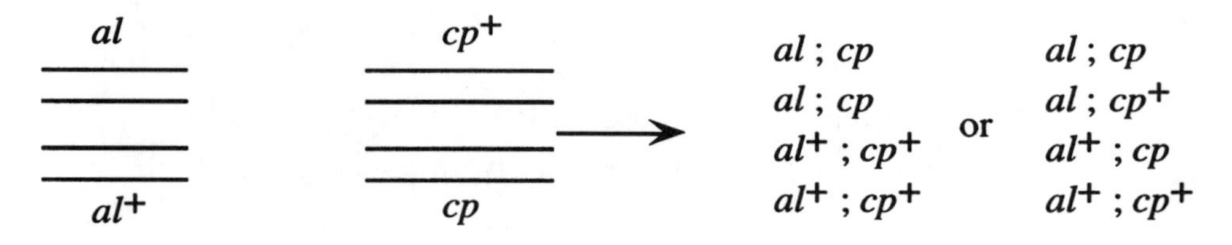

Although NPD asci would be more common for unlinked genes, you cannot make any determination based on just two asci.

37. a. and b. It is likely that the observed abnormalities are the result of mitotic recombination. A crossover between the genes of interest and the centromere in this case will lead to "twin spots" (the adjacent patches of stubby and ebony body observed). On the other hand, a cross-over that occurs between the two genes will lead to "single spots" (the solitary patches of ebony). The position of the two genes with respect to the centromere determines the phenotype of the single spot. When recombination occurs in the region between the genes, the gene more distal to the centromere becomes homozygous while the more proximal gene remains heterozygous. In this problem, since the single patches are ebony, *e* is more distal than *s*. The other product of this event will appear normal and will not be detected among the other "normal" cells.

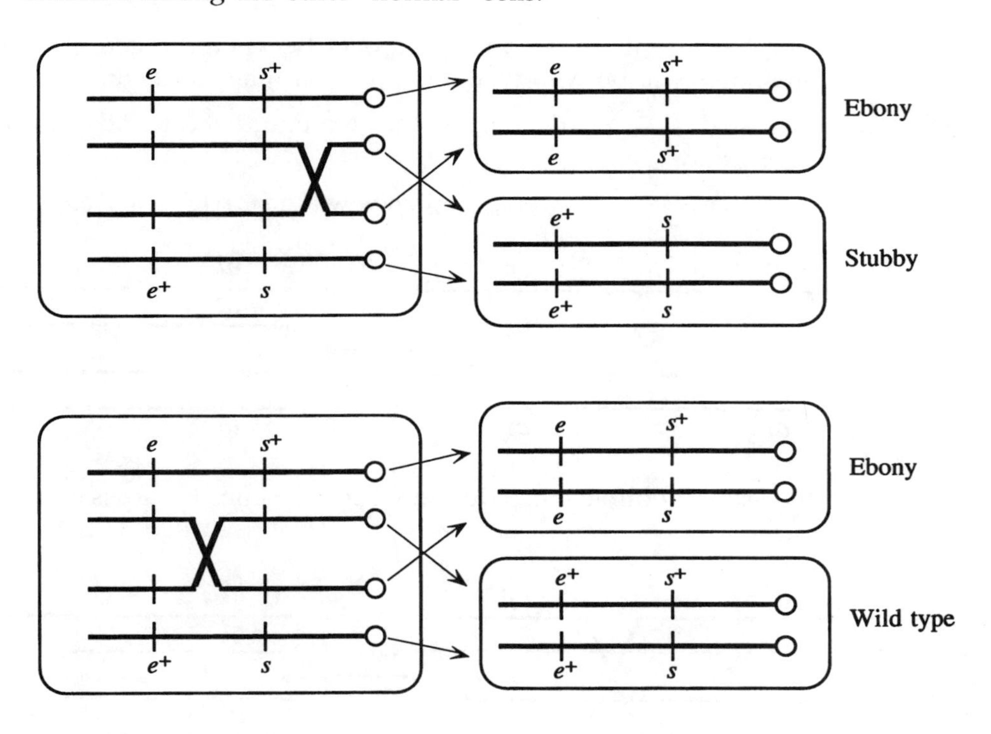

7 Gene Transfer in Bacteria and Their Viruses

1. An Hfr strain has the fertility factor F integrated into the chromosome. An F^+ strain has the fertility factor free in the cytoplasm. An F^- strain lacks the fertility factor.

2. All cultures of F^+ strains have a small proportion of cells in which the F factor is integrated into the bacterial chromosome and are, by definition, Hfr cells. These Hfr cells transfer markers from the host chromosome to a recipient during conjugation.

3. **a.** Hfr cells involved in conjugation transfer host genes in a linear fashion. The genes transferred depend on both the Hfr strain and the length of time during which the transfer occurred. Therefore, a population containing several different Hfr strains will appear to have an almost random transfer of host genes. This is similar to generalized transduction, in which the viral protein coat forms around a specific amount of DNA rather than specific genes. In generalized transduction, any gene can be transferred.

   b. F′ factors arise from improper excision of an Hfr from the bacterial chromosome. They can have only specific bacterial genes on them because the integration site is fixed for each strain. Specialized transduction resembles this in that the viral particle integrates into a specific region of the bacterial chromosome and then, upon improper excision, can take with it only specific bacterial genes. In both cases, the transferred gene exists as a second copy.

4. Generalized transduction occurs with lytic phages that enter a bacterial cell, fragment the bacterial chromosome, and then, while new viral particles are

being assembled, improperly incorporate some bacterial DNA within the viral protein coat. Because the amount of DNA, not the information content of the DNA, is what governs viral particle formation, any bacterial gene can be included within the newly formed virus. In contrast, specialized transduction occurs with improper excision of viral DNA from the host chromosome in lysogenic phages. Because the integration site is fixed, only those bacterial genes very close to the integration site will be included in a newly formed virus.

5. While the interrupted-mating experiments will yield the gene order, it will be relative only to fairly distant markers. Thus, the precise location cannot be pinpointed with this technique. Generalized transduction will yield information with regard to very close markers, which makes it a poor choice for the initial experiments because of the massive amount of screening that would have to be done. Together, the two techniques allow, first, for a localization of the mutant (interrupted-mating) and, second, for precise determination of the location of the mutant (generalized transduction) within the general region.

6. This problem is analogous to forming long gene maps with a series of three-point testcrosses. Arrange the four sequences so that their regions of overlap are aligned:

$\overline{M}$—*Z*—*X*—*W*—*C*
W—*C*—*N*—*A*—*L*
A—*L*—*B*—*R*—*U*
B—*R*—*U*—$\underline{M—Z}$

The regions with the bars above or below are identical in sequence (and "close" the circular chromosome). The correct order of markers on this circular map is

—*M*—*Z*—*X*—*W*—*C*—*N*—*A*—*L*—*B*—*R*—*U*—

7. To interpret the data, the following results are expected:

Cross	Result
F$^+$ × F$^-$	(L) low number of recombinants
Hfr × F$^-$	(M) many recombinants
Hfr × Hfr	(0) no recombinants
Hfr × F$^+$	(0) no recombinants
F$^+$ × F$^+$	(0) no recombinants
F$^-$ × F$^-$	(0) no recombinants

The only strains that show both the (L) and the (M) result when crossed are 2, 3, and 7. These must be F$^-$, since that is the only cell type that can participate in a cross and give either recombination result. Hfr strains will result in only (M) or (0), and F$^+$ will result in only (L) or (0) when crossed. Thus, strains 1 and 8 are F$^+$, and strains 4, 5, and 6 are Hfr.

8. **a.**

Agar type	Selected genes
1	c^+
2	a^+
3	b^+

b. The order of genes is revealed in the sequence of colony appearance. Because colonies first appear on agar type 1, which selects for c^+, *c* must be first. Colonies next appear on agar type 3, which selects for b^+, indicating that *b* follows *c*. Allele a^+ appears last. The gene order is *c b a*. The three genes are roughly equally spaced.

c. In this problem you are looking at the cotransfer of the selected gene with the d^- allele (both from the Hfr). Cells that are d^- do not grow because the medium is lacking D and selecting for those cells that are d^+. Therefore, the farther a gene is from gene *d*, the less likely cotransfer of the selected gene will occur with d^- and the more likely that colonies will grow (remain d^+). From the data, *d* is closest to *b* (only $^8/_{100}$ did not cotransfer d^- with b^+). It is also closer to *a* than it is to *c*. Thus the gene order is *c b d a* (or *a d b c*).

d. With no A or B in the agar, the medium selects for a^+ b^+, and the first colonies should appear at about 17.5 minutes.

9. First, carry out a series of crosses in which you select in a long mating each of the auxotrophic markers. Thus, select for arg^+ T^r. In each case score for penicillin resistance. Although not too informative, these crosses will give the marker that is closest to pen^r by showing which marker has the highest linkage. Then do a second cross concentrating on the two markers on either side of the pen^r locus. Suppose that the markers are *ala* and *glu*. You can first verify the order by taking the cross in which you selected for ala^+, the first entering marker, and scoring the percentage of both pen^r and glu^+. Because of the gradient of transfer, the percentage of pen^r should be higher than the percentage of glu^+ among the selected ala^+ recombinants.

Then, take the mating in which glu^+ was the selected marker. Since this marker enters last, one can use the cross data to determine the map units by determining the percentage of colonies that are ala^+ pen^r, and by the number of ala^- pen^r colonies, as shown in Figure 7-13 of the companion text.

10. **a.** Determine the gene order by comparing arg^+ bio^+ leu^- with arg^+ bio^- leu^+. If the order were *arg leu bio*, four crossovers would be required to get arg^+ leu^- bio^+, while only two would be required to get arg^+ leu^+ bio^-. If the order is arg bio leu, four crossovers would be required to get arg^+ bio^- leu^+, and only two would be required to get arg^+ bio^+ leu^-. There are 8 recombinants that are arg^+ bio^+ leu^- and none that are arg^+ bio^- leu^+. On the basis of the frequencies of these two classes, the gene order is *arg bio leu*.

b. The *arg–bio* distance is determined by calculating the percentage of the exconjugants that are arg^+ bio^- leu^-. These cells would have had a crossing-over event between the *arg* and *bio* genes.

$$\text{RF} = 100\%(48)/376 = 12.76 \text{ m.u.}$$

Similarly, the *bio-leu* distance is estimated by the arg^+ bio^+ leu^- colony type.

$$RF = 100\%(8)/376 = 2.12 \text{ m.u.}$$

11. The most straightforward way would be to pick two Hfr strains that are near the genes in question but are oriented in opposite directions. Then, measure the time of transfer between two specific genes, in one case when they are transferred early and in the other when they are transferred late. For example,

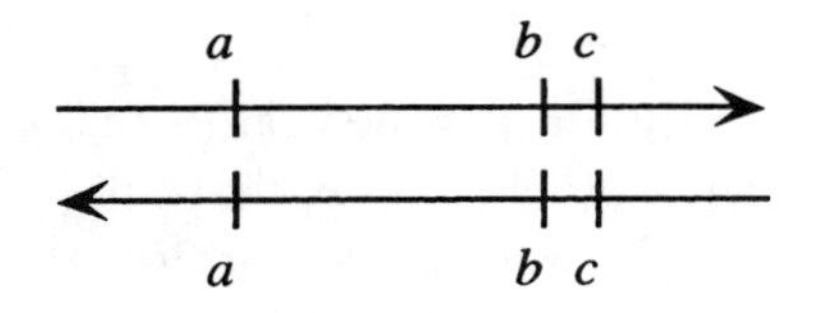

12. a. To survive on the selective medium all cultures must be ery^r. Keep in mind that cells from all 300 colonies were each tested under four separate conditions.

When only erythromycin is added, 263 colonies grew, so these must be arg^+ aro^+ ery^r. The remaining 37 cultures are mutant for one or both genes. One additional colony grew if arginine was also added to the medium (264 – 263 = 1). It must be arg^- aro^+ ery^r. A total of 290 colonies are arg^+ because they grew when erythromycin and aromatic amino acids were added to the medium. Of these, 27 are aro^- (290 – 263 = 27). The genotypes and their frequencies are summarized below:

263	ery^r arg^+ aro^+
27	ery^r arg^+ aro^-
1	ery^r arg^- aro^+
9	ery^r arg^- aro^-
300	

b. Recombination in the *aro–arg* region is represented by two genotypes: aro^+ arg^- and aro^- *arg*+. The frequency of recombination is

$$100\%(1 + 27)/300 = 9.3 \text{ m.u.}$$

Recombination in the ery–arg region is represented by two genotypes: aro^+ arg^- and aro^- arg^-. The frequency of recombination is

$$100\%(1 + 9)/300 = 3.3 \text{ m.u.}$$

Recombination in the *ery–aro* region is represented by three genotypes: arg^+ aro^-, arg^- aro^-, and arg^- aro^+. Recall that the DCO must be counted twice. The frequency of recombination is

$$100\%(27 + 9 + 2)/300 = 12.6 \text{ m.u.}$$

c. The ratio is 28:10, or 2.8:1.0.

13. The best explanation is that the integrated F factor of the Hfr looped out of the bacterial chromosome abnormally and is now an F′ that contains the pro^+ gene. This F′ is rapidly transferred to F^- cells, converting them to pro^+ (and F^+).

14. The high rate of integration and the preference for the same site originally occupied by the F factor suggest that the F′ contains some homology with the original site. The source of homology could be a fragment of the F factor, or more likely, it is homology with the chromosomal copy of the bacterial gene that is also present on the F′.

15. First carry out a cross between the Hfr and F^-, and then select for colonies that are *ala*$^+$ *str*^{r}. If the Hfr donates the *ala* region late, then redo the cross but now interrupt the mating early and select for *ala*$^+$. This selects for an F′, since this Hfr would not have transferred the *ala* gene early.

If the Hfr instead donates this region early, then use a Rec^- strain that cannot incorporate a fragment of the donor chromosome by recombination. Any *ala*$^+$ colonies from the cross should then be used in a second mating to another *ala*$^-$ strain to see whether they can donate the *ala* gene easily, which would indicate that there is F′ *ala*. (This would also require another marker to differentiate the donor and recipient strains. For example, the *ala*$^-$ strain could be tetracycline^r and selection would be for *ala*$^+$ *tet*^{r}.)

16. a. and b.

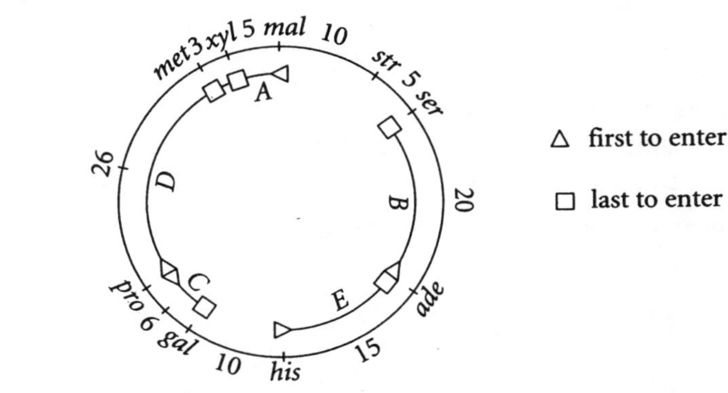

c. A: Select for *mal*$^+$
B: Select for *ade*$^+$
C: Select for *pro*$^+$
D: Select for *pro*$^+$
E: Select for *his*$^+$

17. a. If the two genes are far enough apart to be located on separate DNA fragments, then the frequency of double transformants should be the product of the frequency of the two single transformants, or (4.3%) × (0.40%) = 0.017%. The observed double transformant frequency is 0.17%, a factor of 10 greater than expected. Therefore, the two genes are located close enough together to be cotransformed at a rate of 0.17%.

b. Here, when the two genes must be contained on separate pieces of DNA, the rate of cotransformation is much lower, confirming the conclusion in part (a).

18. a. To determine which genes are close, compare the frequency of double transformants. Pairwise testing gives low values whenever B is involved but fairly high rates when any drug but B is involved. This suggests that the gene for B resistance is not close to the other three genes.

b. To determine the relative order of genes for resistance to A, C, and D, compare the frequency of double and triple transformants. The frequency of resistance to AC is approximately the same as resistance to ACD. This strongly suggests that D is in the middle. Also, the frequency of AD coresistance is higher than AC (suggesting that the gene for A resistance is closer to D than to C), and the frequency of CD is higher than AC (suggesting that C is closer to D than to A).

19. The expected number of double recombinants is (0.01)(0.002)(100,000) = 2. Interference = 1 – (observed DCO/expected DCO) = 1 – 5/2 = –1.5. By definition, the interference is negative.

20. a. The parental genotypes are + + + and *m r tu*. For determining the *m–r* distance, the recombinant progeny are

m + *tu*	162
m + +	520
+ *r tu*	474
+ *r* +	172
	1328

Therefore the map distance is 100%(1328)/10,342 = 12.8 m.u.

Using the same approach, the *r–tu* distance is 100%(2152)/10,342 = 20.8 m.u., and the *m–tu* distance is 100%(2812)/10,342 = 27.2 m.u.

b. Because genes *m* and *tu* are the farthest apart, the gene order must be *m r tu*.

c. The coefficient of coincidence (c.o.c.) compares the actual number of double crossovers with the expected number (where c.o.c. = observed double crossovers/expected double crossovers). For these data, the expected number of double recombinants is (0.128)(0.208)(10,342) = 275. Thus, c.o.c. = (162 + 172)/275 = 1.2. This indicates that there are more double crossover events than predicted and suggests that the occurrence of one crossover makes a second crossover between the same DNA molecules more likely to occur.

21. a. I: minimal plus proline and histidine
II: minimal plus purines and histidine
III: minimal plus purines and proline

b. The order can be deduced from cotransfer rates. It is *pur–his–pro*.

c. The closer the two genes, the higher the rate of cotransfer; *his* and *pro* are closest.

d. Transduction of pro^+ requires a crossover on both sides of the *pro* gene. Because *his* is closer to *pro* than *pur*, you get the following:

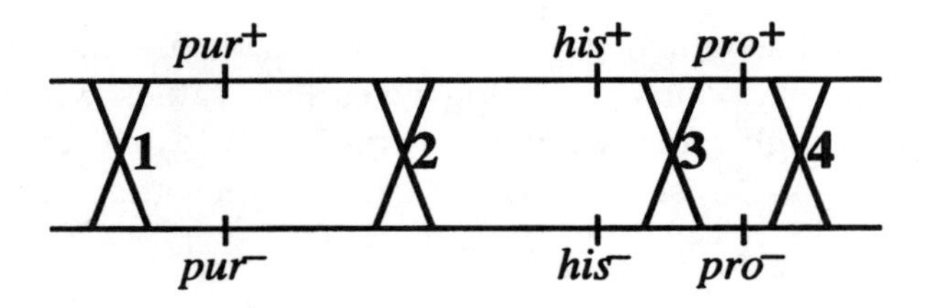

Genotypes	Frequency	Crossovers
pur⁻ his⁻	43%	4 and 3
pur⁺ his⁻	0%	4 and 3 and 2 and 1
pur⁻ his⁺	55%	4 and 2
pur⁺ his⁺	2%	4 and 1

As can be seen, a *pur⁺ his⁻ pro⁺* genotype requires four crossovers and, as expected, would occur less frequently (in this example, 0%).

22. In a small percent of the cases, *gal⁺* transductants can arise by recombination between the *gal⁺* DNA of the λdgal transducing phage and the *gal⁻* gene on the chromosome. This will generate *gal⁺* transductants without phage integration.

23. a. This appears to be specialized transduction. It is characterized by the transduction of specific markers based on the position of the integration of the prophage. Only those genes near the integration site are possible candidates for misincorporation into phage particles that then deliver this DNA to recipient bacteria.

b. The only media that supported colony growth were those lacking either cysteine or leucine. These selected for *cys⁺* or *leu⁺* transductants and indicate that the prophage is located in the *cys-leu* region.

24. a. and b.

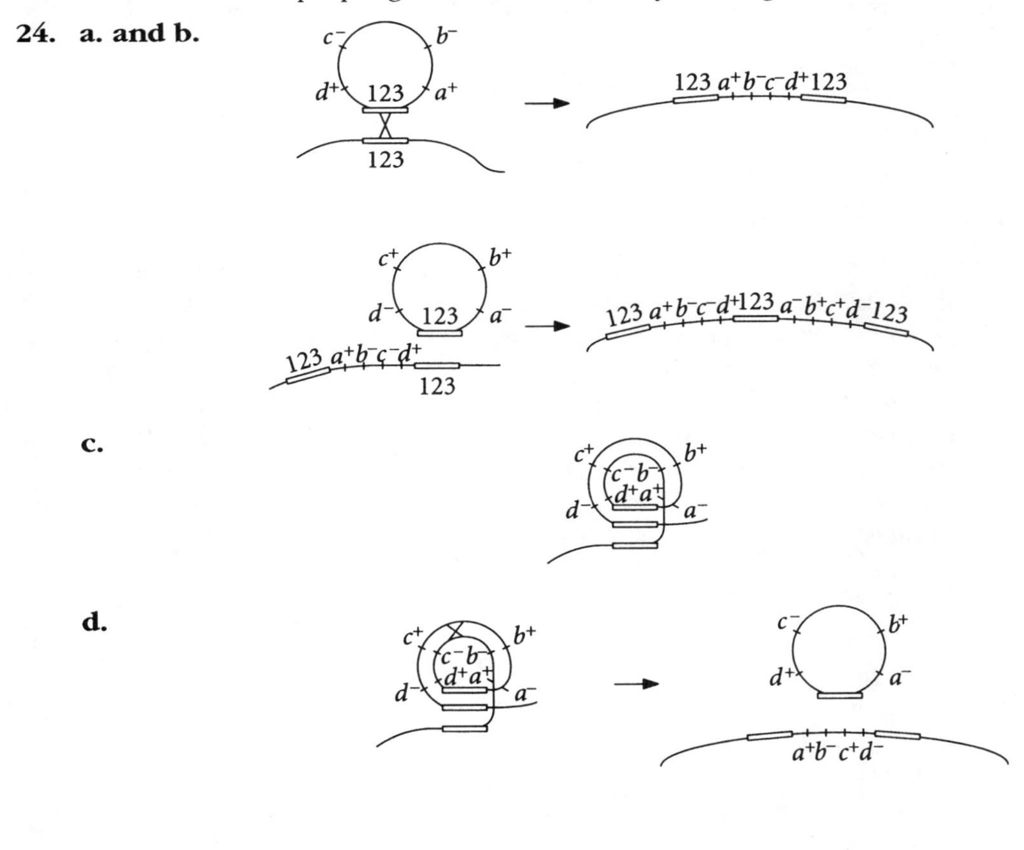

25. If *trp*1 and *trp*2 are alleles of the same locus, then a cross between strains A and B will not result in Trp^+ cells; if they are not allelic, strain B cells that have received the F′ from strain A, will be Trp^+ (complementation).

26. If a compound is not added and growth occurs, the *E. coli* recipient cell must have received the wild-type genes for production of those nutrients by transduction. Thus, the BCE culture selects for cells that are now a^+ and d^+, the BCD culture selects for cells that are a+ and e+, and the ABD culture selects for cells that are c^+ and e^+. These genes can be aligned, see below, to give the map order of *d a e c*. (Notice that *b* is never cotransduced and is therefore distant from this group of genes.)

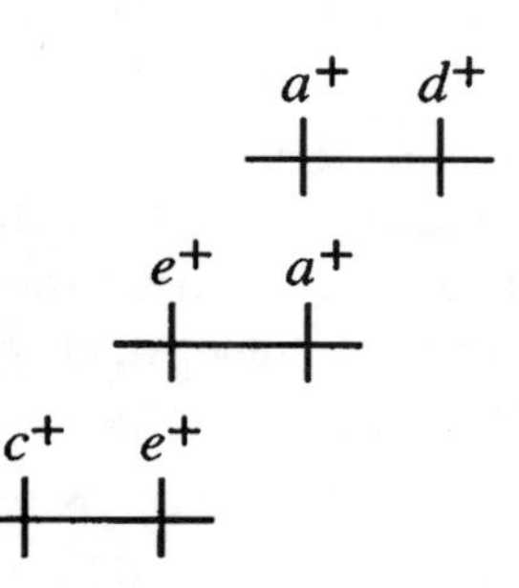

27. a. This is simply calculated as the percentage of pur^+ colonies that are also nad^+:

$$= 100\%(3 + 10)/50 = 26\%$$

b. This is calculated as the percentage of pur^+ colonies that are also pdx^-:

$$= 100\%(10 + 13)/50 = 46\%$$

c. *pdx* is closer, as determined by cotransduction rates.

d. From the cotransduction frequencies, you know that *pdx* is closer to *pur* than *nad* is, so there are two gene orders possible: *pur pdx nad* or *pdx pur nad*. Now, consider how a bacterial chromosome that is pur^+ pdx^+ nad^+ might be generated given the two gene orders: if *pdx* is in the middle, 4 crossovers are required to get pur^+ pdx^+ nad^+; if pur is in the middle, only 2 crossovers are required (see next page). The results indicate that there are fewer pur^+ pdx^+ nad^+ transductants than any other class, suggesting that this class is "harder" to generate than the others. This implies that *pdx* is in the middle and the gene order is *pur pdx nad*.

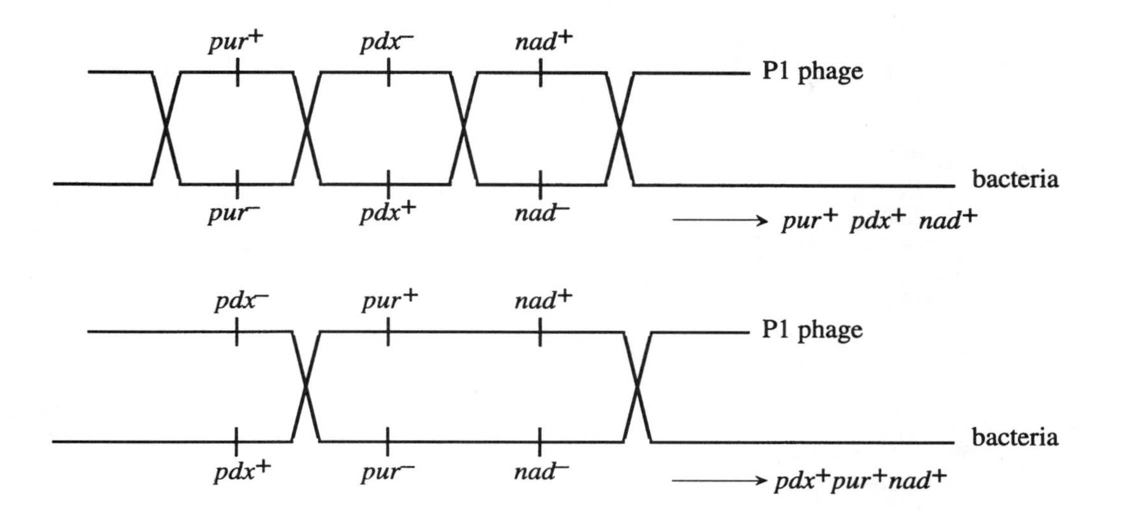

28. **a.** Owing to the medium used, all colonies are *cys*$^+$ but either + or – for the other two genes.

b. (1) *cys*$^+$ *leu*$^+$ *thr*$^+$ and *cys*$^+$ *leu*$^+$ *thr*$^-$ (supplemented with threonine)

(2) *cys*$^+$ *leu*$^+$ *thr*$^+$ and *cys*$^+$ *leu*$^-$ *thr*$^+$ (supplemented with leucine)

(3) *cys*$^+$ *leu*$^+$ *thr*$^+$ (no supplements)

c. Because none grew on minimal medium, no colony was *leu*$^+$ *thr*$^+$. Therefore, medium (1) had *cys*$^+$ *leu*$^+$ *thr*$^-$, and medium (2) had *cys*$^+$ *leu*$^-$ *thr*$^+$. The remaining cultures were *cys*$^+$ *leu*$^-$ *thr*$^-$, and this genotype occurred in 100% – 56% – 5% = 39% of the colonies.

d. *cys* and *leu* are cotransduced 56% of the time, while *cys* and *thr* are cotransduced only 5% of the time. This indicates that *cys* is closer to *leu* than it is to *thr*. Since no *leu*$^+$ *cys*$^+$ *thr*$^+$ cotransductants are found, it indicates that *cys* is in the middle.

29. To isolate the specialized transducing particles of phage ϕ80 that carried *lac*$^+$, the researchers would have had to lysogenize the strain with ϕ80, induce the phage with UV, and then use these lysates to transduce a Lac$^-$ strain to Lac$^+$. Lac$^+$ colonies would then be used to make a new lysate, which should be highly enriched for the *lac*$^+$ transducing phage.

8 The Structure and Replication of DNA

1. The DNA double helix is held together by two types of bonds, covalent and hydrogen. Covalent bonds occur within each linear strand and strongly bond the bases, sugars, and phosphate groups (both within each component and between components). Hydrogen bonds occur between the two strands and involve a base from one strand with a base from the second in complementary pairing. These hydrogen bonds are individually weak but collectively quite strong.

2. Conservative replication is a form of DNA synthesis in which the two template strands remain together but dictate the synthesis of two new DNA strands, which then form a second DNA helix. The end point is two double helices, one containing only old DNA and one containing only new DNA. This hypothesis was found not to be correct. Semiconservative replication is a form of DNA synthesis in which the two template strands separate and each dictates the synthesis of a new strand. The end point is two double helices, both containing one new and one old strand of DNA. This hypothesis was found to be correct.

3. A primer is a short segment of RNA that is synthesized by primase using DNA as a template during DNA replication. Once the primer is synthesized, DNA polymerase then adds DNA to the 3′ end of the RNA. Primers are required because the major DNA polymerase involved with DNA replication is unable to initiate DNA synthesis and, rather, requires a 3′ end. The RNA is subsequently removed and replaced with DNA so that no gaps exist in the final product.

4. Helicases are enzymes that disrupt the hydrogen bonds which hold the two DNA strands together in a double helix. This breakage is required for both RNA and DNA synthesis. Topoisomerases are enzymes that create and relax supercoiling in the DNA double helix. The supercoiling itself is a result of the twisting of the DNA helix that occurs when the two strands separate.

5. Because the DNA polymerase is capable of adding new nucleotides only at the 3′ end of a DNA strand, and because the two strands are antiparallel, at least two molecules of DNA polymerase must be involved in the replication of any specific region of DNA. When a region becomes single-stranded, the two strands have an opposite orientation. Imagine a single-stranded region that runs from right to left. The 5′ end is at the right, with the 3′ end pointing to the left; synthesis can initiate and continue uninterrupted toward the right end of this strand. Remember: new nucleotides are added in a 5′ → 3′ direction, so the template must be copied from its 3′ end. The other strand has a 5′ end at the left with the 3′ end pointing right. Thus, the two strands are oriented in opposite directions (antiparallel), and synthesis (which is 5′ → 3′) must proceed in opposite directions. For the leading strand (say, the top strand) replication is to the right, following the replication fork. It is continuous and may be thought of as moving "downstream." Replication on the bottom strand cannot move in the direction of the fork (to the right), since, for this strand, that would mean adding nucleotides to its 5′ end. Therefore, this strand must replicate discontinuously: as the fork creates a new single-stranded stretch of DNA, this is replicated *to the left* (away from the direction of fork movement). For this lagging strand, the replication fork is always opening new single-stranded DNA for replication *upstream* of the previously replicated stretch, and a new fragment of DNA is replicated back to the previously created fragment. Thus, one (Okazaki) fragment follows the other in the direction of the replication fork, but each fragment is created in the opposite direction.

6. If the DNA is double stranded, A = T and G = C and A + T + C + G = 100%. If T = 15% then C = [100 – 15(2)]/2 = 35%.

7. If the DNA is double stranded, G = C = 24% and A = T = 26% .

8. 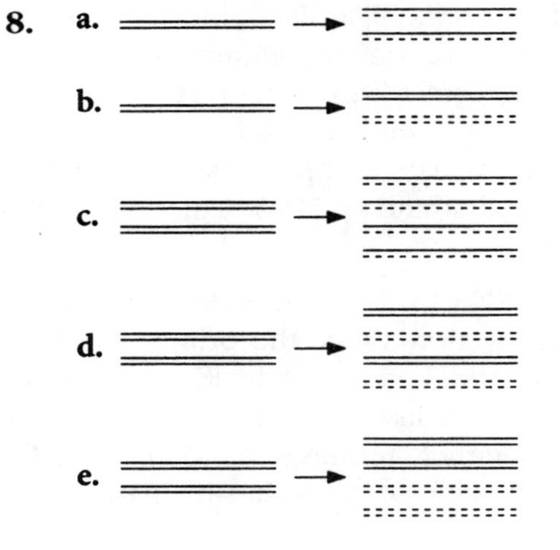

f. Models b and e are ruled out by the experiment. The results were compatible with semiconservative replication, but the exact structure could not be predicted from the results. In this experiment, it was proved that DNA replication does not occur conservatively at either the DNA or chromosomal level.

9. The results suggest that the DNA is replicated in short segments that are subsequently joined by enzymatic action (DNA ligase). Because DNA replication is bidirectional, because there are multiple points along the DNA where replication is initiated, and because DNA polymerases work only in a $5' \rightarrow 3'$ direction, one strand of the DNA is always in the wrong orientation for the enzyme. This requires synthesis in fragments.

10. Replication requires that the enzymes and initiation factors of replication have access to the DNA. One possible answer is that the heterochromatic regions, being more condensed than euchromatin, require a longer period to decondense to the point where replication can be initiated. A second possibility is that the heterochromatic regions, which are generally located at centromeres and telomeres in many organisms, "anchor" the chromosome to the nuclear membrane. Embedded in part in the nuclear membrane, these regions could require extra time to disentangle so that replication may proceed. A third possibility involves the mechanism of heterochromatization. Genes that are inactive in a given cell type are generally heterochromatic. This mechanism may also specifically delay replication. Many other possibilities can be proposed.

11. **a.** A very plausible model is of a triple helix, which would look like a braid, with each strand interacting by hydrogen bonding to the other two.

b. Replication would have to be terticonservative. The three strands would separate, and each strand would dictate the synthesis of the other two strands.

c. The reductional division would have to result in three daughter cells, and the equational would have to result in two daughter cells, in either order. Thus, meiosis would yield six gametes.

12. Chargaff's rules are that A = T and G = C. Because this is not observed, the most likely interpretation is that the DNA is single-stranded. The phage would first have to synthesize a complementary strand before it could begin to make multiple copies of itself.

13. Remember that there are two hydrogen bonds between A and T, while there are three hydrogen bonds between G and C. Denaturation involves the breaking of these bonds, which requires energy. The more bonds that need to be broken, the more energy that must be supplied. Thus the temperature at which a given DNA molecule denatures is a function of its base composition. The higher the temperature of denaturation, the higher the percentage of G–C pairs.

14. Deletions and duplications at the DNA level would look exactly like the same events during homologous pairing of chromosomes at the duplex DNA level:

Inversions would have two possibilities, depending on the relative length of the inverted and noninverted regions:

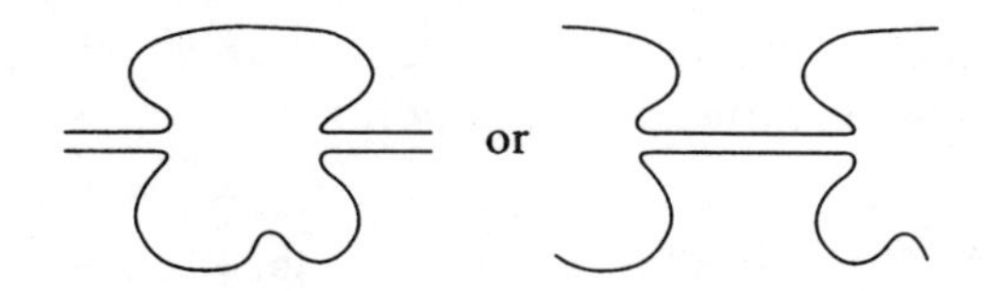

15. a. The first shoulder appears before strand interaction takes place, suggesting that the complementary regions are in the same molecule. This is called a palindrome:

A T G C A T G G C C A————————T G G C C A T G C A T

T A C G T A C C G G T————————A C C G G T A C G T A

When the strands separate, each strand can base-pair with itself:

b. The second shoulder represents sequences that are present in many copies in the genome (repeated sequences). Because they are at a higher concentration than are sequences present in only one copy per haploid genome (unique sequences), they have a higher probability of encountering one another during a given time period.

16. First realize that the repeated and unique sequences could be completely interspersed, partially interspersed, or completely segregated from one another in the chromosomal DNA. The distribution of these two types of sequences will affect the reannealing curves observed.

Extract the DNA from the cells. Do separate renaturation curves for unsheared DNA and DNA sheared to fragments of just a few kilobases. If there is generalized interspersion, then the part of the curve corresponding to the annealing of unique sequences will be farther to the left for the unsheared sample compared with the sheared sample. This is because the interspersed repeated sequences in the unsheared sample will facilitate pairing.

You might also examine the rapidly annealing DNA from the unsheared sample by electron microscopy. Interspersion would result in partial duplexes, with unmatched single-stranded regions (unique sequence DNA) interspersed.

17. This observation suggests that the mouse cancer cells have copies of the virus genome integrated into them. Thus, it may be that the viral genes somehow alter cell function, triggering malignancy. Alternatively, the viral genome may carry one or more genes that directly result in malignancy. In either case, viral infection may also be the mechanism by which human malignancy is triggered in some instances.

18. The data suggest that each chromosome is composed of one continuous molecule of DNA and that translocations can alter their size. In part (c), it appears that part of the longest chromosome has been translocated to the shortest.

19. Let the broken line indicate DNA that has incorporated bromodeoxyuridine and the unbroken line indicate normal DNA.

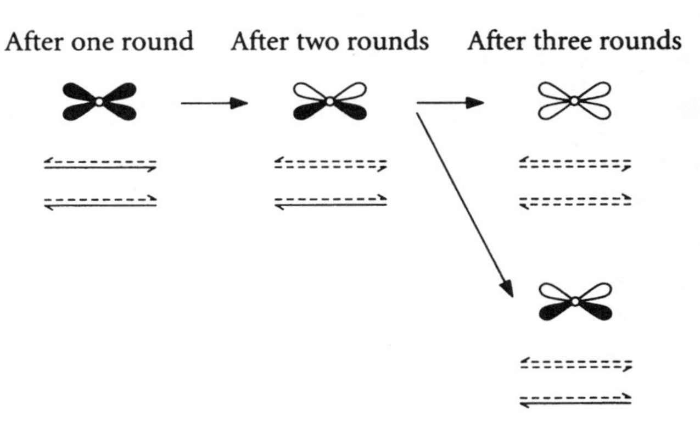

9 Genetics of DNA Function

1. The primary structure of a protein is the sequence of amino acids along its length. It is held together by covalent bonds. The secondary structure of a protein is caused by hydrogen bonding between CO and NH groups of the polypeptide backbone. Common secondary structures are the α helix and the ß pleated sheet. The tertiary structure of a protein is its final three-dimensional shape. If the functional protein is composed of more than one polypeptide, then the protein also has quaternary structure. For example, hemoglobin is composed of four polypeptides — two α-globins and two ß-globins. Each globin has a primary and tertiary structure (as well as some secondary structure), but only when associated in hemoglobin's tetrameric structure is quaternary structure present.

2. The one-gene–one-enzyme hypothesis suggested that genes were responsible for the function of enzymes and that each gene controlled one specific enzyme.

3. An auxotroph is a strain that requires at least one nutrient for growth beyond that normally required for the wild-type organism.

4. Sickle-cell anemia results from the replacement of a valine for a glutamic acid at position 6 in the ß chain of hemoglobin.

5. Yanofsky analyzed mutations in the *trpA* gene and ordered them using transduction. He also determined the amino acid sequence of the altered gene products. By this, he demonstrated an exact correlation between the sequence of mutation sites and the sequence of altered amino acids.

6. Complementation is the cooperation of the products of two or more genes to produce a nonmutant phenotype, while recombination is the exchange of DNA segments between chromosomes. Recombination can occur both within and between genes, while complementation occurs between gene products.

7.

	B	K
rII	large plaques	no plaques
rII^+	small plaques	plaques

8. The defective enzyme that results in albinism may not be able to detoxify a chemical component of Saint-John's-wort that the wild-type enzyme can detoxify. In fact, the plant contains a chemical that is made toxic by light (a phototoxin).

9. Lactose is composed of one molecule of galactose and one molecule of glucose. A secondary cure would result if all galactose and lactose were removed from the diet. The disorder would be expected not to be dominant, because one good copy of the gene should allow for at least some, if not all, breakdown of galactose. In fact, the disorder is recessive.

10. Amniocentesis can be used to detect chromosomal abnormalities and any single-gene abnormalities for which there is a test. Thus, Down syndrome, Turner syndrome, and other chromosomal abnormalities can be detected. Furthermore, biochemical disorders such as galactosemia, Tay-Sachs disease, sickle-cell anemia, and phenylketonuria can be detected. Amniocentesis would thus be useful whenever there is a family history of chromosomal or biochemical disorders and whenever there is a history of parental exposure to mutagens (X rays or chemicals). It would also be useful when the maternal age is over 35, because the rate of nondisjunction is elevated in this population.

11. **a.** The main use is in detecting carrier parents and in diagnosing the disorder in the fetus.

b. Because the values for normal individuals and carriers overlap for galactosemia, there is ambiguity if a person has 25 to 30 units. That person could be either a carrier or normal.

c. These wild-type genes are phenotypically dominant but are incompletely dominant at the molecular level. A minimal level of enzyme activity apparently is enough to ensure normal function and phenotype.

12. One less likely possibility is a germ-line mutation. More likely is that each parent was blocked at a different point in a metabolic pathway. If one were $A/A \cdot b/b$ and the other were $a/a \cdot B/B$, then the child would be $A/a \cdot B/b$ and would have sufficient levels of both resulting enzymes to produce pigment.

13. Assuming homozygosity for the normal gene, the mating is $A/A \cdot b/b \times a/a \cdot B/B$. The children would be normal, $A/a \cdot B/b$ (see Problem 12).

14. **a.** Complementation refers to gene products within a cell, which is not what is happening here. Most likely, what is known as *cross-feeding* is occurring, whereby a product made by one strain diffuses to another strain and allows growth of the second strain. This is equivalent to supplementing the medium. Because cross-feeding seems to be taking place, the suggestion is that the strains are blocked at different points in the metabolic pathway.

b. For cross-feeding to occur, the growing strain must have a block that occurs earlier in the metabolic pathway than does the block in the strain from which it is obtaining the product for growth.

c. The *trpE* strain grows when cross-fed by either *trpD* or *trpB* but the converse is not true (placing *trpE* earlier in the pathway than either *trpD* or *trpB*), and *trpD* grows when cross-fed by *trpB* (placing *trpD* prior to *trpB*). This suggests that the metabolic pathway is

$$trpE \rightarrow trpD \rightarrow trpB$$

d. Without some tryptophan, no growth at all would occur, and the cells would not have lived long enough to produce a product that could diffuse.

15.

Experiment	Result	Interpretation
v into *v* hosts	scarlet	defects in same gene
cn into *cn* hosts	scarlet	defects in same gene
v into wild-type hosts	wild type	wild type provides *v* product
cn into wild-type hosts	wild type	wild type provides *cn* product
cn into *v* hosts	scarlet	*v* cannot provide *cn* product; *cn* later than *v* in metabolic pathway
v into *cn* hosts	wild type	*cn* provides *v* product; *v* defect earlier than *cn*

A simple test would be to grind up *cn* animals, inject *v* larvae with the material, and look for wild-type development.

16.

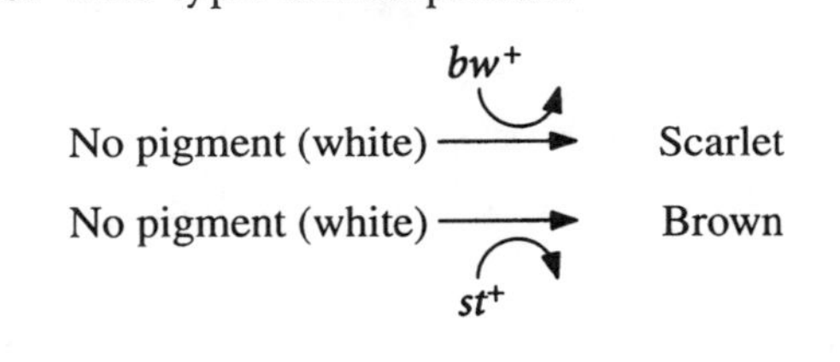

Scarlet plus brown results in red.

17. Growth will be supported by a particular compound if it is later in the pathway than the enzymatic step blocked in the mutant. Restated, the more mutants a compound supports, the later in the pathway it must be. In this example, compound G supports growth of all mutants and can be considered the end product of the pathway. Alternatively, compound E does not support the growth of any mutant and can be considered the starting substrate for the pathway. The data indicate the following:

a. and b. E —|→ A —|→ C —|→ B —|→ D —|→ G
(blocks: 5, 4, 2, 1, 3)

vertical lines indicate step where each mutant is blocked

18. The formula that is needed to calculate the proportion of meiocytes with different numbers of crossovers is $f(i) = e^{-m}m^i/i!$, where m is the mean number of crossovers. In this case, you are told that $m = 0.5$.

a. The probability of no crossover is

$$f(0) = e^{-0.5}(0.5)^0/0! = 0.6 \text{ or } 60\%$$

b. Because RF = $(1 - e^{-m})/2$, RF = $^1/_2(1 - 0.60) = 0.20$. That is, there are 20 m.u. between the two genes.

c. The recombinants will be $^1/_2$(*val-1 val-2*) and $^1/_2$($val\text{-}1^+$ $val\text{-}2^+$).

Therefore, $^1/_2$RF = p($val\text{-}1^+$ $val\text{-}2^+$) = 10 percent.

d. Remember that accumulation of substance x means the gene responsible for converting substance x to the next metabolic substance is defective. Also, if substance y permits growth, it is beyond a block in a metabolic pathway.

19. a. If enzyme A was defective or missing (m_2/m_2), red pigment would still be made and the petals would be red.

b. Purple, because it has a wild-type allele for each gene and you are told that the mutations are recessive.

c.

9	$M_1/- \,;\, M_2/-$	purple
3	$m_1/m_1 \,;\, M_2/-$	blue
3	$M_1/- \,;\, m_2/m_2$	red
1	$m_1/m_1 \,;\, m_2/m_2$	white

d. The mutant alleles do not produce functional enzyme. However, enough functional enzyme must be produced by the single wild-type allele of each gene to synthesize normal levels of pigment.

20. a. If enzyme B is missing, a white intermediate will accumulate and the petals will be white.

b. If enzyme D is missing, a blue intermediate will accumulate and the petals will be blue.

c. purple (*B/b* ; *D/d*)

d. P   *b/b* ; *D/D* × *B/B* ; *d/d*

F_1 *B/b* ; *D/d* × *B/b* ; *D/d*

F_2	9	*B/–* ; *D/–*	purple
	3	*b/b* ; *D/–*	white
	3	*B/–* ; *d/d*	blue
	1	*b/b* ; *d/d*	white

The ratio of purple:blue:white would be 9:3:4.

21. The cis and trans burst size should be the same if the mutants are in different cistrons, and if they are in the same cistron, the trans burst size should be zero. Therefore, assuming *rV* is in *A*, *rW* also is in *A*, and *rU*, *rX*, *rY*, and *rZ* are in *B*.

22. The cross is *pan2x pan2y*$^+$ × *pan2x*$^+$ *pan2y*.

a. If one centromere precociously divides, that will put three chromatids in one daughter cell and one in the other.

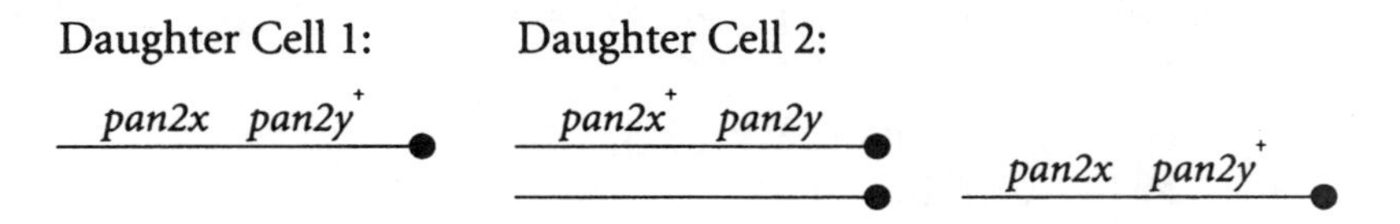

After meiosis II and mitosis, the first daughter cell would give rise to two pale (*pan2x pan2y*$^+$) and two white, aborted (nullisomic) ascospores, while the second daughter cell would give rise to two black (*pan2x pan2y*$^+$/*pan2x*$^+$ *pan2y*) and two pale (*pan2x*$^+$ *pan2y*) ascospores. The same result (4 pale:2 colorless:2 black) would occur if only the other centromere divided early.

b. If both centromeres divided precociously, each daughter cell would be *pan2x pan2y*$^+$/*pan2x*$^+$ *pan2y*. This would lead to 4 colorless (nullisomic) and 4 black (disomic) ascospores.

23. a. The mutant fails to complement any other mutant. One interpretation is that it is dominant. Alternatively, it could be a deletion that spans the other genes represented by these mutants.

b. Complementation groups do not complement within a group but do complement between groups (+). Notice that mutants 1, 5, 8, and 9 complement all others but do not complement within the group. The same holds for mutants 2, 3, 4, and 12 as a group and mutants 6, 7, 10, 11, and 13 as a group. These are the three complementation groups.

c. Mutants 1 and 2 are in different genes, so the cross can be written $1 \cdot 2^+ \times 1^+ \cdot 2$. Assuming independent assortment, the progeny would be

$1/4$	$1\ ;\ 2^+$	eye$^-$
$1/4$	$1\ ;\ 2$	eye$^-$
$1/4$	$1^+\ ;\ 2$	eye$^-$
$1/4$	$1^+\ ;\ 2^+$	eye$^+$

or 3 eye$^-$:1 eye$^+$

Mutants 2 × 6 also complement each other. If independent assortment existed, a 3:1 ratio would be observed. Because the ratio is 113:5, there is no independent assortment and the genes are linked. Only one of the two recombinant classes can be distinguished: the eye$^+$ class.

Because the recombinants should be of equal frequency, the total number of recombinants is 10 out of 118, which leads to 100% (10/118) = 8.47 m.u. distance between the genes.

Mutant 14 fails to complement mutant 1 and at least for this sample set, recombination is also not observed.

d. There are three complementation groups; therefore, there are three loci, plus mutant 14. Because two of the groups are independently assorting, mutant 14 is either a dominant mutation or a very large deletion spanning the three loci (and they are therefore on the same chromosome), or it is a separate fourth locus that in some fashion controls the expression of the other three loci, or it is a double mutant with a point mutation within one complementation group (1, 5, 8, 9) and a deletion spanning the two linked complementation groups.

e. Two groups are linked, with 8.5 m.u. between them (2, 3, 4, 12 and 6, 7, 10, 11, 13), and the third group is either on a separate chromosome or more than 50 m.u. from the two other groups.

24. The best interpretation is that *rW* is a deletion spanning *rZ* and *rD*. Alternatively, it could be a double point mutation in *rZ* and *rD*. If it is a double point mutation, then a cross of *rW* with a mutant between *rZ* and *rD* should yield wild type at a low frequency (double crossover). If it is a deletion, no wild type will be observed.

25. a. A point mutation within a deletion yields no growth (–), while a point mutation external to a deletion yields growth (+). Only mutant *c* fails to complement deletion 1, so it is within the deletion and the other mutants are not. Mutant *d* is in the overlapping region of deletions 2 and 3, while mutant *e* is in the small region of deletion 2 that is not overlapped by other deletions. The partial order is therefore *c e d*. Mutant *a* is within the nonoverlapped deletion 3, and mutant *b* is within deletions 3 and 4. The final map is

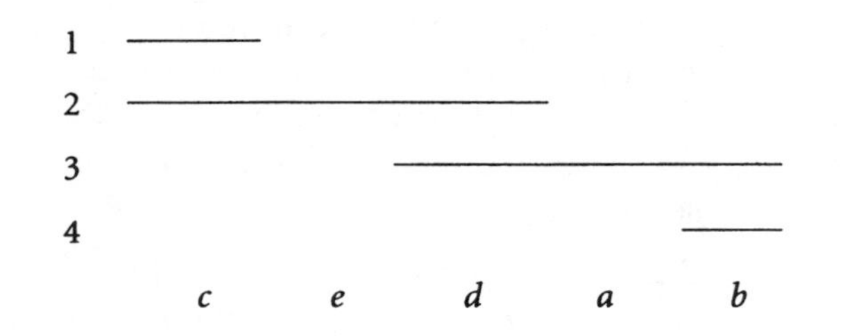

b. The suggestion is that deletion 4 spans both genes. This does not affect the conclusions from part a.

26. a. Here, + indicates nonoverlapping and – indicates overlapping. Therefore, 1 overlaps 3 and 5, 2 overlaps 5 only, 3 overlaps all but 2, 4 overlaps 3 only, and 5 overlaps all but 4. Putting all these pieces together yields the deletion map

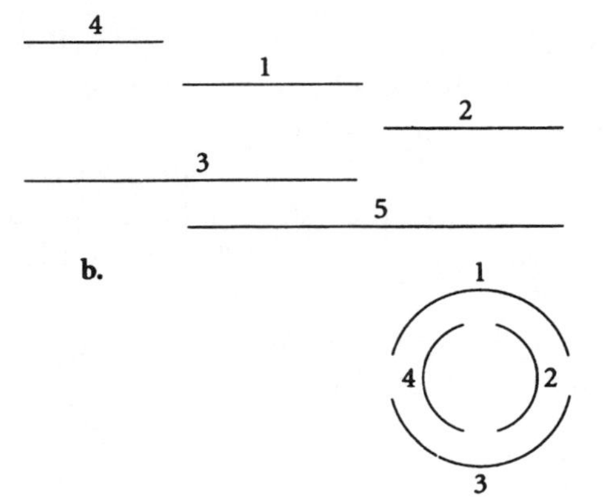

27. a. The values equal half the recombinants. Because the problem asks for relative frequency, however, it is not necessary to convert to real map units. Therefore, the distances are

1–2: 14	2–3: 12
1–3: 2	2–4: 6
1–4: 20	3–4: 18

The only map possible is

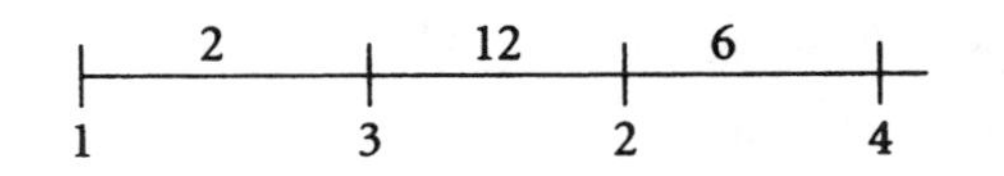

b. No. This might have been observed in the same mutant × same mutant matings (e.g., 1 × 1 or 2 × 2), but because these all gave 0 prototrophs, evidence of reversion was not observed.

28. This is like any other mapping problem when the reciprocal cannot be detected: the detected class is multiplied by 2. Rewrite the crosses and results, using parentheses to indicate unknown order, so that it is clear what the results mean:

Cross 1: $his\text{-}2^{+}$ (a, b^{+}) $nic\text{-}2^{+}$ × $his\text{-}2$ (a^{+}, b) $nic\text{-}2$

Cross 2: $his\text{-}2^{+}$ (a, c^{+}) $nic\text{-}2$ × $his\text{-}2$ (a^{+}, c) $nic\text{-}2^{+}$

Cross 3: $his\text{-}2$ (b, c^{+}) $nic\text{-}2$ × $his\text{-}2^{+}$ (b^{+}, c) $nic\text{-}2^{+}$

Results:

Genotype	Cross 1	Cross 2	Cross 3
his-2 nic-2	DCO between *a* and *b*, and *b* and *nic*	1 CO between *a* and *c*	DCO between *b* and *c*, and *b* and *nic*
$his\text{-}2^{+}$ $nic\text{-}2^{+}$	DCO between *a* and *b*, and *his* and *a*	as above	DCO between *b* and *c*, and *his* and *c*
his-2 $nic\text{-}2^{+}$	1 CO between *a* and *b*	DCO between *a* and *c*, and *nic* and *c*	1 CO between *b* and *c*
$his\text{-}2^{+}$ *nic-2*	3 CO between *a* and *b*, *his* and *a*, and *b* and *nic*	DCO between *a* and *c*, *his* and *a*	3 CO between *b* and *c*, *his* and *c*, and *b* and *nic*

Putting all this information together, the only gene sequence possible is *his-2 a c b nic-2*. To calculate the genetic distances, it is necessary to multiply the frequency of recombinants by 2 because reciprocals are not seen:

a–b: 100%(2)(15)/41,236 = 0.072 m.u.

a–c: 100%(2)(6)/38,421 = 0.031 m.u.

b–c: 100%(2)(5)/43,600 = 0.023 m.u.

The final map is

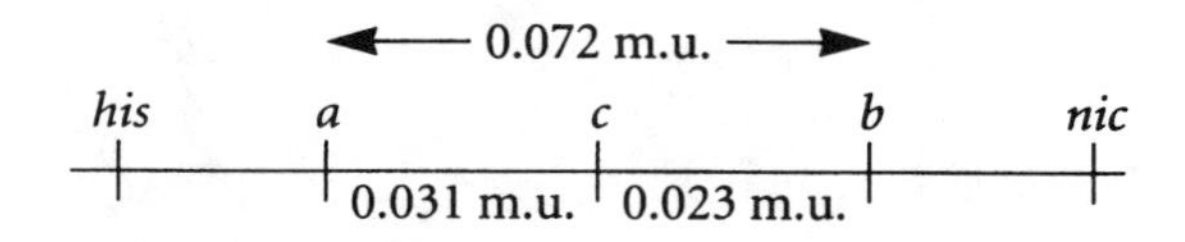

29. **a.** The allele s^n will show dominance over s^f because there will be only 40 units of square factor in the heterozygote.

b. Here, the functional allele is recessive.

c. The allele s^f may become dominant over time in two ways: (1) it could mutate slightly, so that it produces more than 50 units, or (2) other modifying genes may mutate to increase the production of s^f.

30. Benzer used the Poisson distribution to make his calculations. While he knew the number of 1 and 2 occurrence sites, he did not know the total number of occurrence sites. Therefore, he had to assume that the Poisson distribution was applicable.

The equation is

$$f(i) = e^{-m}m^i/i!$$

The terms:

$$f(0) = e^{-m}m^0/0! = e^{-m}$$

$$f(1) = e^{-m}m^1/1! = e^{-m} = 117 \text{ (from Figure 9-27)}$$

$$f(2) = e^{-m}m^2/2! = e^{-m}m^2/2 = 53 \text{ (from Figure 9-27)}$$

These equations can be used to determine the value of m:

$$f(2)/f(1) = (e^{-m}m^2/2)/(e^{-m}m^1) = 53/117$$

$$m/2 = 53/117$$

$$m = 0.9059$$

The number of 0 occurrences is

$$f(0)/f(1) = e^{-m}/e^{-m}m$$

$$f(0) = f(1)/m = 117/0.9059 = 129.15$$

31. **a.** Cross 1 × 2: All purple F_1 indicates that two genes are involved. Call the defect in 1 *a/a* and the defect in 2 *b/b*. The cross is

P $\quad a/a \cdot B/B \times A/A \cdot b/b$

F_1 $\quad A/a \cdot B/b$

If the two genes assort independently, a 9:7 ratio of purple:white would be seen. A 1:1 ratio indicates tight linkage. The cross above now needs to be rewritten

P $\quad a\ B/a\ B \times A\ b/A\ b$

F_1 $\quad a\ B/A\ b$

F_2 $\quad 1\ a\ B/a\ B$ (white):2 $a\ B/A\ b$ (purple):1 $A\ b/A\ b$ (white)

Cross 1 × 3: Again, an F_1 of all purple indicates two genes. The 9:7 F_2 indicates independent assortment. Therefore, let the strain 3 defect be symbolized by *d*:

P a/a ; D/D × A/A ; d/d

F_1 A/a ; D/d × A/a ; D/d

F_2 9 $A/–$; $D/–$ (purple)
3 a/a ; $D/–$ (white)
3 $A/–$; d/d (white)
1 a/a ; d/d (white)

Cross 1 × 4: All white F_1 and F_2 indicates that the two mutations are in the same gene. The cross is

P a_1/a_1 × a_2/a_2

F_1 a_1/a_2

F_2 1 a_1/a_1 : 2 a_1/a_2 : 1 a_2/a_2

b. Cross 2 × 3: (because 2 is tightly linked to 1, this cross is similar to 1 × 3)

P b/b ; D/D × B/B ; d/d

F_1 B/b ; D/d (purple)

F_2 9 $B/–$; $D/–$ (purple)
3   b/b ; $D/–$ (white)
3    $B/–$; d/d (white)
1   b/b ; d/d (white)

Cross 2 × 4: (because 1 and 4 are mutant for the same gene, this cross is the same as 1 × 2)

32. Mutants *a* and *e* have point mutations within the same gene. The other point mutations are in different genes. There are at least four genes involved with leucine synthesis. With the exception of two crosses (*a* × *e*, *b* × *d*), the frequency of prototrophic progeny is approximately 25 percent. This indicates independent assortment of *a* (and *e*) with *b*, *d*, and *c*, and *c* with *b* and *d*. Cistrons *b* and *d* are linked: RF = 100%(4)/500 = 0.8 m.u. (The recombinants are doubled because only one class can be observed.)

33. a. There are three genes:

Gene 1: mutants 1, 3, and 4
Gene 2: mutants 2 and 5
Gene 3: mutant 6

b. Diagram the heterozygotes that yield *star*$^+$ gametes, using parentheses to indicate unknown order:

1–6: A (1^+, 6) B/a (1, 6^+) $b \rightarrow a$ (1^+, 6^+) B
2–4: A (2^+, 4) B/a (2, 4^+) $b \rightarrow a$ (2^+, 4^+) B

To determine gene order within the 1–6 cross, note that if the order is *A 1 6 B*, three crossovers are required to generate the necessary gamete, while the order *A 6 1 B* requires only one crossover. Therefore, the order is more likely *A 6 1 B*.

Similarly, to determine gene order from the 2–4 cross, note that the order *A 2 4 B* would require three crossovers, while the order *A 4 2 B* would require one crossover. The order is more likely *A 4 2 B*. The final order is *A 6 (1,3,4) (2,5) B*.

34. *Unpacking the Problem*

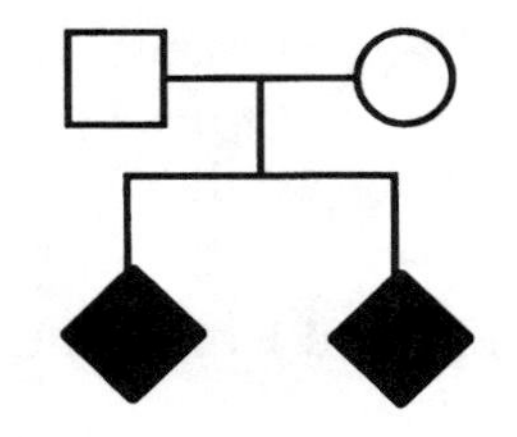

1. Because the trait is recessive, possible siblings could be unaffected.
2. By age 10, most of the brain damage caused by an excess level of phenylalanine has already occurred.
3. The bacteria were needed to convert phenylalanine, for which no test existed, to phenylpyruvic acid, for which a test existed.
4. The level of phenylalanine in the urine is only an estimate of the level in the blood. It is the blood level that affects brain development, not the urine level.
5. The phenylpyruvic acid in urine was formed from excess phenylalanine in the blood.
6. Blood concentrations would be expected to be much higher than urine concentrations.
7. The green substance in the ferric chloride test was not in itself important. However, the presence of the colored substance was a definitive indicator of the presence of phenylpyruvic acid.
8. Both phenylalanine and phenylpyruvic acid would be expected to be very low in the blood and urine of unaffected children because both substances are intermediates in a biochemical pathway.
9. The odor in the urine was due to the presence of phenylpyruvic acid.
10. Both parents were heterozygotes: *A/a*.
11. Because the allele is rare, the genotype is also rare. It would be very rare to have both parents of a family heterozygous.
12. The progeny of heterozygous parents would be expected to occur as follows: 1 *A/A*:2 *A/a*:1 *a/a*.
13. Most families in the population would have parents who were *A/A*.
14. Inheritance was inferred because of the observation that in families with retardation associated with green-staining urine, the affected:unaffected ratio was 1:3. This is the expected ratio in a heterozygous-by-heterozygous cross.

15. The disorder was inferred to be a Mendelian recessive because both parents were free of the disorder but had children with the disorder.
16. Most of the nervous system develops before birth.
17. Adults with PKU do not need to remain on the diet because their nervous systems are completely developed.
18. The maternal blood circulatory system contains much more fluid than the fetal circulatory system.
19. Macromolecules and small molecules can pass through the placental barrier.
20. Phenylalanine and phenylpyruvic acid pass from mother to child.
21. An essential amino acid is an amino acid that must be derived from a dietary source because an organism is incapable of making it, or it is an amino acid that is made by the organism in insufficient amounts for its stage of development.
22. Phenylalanine is an essential amino acid for humans.
23. Phenylalanine blood concentration increases in the absence of the enzyme that converts it to the next product of the metabolic pathway (tyrosine).
24. PKU is recessive because one normal copy of the gene produces enough enzyme to convert all excess phenylalanine to tyrosine.
25. PKU is the result of a blocked biochemical pathway.
26. See Figure 9-21.
27.

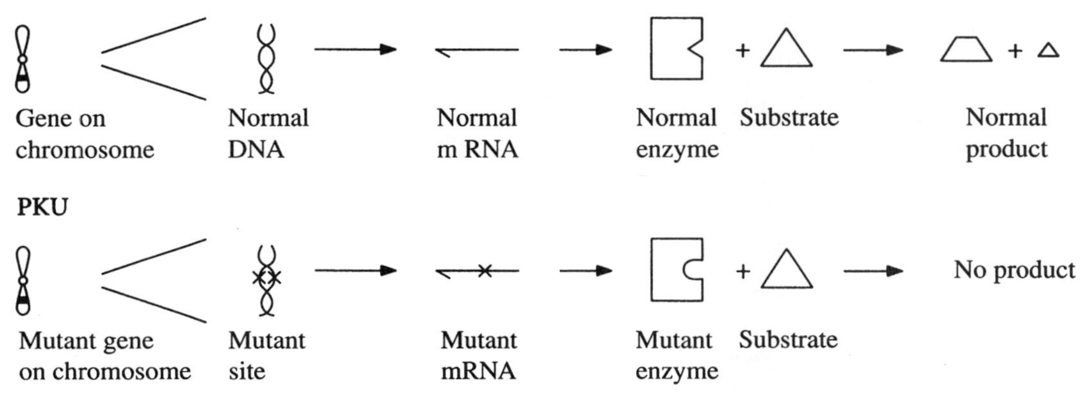

Solution to the Problem

a. The mothers had an excess of phenylalanine in their blood, and that excess was passed through the placenta into the fetal circulatory system, where it caused brain damage prior to birth.

b. A fetus with two mutant copies of the allele that causes PKU makes no functional enzyme. However, the mother of such a child is heterozygous and makes enough enzyme to block any brain damage; the excess phenylalanine in the fetal circulatory system enters the maternal circulatory system and is processed by the maternal gene product. After birth, which is when PKU damage occurs in a PKU child, dietary restrictions block a buildup of phenylalanine in the circulatory system until brain development is completed.

The fetus of a PKU mother is exposed to the very high level of phenylalanine in its circulatory system during the time of major brain development. Therefore, brain damage occurs before birth, and no dietary restrictions after birth can repair that damage.

c. The obvious solution to the brain damage seen in the babies of PKU mothers is to return the mother to a restricted diet during pregnancy in order to block high levels of exposure to her child.

d. PKU is characterized as a rare recessive disorder. A child with PKU has two parents that carry a mutant allele for the metabolism of phenylalanine. When two individuals who are heterozygous for PKU have a child, the risk that the child will have PKU is 25%. A PKU child is unable to make a functional enzyme that converts phenylalanine to tyrosine. As a result, an excess level of phenylalanine is found in the blood, and the excess is detected as an increase in phenylpyruvic acid in both the blood and the urine. The excess phenylpyruvic acid blocks normal development of the brain, resulting in retardation.

10 Molecular Biology of Gene Function

1. Because RNA can hybridize to both strands, the RNA must be transcribed from both strands. This does not mean, however, that both strands are used as a template *within each gene*. The expectation is that only one strand is used within a gene but that different genes are transcribed in different directions along the DNA. The most direct test would be to purify a specific RNA coding for a specific protein and then hybridize it to the λ genome. Only one strand should hybridize to the purified RNA.

2. **a.** The data cannot indicate whether one or both strands are used for transcription. You do not know how much of the DNA is transcribed nor which regions of DNA are transcribed. Only when the purine/pyrimidine ratio is not unity can you deduce that only one strand is used as template.

b. If the RNA is double-stranded, the percentage of purines (A + G) would equal the percentage of pyrimidines (U + C) and the (A +G)/(U + C) ratio would be 1.0. This is clearly not the case for *E. coli*, which has a ratio of 0.80. The ratio for *B. subtilis* is 1.02. This is consistent with the RNA being double-stranded but does not rule out single-stranded if there are an equal number of purines and pyrimidines in the strand.

3. A single nucleotide change should result in three adjacent amino acid changes in a protein. One and two adjacent amino acid changes would be expected to be much rarer than the three changes. This is directly the opposite of what is observed in proteins. Also, given any triplet coding for an amino acid, the next triplet could be only one of four. For example, if the first is GGG, then the next must be GGN (N = any base). This puts severe limits on which amino acids could be adjacent to each other. You could check amino acid sequences of various proteins to show that this is not the case.

4. It suggests very little evolutionary change between *E. coli* and humans with regard to the translational apparatus. The code is universal, the ribosomes are interchangeable, the tRNAs are interchangeable, and the enzymes involved are interchangeable.

5. There are three codons for isoleucine: 5′ AUU 3′, 5′ AUC 3′, and 5′ AUA 3′. Possible anticodons are 3′ UAA 5′ (complementary), 3′ UAG 5′ (complementary), and 3′ UAI 5′ (wobble). 5′ UAU 3′, although complementary, would also base-pair with 5′ AUG 3′ (methionine) owing to wobble and therefore would not be an acceptable alternative.

6. **a.** By studying the genetic code table provided in the textbook, you will discover that there are eight cases in which knowing the first two nucleotides does not tell you the specific amino acid.

b. If you knew the amino acid, you would not know the first two nucleotides in the cases of Arg, Ser, and Leu.

7. The codon for amber is UAG. Listed below are the amino acids that would have needed to be inserted to continue the wild-type chain and their codons:

glutamine	CAA, CAG*
lysine	AAA, AAG*
glutamic acid	GAA, GAG*
tyrosine	UAU*, UAC*
tryptophan	UGG*
serine	AGU, AGC, UCU, UCC, UCA, UCG*

In each case, the codon marked by an asterisk would require a single base change to become UAG.

8. **a.** The codons for phenylalanine are UUU and UUC. Only the UUU codon can exist with randomly positioned A and U. Therefore, the chance of UUU is $(^1/_2)(^1/_2)(^1/_2) = {^1/_8}$.

b. The codons for isoleucine are AUU, AUC, and AUA. AUC cannot exist. The probability of AUU is $(^1/_2)(^1/_2)(^1/_2) = {^1/_8}$, and the probability of AUA is $(^1/_2)(^1/_2)(^1/_2) = {^1/_8}$. The total probability is thus $^1/_4$.

c. The codons for leucine are UUA, UUG, CUU, CUC, CUA, and CUG, of which only UUA can exist. It has a probability of $(^1/_2)(^1/_2)(^1/_2) = {^1/_8}$.

d. The codons for tyrosine are UAU and UAC, of which only UAU can exist. It has a probability of $(^1/_2)(^1/_2)(^1/_2) = {^1/_8}$.

9. **a.** 1 U:5 C — The probability of a U is $^1/_6$, and the probability of a C is $^5/_6$.

Codon	Amino acid	Probability	Sum
UUU	Phe	$(^1/_6)(^1/_6)(^1/_6) = 0.005$	Phe = 0.028
UUC	Phe	$(^1/_6)(^1/_6)(^5/_6) = 0.023$	
CCC	Pro	$(^5/_6)(^5/_6)(^5/_6) = 0.578$	Pro = 0.694
CCU	Pro	$(^5/_6)(^5/_6)(^1/_6) = 0.116$	
UCC	Ser	$(^1/_6)(^5/_6)(^5/_6) = 0.116$	Ser = 0.139
UCU	Ser	$(^1/_6)(^5/_6)(^1/_6) = 0.023$	
CUC	Leu	$(^5/_6)(^1/_6)(^5/_6) = 0.116$	Leu = 0.139
CUU	Leu	$(^5/_6)(^1/_6)(^1/_6) = 0.023$	

1 Phe:25 Pro:5 Ser:5 Leu

b. Using the same method as above, the final answer is 4 stop:80 Phe:40 Leu:24 Ile:24 Ser:20 Tyr:6 Pro:6 Thr:5 Asn:5 His: 1 Lys: 1 Gln.

c. All amino acids are found in the proportions seen in the code table.

10. a. $(GAU)_n$ codes for Asp_n $(GAU)_n$, Met_n $(AUG)_n$, and $stop_n$ $(UGA)_n$. $(GUA)_n$ codes for Val_n $(GUA)_n$, Ser_n $(AGU)_n$, and $stop_n$ $(UAG)_n$. One reading frame in each contains a stop codon.

b. Each of the three reading frames contains a stop codon.

c. The way to approach this problem is to focus initially on one amino acid at a time. For instance, line 4 indicates that the codon for Arg might be AGA or GAG. Line 7 indicates it might be AAG, AGA, or GAA. Therefore, Arg is at least AGA. That also means that Glu is GAG (line 4). Lys and Glu can be AAG or GAA (line 7). Because no other combinations except the ones already mentioned result in either Lys or Glu, no further decision can be made with respect to them. However, taking wobble into consideration, Glu may also be GAA, which leaves Lys as AAG.

Next, focus on lines 1 and 5. Ser and Leu can be UCU and CUC. Ser, Leu, and Phe can be UUC, UCU, and CUU. Phe is not UCU, which is seen in both lines. From line 14, CUU is Leu. Therefore, UUC is Phe, and UCU is Ser.

The footnote states that lines 13 and 14 are in the correct order. In line 13, if UCU is Ser (see above), then Ile is AUC, Tyr is UAU, and Leu is CUA.

Continued application of this approach will allow the assignment of an amino acid to each codon.

11. Mutant 1: A simple substitution of Arg for Ser exists, suggesting a nucleotide change. Two codons for Arg are AGA and AGG, and one codon for Ser is AGU. The U of the Ser codon could have been replaced by either an A or a G.

Mutant 2: The Trp codon (UGG) changed to a stop codon (UGA or UAG).

Mutant 3: Two frameshift mutations occurred:

5′—GCN CCN (–U)GGA GUG AAA AA(+U or C) UGU(or C) CAU(or C)—3′.

Mutant 4: An inversion occurred after Trp and before Cys. The DNA original sequence (with both strands shown for the area of inversion) was

3′—CGN GGN ACC TCA CTT TTT ACA(or G) GTA(or G)—5′
5′— —AGT GAA AAA— —3′

Therefore, the complementary RNA sequence was

5′—GCN CCN UGG AGU GAA AAA UGU/C CAU/C—3′

The DNA inverted sequence became

3′—CGN GGN ACC AAA AAG TGA ACA/G GTA/G—5′

Therefore the complementary RNA sequence was

5′—GCN CCN UGG UUU UUC ACU UGU/C CAU/C—3′

12. **a. and b.** The goal of this type of problem is to align the two sequences. You are told that there is a single nucleotide addition and single nucleotide deletion, so look for single base differences that effect this alignment. These should be located where the protein sequence changes (i.e., between Lys-Ser and Asn-Ala). Remember also that the genetic code is redundant. (N = any base)

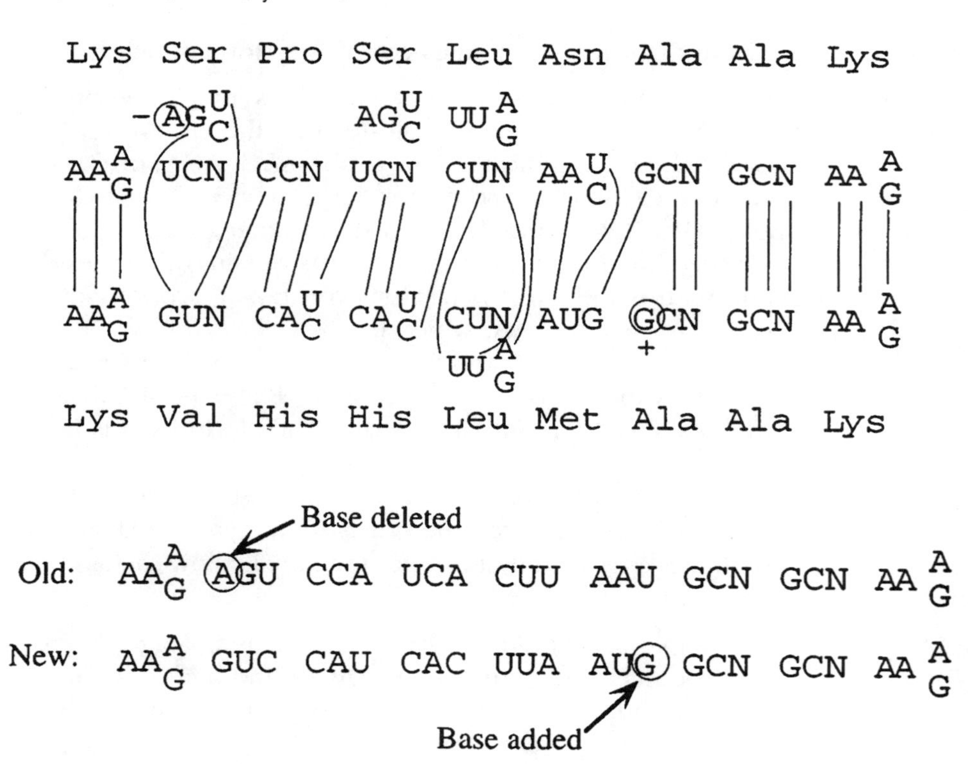

13. If the anticodon on a tRNA molecule also was altered by mutation to be four bases long, with the fourth base on the 5′ side of the anticodon, it would suppress the insertion. Alterations in the ribosome can also induce frameshifting.

14.

3′ CGT	ACC	ACT	GCA 5′	DNA double helix (transcribed strand)
5′ GCA	TGG	TGA	CGT 3′	DNA double helix
5′ GCA	UGG	UGA	CGU 3′	mRNA transcribed
3′ CGU	ACC	ACU	GCA 5′	Appropriate tRNA anticodon
NH_3 - Ala -	Trp	- (stop) - COOH		Amino acids incorporated

15. (5) With an insertion, the reading frame is disrupted. This will result in a drastically altered protein from the insertion to the end of the protein (which may be much shorter or longer than wild type because of altered stop signals).

16. f, d, j, e, c, i, b, h, a, g

17. Cells in long-established culture lines usually are not fully diploid. For reasons that are currently unknown, adaptation to culture frequently results in both karyotypic and gene dosage changes. This can result in hemizygosity for some genes, which allows for the expression of previously hidden recessive alleles.

18. **a. and b.** The sequence of double-stranded DNA is as follows:

5′—TAC ATG ATC ATT TCA CGG AAT TTC TAG CAT GTA—3′
3′—ATG TAC TAG TAA AGT GCC TTA AAG ATC GTA CAT—5′

First look for stop codons. Next look for the initiating codon, AUG (3′—TAC—5′ in DNA). Only the upper strand contains a code exactly five amino acids long:

DNA	3′ TAC	GAT	CTT	TAA	GGC	ACT 5′
RNA	5′ AUG	CUA	GAA	AUU	CCG	UGA 3′
protein	Met	Leu	Glu	Ile	Pro	stop

The DNA strand is read from right to left as written in your text and is written above in reverse order from your text.

c. Remember that polarity must be taken into account. The inversion is

DNA	5′ TAC ATG	CTA	GAA	ATT	CCG	TGA	AAT	GAT	CAT GTA 3′		
RNA	3′	—GAU	CUU	UAA	GGC	ACU	UUA	CUA	GUA—	5′	
amino acids		HOOC 7	6	5	4	3	2	1	—NH_3		

d.

DNA	3′ATG TAC	TAG	TAA	AGT	GCC	TTA	AAG	ATC	GTA CAT 5′	
mRNA	5′UAC AUG	AUC	AUU	UCA	CGG	AAU	UUC	UAG	3′	
	1	2	3	4	5	6	7	stop		

Codon 4 is 5′—UCA—3′, which codes for Ser. Anticodon 4 would be 3′—AGU—5′.

11 Regulation of Gene Transcription

1. The I gene determines the synthesis of a repressor molecule, which blocks expression of the *lac* operon and which is inactivated by the inducer. The presence of the repressor I^+ will be dominant to the absence of a repressor I^-. I^S mutants are unresponsive to an inducer. For this reason, the gene product cannot be stopped from interacting with the operator and blocking the *lac* operon. Therefore, I^S is dominant to I^+.

2. O^C mutants are changes in the DNA sequence of the operator that impair the binding of the *lac* repressor. Therefore, the *lac* operon associated with the O^C operator cannot be turned off. Because an operator controls only the genes on the same DNA strand, it is cis (on the same strand) and dominant (cannot be turned off).

3. **a.** You are told that a, b, and c represent *lacI*, *lacO*, and *lacZ*, but you do not know which is which. Both a^- and c^- have constitutive phenotypes (lines 1 and 2) and therefore must represent mutations in either the operator (*lacO*) or the repressor (*lac I*). b^- (line 3) shows no ß-gal activity and by elimination must represent the *lacZ* gene.

   Mutations in the operator will be cis-dominant and will cause constitutive expression of the *lacZ* gene only if it's on the same chromosome. Line 6 has c^- on the same chromosome as b^+, but the phenotype is still inducible (owing to c^+ in trans). Line 7 has a^- on the same chromosome as b^+ and is constitutive even though the other chromosome is a^+. Therefore a is *lacO*, c is *lacI*, and b is *lacZ*.

b. Another way of labeling mutants of the operator is to denote that they lead to a constitutive phenotype; $lacO^-$ (or a^-) can also be written as $lacO^C$. There are also mutations of the repressor that fail to bind inducer (allolactose) as opposed to fail to bind DNA. These two classes have quite different phenotypes and are distinguished by $lacI^S$ (fails to bind allolactose and leads to a dominant uninducible phenotype in the presence of a wild-type operator) and $lacI^-$ (fails to bind DNA and is recessive). It is possible that line 3, line 4, and line 7 have $lacI^S$ mutations (because dominance cannot be ascertained in a cell that is also $lacO^C$), but the other c^- alleles must be $lacI^-$.

4.

	ß-Galactosidase		**Permease**	
Part	No lactose	Lactose	No lactose	Lactose
a	+	+	–	+
b	+	+	–	–
c	–	–	–	–
d	–	–	–	–
e	+	+	+	+
f	+	+	–	–
g	–	+	–	+

5 **a.** A lack of only E_1 or only E_2 function indicates that both genes have enzyme products that are responsible for a conversion reaction. Because the two genes are in different linkage groups, they cannot be regulated by a single operator and promoter like the *Z* and *Y* genes of the *lac* operon. Type 3 mutants must be mutants of a site that produces a diffusible regulator of the E_1 and E_2 genes. The type 3 mutants identify a site that produces either a repressor (like *I* in the *lac* operon) or an activator (analogous to CAP) of the other two genes.

b. Separate operator and promoter mutants might be found for each gene.

6. If there is an operon governing both genes, then a frameshift mutation could cause the stop codon separating the two genes to be read as a sense codon. Therefore, the second gene product will be incorrect for almost all amino acids. However, there are no known polycistronic messages in eukaryotes. The alternative, and better, explanation is that both enzymatic functions are performed by the same gene product. Here, a frameshift mutation beyond the first function, carbamyl phosphate synthetase, will result in the second half of the protein molecule being nonfunctional.

7. Nonpolar Z^- mutants cannot convert lactose to allolactose, and thus, the operon is never induced.

8. Because very small amounts of the repressor are made, the system as a whole is quite responsive to changes in lactose concentration. In the heterodiploids, repressor tetramers may form by association of polypeptides encoded by I^- and I^+. The operator binding site binds two subunits at a time. Therefore, the repressors produced may reduce operator binding, which in turn would result in some expression of the *lac* genes in the absence of lactose.

9. A gene is turned off or inactivated by the "modulator" (usually called a *repressor*) in negative control, and the repressor must be removed for transcription to occur. A gene is turned on by the "modulator" (usually called an *activator*) in positive control, and the activator must be added or converted to an active form for transcription to occur.

10. The *lacY* gene produces a permease that transports lactose into the cell. A cell containing a *lacY*$^{-}$ mutation cannot transport lactose into the cell, so ß-galactosidase will not be induced.

11. Activation of gene expression by trans-acting factors occurs in both prokaryotes and eukaryotes. In both cases, the transacting factors interact with specific DNA sequences that control expression of cis genes.

In prokaryotes, proteins bind to specific DNA sequences, which in turn regulate one or more downstream genes.

In eukaryotes, highly conserved sequences such as CCAAT and various enhancers in conjunction with trans-acting binding proteins increase transcription controlled by the downstream TATA box promoter. Several proteins have been found that bind to the CCAAT sequence, upstream GC boxes, and the TATA sequence in *Drosophila*, yeast, and other organisms. Specifically, the Sp1 protein recognizes the upstream GC boxes of the SV40 promoter and many other genes; GCN4 and GAL4 proteins recognize upstream sequences in yeast; and many hormone receptors bind to specific sites on the DNA (e.g., estrogen complexed to its receptor binding to a sequence upstream of the ovalbumin gene in chicken oviduct cells). Additionally, the structure of some of these trans-acting DNA-binding proteins is quite similar to the structure of binding proteins seen in prokaryotes. Further, protein-protein interactions are important in both prokaryotes and eukaryotes. For the above reasons, eukaryotic regulation is now thought to be very close to the model for regulation of the bacterial *ara* operon.

12. Bacterial operons contain a promoter region that extends approximately 35 bases upstream of the site where transcription is initiated. Within this region is the promoter. Activators and repressors, both of which are trans-acting proteins that bind to the promoter region, regulate transcription of associated genes in cis only.

The eukaryotic gene has the same basic organization. However, the promoter region is somewhat larger. Also, enhancers up to several thousand nucleotides upstream or downstream can influence the rate of transcription. A major difference is that eukaryotes have not been demonstrated to have polycistronic messages.

13. The *araC* product has two conformations, which are determined by the presence and absence of arabinose. When it has bound arabinose, the *araC* product can bind to the initiator site (*araI*) and activate transcription. When it is not bound to arabinose, the *araC* product binds to both the initiator (*araI*) and the operator (*araO*) sites, forming a loop of the intermediary DNA. When both sites are bound to the *araC* product, transcription is inhibited. The *araC* product is trans-acting.

Many eukaryotic transacting protein factors also bind to promoters, enhancers, or both that are upstream from the protein-encoding gene. These factors are required for the initiation of transcription. Additionally, some bind to other proteins, such as RNA polymerase II, in order to initiate transcription. Like their counterparts in the *ara* operon, the eukaryotic transacting protein factors can bind DNA at two sites, with the intermediary DNA forming a loop between the binding sites.

14. A reasonable model is that one dimer portion of the tetramer binds O_1, and one dimer portion of the tetramer binds O_2. The two dimers then bend the DNA when forming the tetramer complex, which results in a blocking of transcription.

15. Normally, the repressor searches for the operator by rapidly binding and dissociating from nonoperator sequences. Even for sequences that mimic the true operator, the dissociation time is only a few seconds or less. Therefore, it is easy for the repressor to find new operators as new strands of DNA are synthesized. However, when the affinity of the repressor for DNA and operator is increased, it takes too long for the repressor to dissociate from sequences on the chromosome that mimic the true operator, and as the cell divides and new operators are synthesized, the repressor never quite finds all of them in time, leading to a partial synthesis of ß-galactosidase. This explains why in the absence of IPTG there is some elevated ß-galactosidase synthesis. When IPTG binds to the repressors with increased affinity, it lowers the affinity back to that of the normal repressor (without IPTG bound). Then, the repressor can rapidly dissociate from sequences in the chromosome that mimic the operator and find the true operator. Thus, ß-galactosidase is repressed in the presence of IPTG in strains with repressors that have greatly increased affinity for operator. In summary, because of a kinetic phenomenon, we see a reverse induction curve.

16. Construct a set of reporter genes with the promoter region, the introns, and the region 3′ to the transcription unit of the gene in question containing different alterations that do not disrupt transcription or processing. Use these reporter genes to make transgenic animals by germ-line transformation. Assay for expression of the reporter gene in various tissues and the kidney of both sexes.

12 Recombinant DNA Technology

1. *Unpacking the Problem*

1. Of the two discussed in the text, pBR322 is the closer.
2. The single *Hin*dIII site in pBP1 allows for a simple opening up of the plasmid so that a DNA fragment made with *Hin*dIII can be inserted.
3. It is important because it allows for screening for insertions of DNA into the plasmid. If the plasmid simply recircularizes, the transformed bacteria will be tetracycline-resistant. If the plasmid contains "foreign" DNA, the *tet* gene will be disrupted and the strain will be tetracycline-sensitive.
4. Insertion of donor DNA into the plasmid disrupts the *tet* gene. It is not relevant to the problem but was important in the construction of the library.
5. A library is a large collection of cloned DNA maintained within easily cultured vectors. For this question, the source of donor DNA was *Hin*dIII-digested fruit fly genomic DNA, and the vector was pBP1. Although it is not relevant to this question, the source of donor DNA is often key to the type of research being conducted.
6. The gene of interest would have been "found" by using a probe composed of the gene's sequence (typically just a small region is required). This could have been synthesized by "guessing" the DNA sequence on the basis of the gene product's amino acid sequence or by homology to a similar gene from another organism, etc. For this particular question, how this clone was identified does not matter.

7. An electrophoretic gel is an apparatus to separate fragments of DNA by their size. Generally, the mix of DNA fragments is forced to migrate through an agarose gel by an electric field that is negative at the end where the DNA is placed and positive at the far end. Because DNA is negatively charged, it will move to the far end but at rates that are inversely proportional to its size: small fragments will move more rapidly than large.
8. Ethidium bromide binds to DNA and fluoresces when exposed to UV light. It is used to visualize the location of the various DNA fragments within the gel.
9. The DNA from this gel is not "blotted" onto filter paper in this problem. If it had been, it would have been a Southern blot (because DNA was in the gel).
10. In this gel, DNA molecules of different sizes bound to ethidium bromide are visible under UV illumination.
11. There is only one linear fragment generated when a circle is cut once.
12. If cut twice, two linear fragments are generated.
13. There is a one-to-one relationship between the number of sites cut in a circular plasmid and the number of fragments generated.
14. Because the two enzymes will cut the DNA independently, the total number of fragments will be $n + m$.
15. They were loaded into the wells located at the top of the diagram.
16. Smaller fragments move more rapidly and travel farther per unit time than larger fragments.
17. All the control lanes contain 5 kb of DNA, the size of the plasmid. Both *Hin*dIII and *Eco*RV cut the plasmid once but at separate locations, as seen in the lane of the double digest (both single digests generate a single band while the double digest generates two). The lanes with the clone 15–containing plasmid always add up to 7.5 kb, indicating that the donor DNA is 2.5 kb.
18. No. The 5-kb plasmid is cut twice, and the resulting fragments must add up to the total length.
19. No. They represent the cloned DNA cut out from the plasmid by *Hin*dIII and then cut once again by *Eco*RV. The sum of these fragments must equal the whole.
20. It tells you that the fragment that disappeared also contains a restriction site for the second enzyme.
21. A probe will hybridize to any fragments to which it is complementary in sequence.
22. If the two vectors are nonhomologous, the only hybridization observed will be because the gene of interest from the one species is complementary to the gene of interest in the other.

Solution to the problem

a.

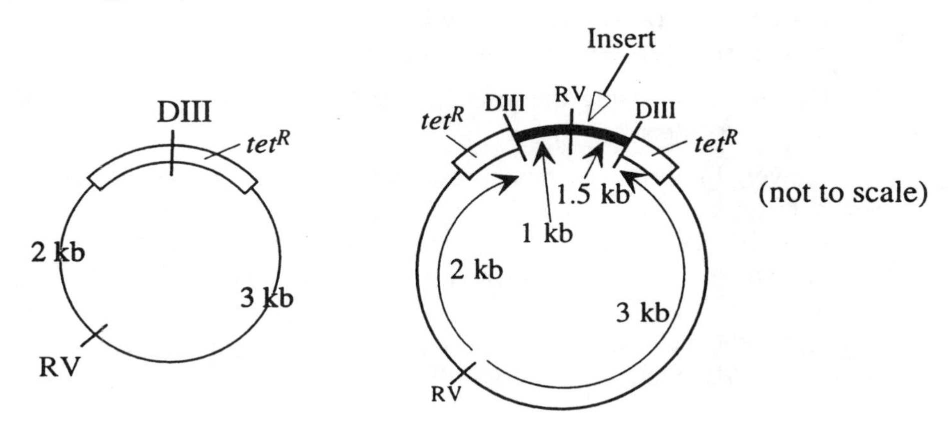

b. The tet^R gene used as a probe will detect only those bands that contain tet^R DNA. Thus, all bands in control lanes will have sequences complementary to the probe. For the clone 15 digests, the *Hin*dIII 5-kb band will be radioactive and both *Eco*RV bands will be radioactive. For the *Hin*dIII + *Eco*RV double digest, the 3-kb and 2-kb bands will be radioactive.

c. The homologous gene used as a probe will detect only those fragments containing the gene of interest. Thus, no bands will be radioactive in the control lanes, and the clone 15 lanes will all have at least one radioactive band. For *Hin*dIII, the 2.5-kb band (the insert) will be radioactive. For *Eco*RV, the 4.5-kb and 3.0-kb bands will be radioactive. For the *Hin*dIII + *Eco*RV double digest, the 1.5-kb and 1-kb bands will be radioactive.

2. This problem assumes a random and equal distribution of nucleotides. Thus, a specific sequence of length n will occur on average once in every 4^n base pairs. GTTAAC occurs, on average, every 4^6 bases, which is 4.096 kb. GGCC occurs, on average, every 4^4 bases, which is 0.256 kb.

3. The data indicate that *Eco*RI fragments 1 and 4 contain no *Hin*dII sites, fragment 3 contains one *Hin*dII site, and fragment 2 contains two sites. Conversely, *Hin*dII fragments A, B, and D all contain one *Eco*RI site, and fragment C contains none. Fragment D contains fragments 1 and 3_1; fragment A contains fragments 3_2 and 2_1; fragment C is the same as fragment 2_2; and fragment B contains fragments 2_3 and 4. The only map consistent with these data is

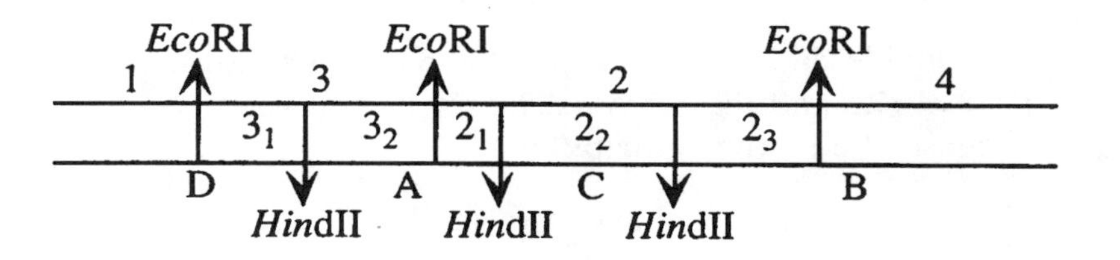

4. **a.** Because the actin protein sequence is known, a probe could be synthesized by "guessing" the DNA sequence based on the amino acid sequence. (This works best if there is a region of amino acids that can be coded with minimal redundancy.) Alternatively, the gene for actin cloned in another species can be used as a probe to find the homologous gene in *Drosophila*. If an expression vector was used, it might also be possible to detect a clone coding for actin by screening with actin antibodies.

b. Hybridization using the specific tRNA as a probe could identify a clone coding for itself.

5. To answer this problem, you must realize what is being visualized. The 8.5 *Eco*RI fragment is radioactive only at one 5′ end, and only fragments containing that end will be seen by autoradiography. When this fragment is cleaved by other restriction enzymes, the longest fragments will have been cut at sites farthest from the radioactive end. In the following figure, if cut at position labeled 2, the fragment will be longer than any fragment cut at 3, 4, or 5 and shorter than any cut at position 1.

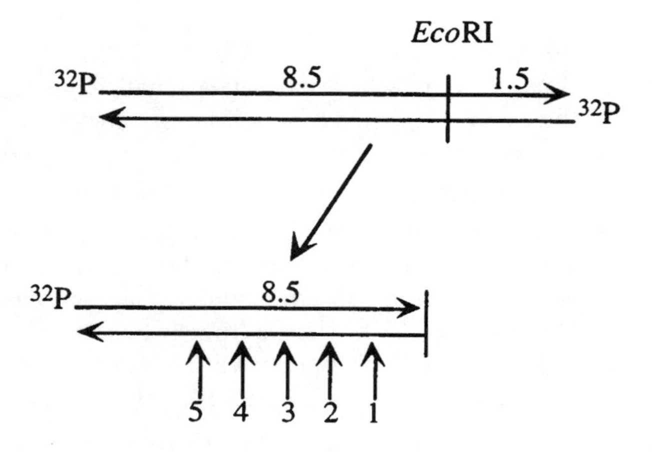

Using this logic, a relative map of the restriction sites of *Hin*dII and *Hae*III for this fragment can be generated.

Reading the gels from the top (and from the farthest to the nearest to the labeled end) the order is (*Eco*RI) - *Hin*dII - *Hae*III - *Hae*III - *Hin*dII - *Hae*III - *Hae*III - *Hae*III - *Hin*dII - *Hin*dII - *Hae*III - *Hin*dII - *Hae*III - labeled end

6. This problem assumes a random and equal distribution of nucleotides.

*Alu*I $(^1/_4)^4$ = on average, once in every 256 nucleotide pairs

*Eco*RI $(^1/_4)^6$ = on average, once in every 4096 nucleotide pairs

*Acy*I $(^1/_4)^4(^1/_2)^2$ = on average, once in every 1024 nucleotide pairs

7. **a.** The double digest indicates that the 5.0-kb *Hin*dIII fragment also contains a *Sma*I site and the 5.5 *Sma*I fragment also contains a *Hin*dIII site. This suggests the following map:

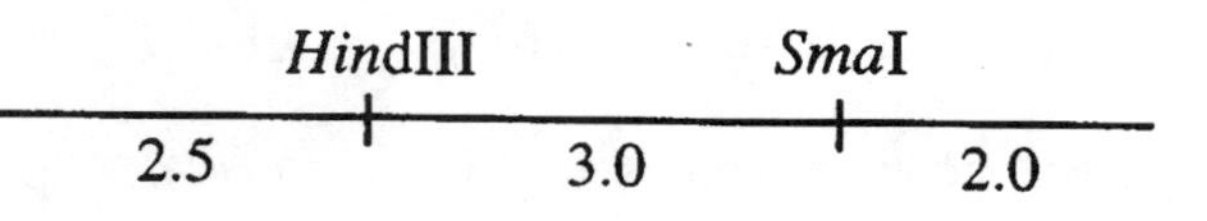

b. Because the only band to disappear is the 3.0-kb fragment, it is the only one that also contains an *Eco*RI site. The appearance of a new 1.5-kb fragment suggests the following:

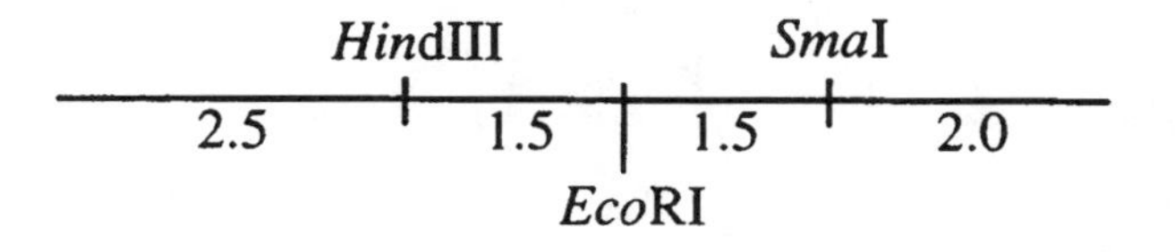

8. **a.** Because the protein is present only after the mRNA has been processed, the mature mRNA must be 1200 nucleotides.

b. The autoradiogram will show only the radioactive RNA molecules. At 2 hours the cDNA does not protect the entire RNA from RNase, but at 10 hours it does. Because the cDNA would not contain sequences complementary to introns, there must be an intron present in the pre-mRNA at 2 hours that is spliced out by 10 hours.

At 2 hours the viral transcript contains an intron sequence that does not hybridize to the cDNA. The RNase removes the single-stranded sequence, leaving behind 500- and 700-base fragments:

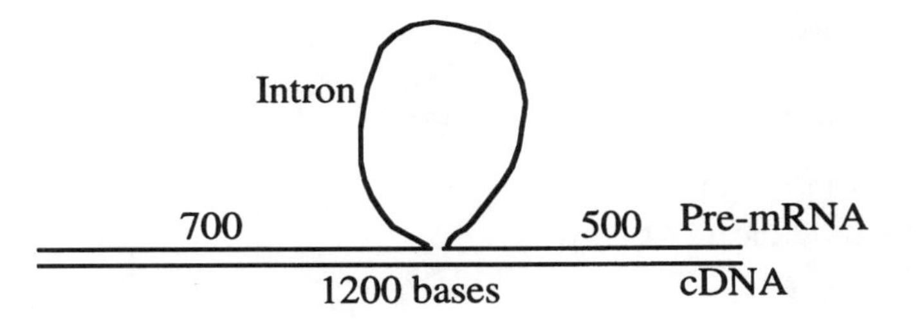

By 10 hours, the intron has been spliced out, and a perfect hybrid forms between the 1200-base viral mRNA and the cDNA:

1200 bases mRNA

1200 bases cDNA

c. It takes a minimum of 2 hours to transcribe and splice the mRNA and then translate it into protein.

9. Create a library of *Podospora* DNA. This would be accomplished by isolating DNA from *Podospora*, cutting the DNA with *Hae*II, mixing with the vector pBR also cut with *Hae*II, ligating the mixture, and transforming *E. coli.* Only those bacteria that contain the plasmid will be tet^R. Of these, those that are kan^S contain plasmids with inserts.

Assuming that the same genes from different species have approximately the same base sequence, the ß-tubulin gene cloned from *Neurospora* can be used as a probe to isolate the ß-tubulin gene from *Podospora.* Identify which clone or clones in the library contain the desired sequence by colony hybridization using the cloned *Neurospora* actin gene as a probe.

10. a. There is one *Bgl*II site, and the plasmid is 14 kb.

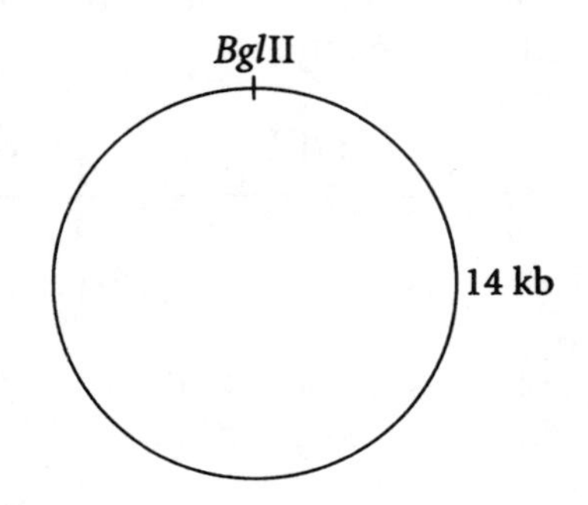

b. There are two *Eco*RV sites.

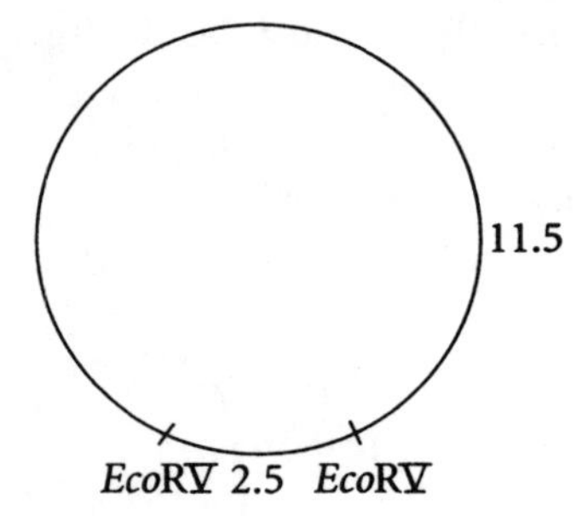

c. The 11.5 kb RV fragment is cut by *Bgl*II. The arrangement of the sites must be as indicated below.

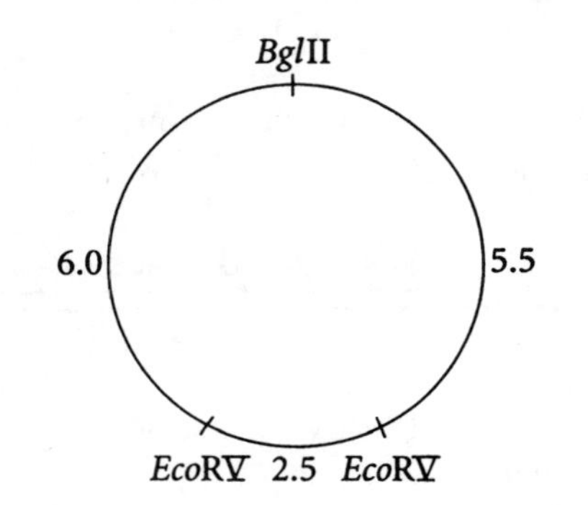

d. The *Bgl*II site must be within the *tet* gene.

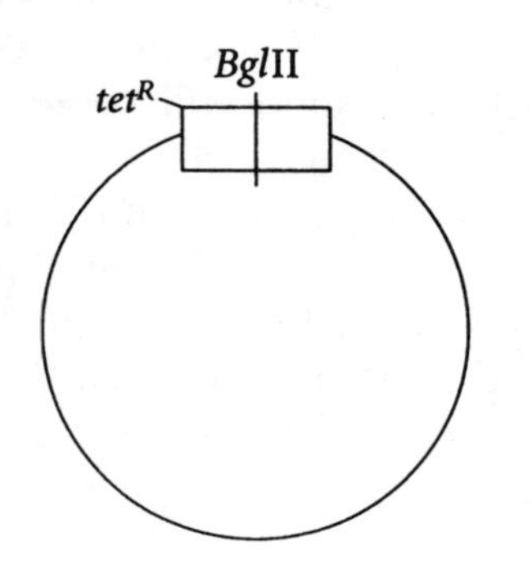

e. There was an insert of 4 kb.

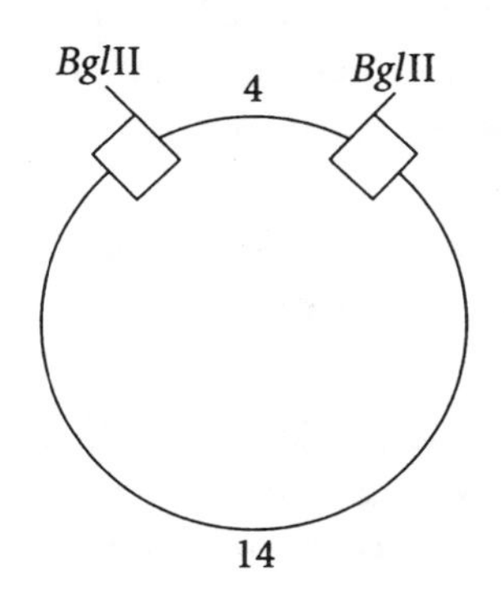

f. There was an *Eco*RV site within the insert.

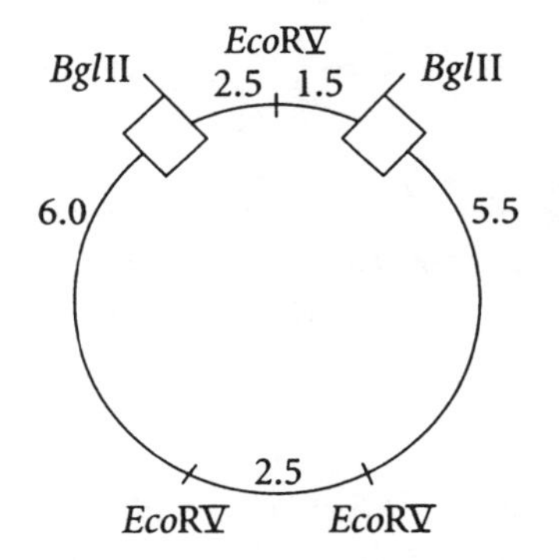

11. a. The restriction map of pBR322 with the mouse fragment inserted is shown below. The 2.5-kb and 3.5-kb fragments would hybridize to the pBR322 probe.

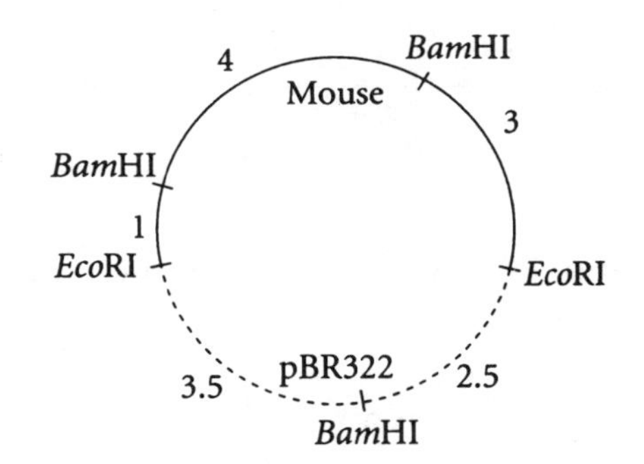

b. A protein 400 amino acids long requires a minimum of 1200 nucleotide bases. Only fragment 3 is long enough (3000 bp) to contain two or more copies of the gene. However, nothing can be said about their orientation.

12. a. To ensure that a colony is not, in fact, a prototrophic contaminant, the prototrophic line should be sensitive to a drug to which the recipient is resistant. A simple additional marker would also achieve the same end.

b. Use a nonrevertible auxotroph as the recipient (such as one containing a deletion).

13. **a.** The transformed phenotype would map to the same locus. If gene replacement occurred by a double crossing-over event, the transformed cells would not contain vector DNA. If a single crossing-over took place, the entire vector would now be part of the linear *Neurospora* chromosome.

b. The transformed phenotype would map to a different locus than that of the auxotroph if the transforming gene was inserted ectopically (i.e., at another location).

14. Size, translocations between known chromosomes, and hybridization to probes of known location can all be useful in identifying which band on a PFGE gel corresponds to a particular chromosome.

15. **a. and b.**

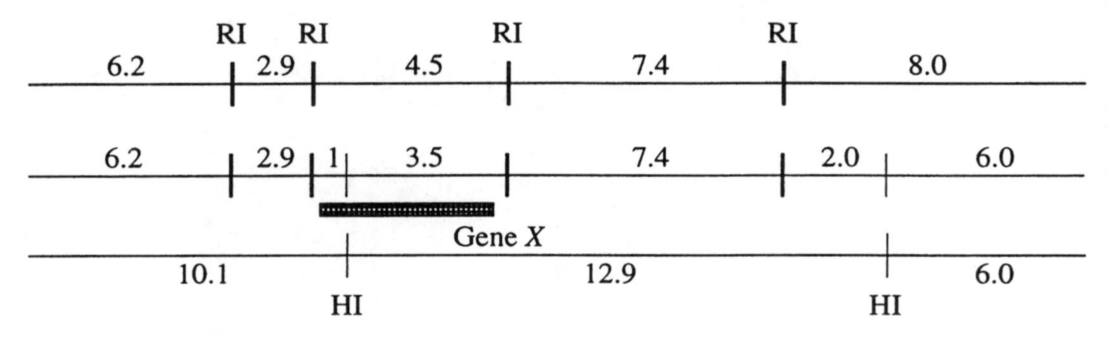

16. **a.** The total weight of the bands is 2.3 kilobases. The combined digest results in four bands, and *Taq*I results in two bands. This suggests that *Taq*I cuts once in the linearized fragment and that *Hae*III cuts once within each *Taq*I fragment. Note that 0.8 + 0.4 = 1.2, and that 0.6 + 0.5 = 1.1. The map of all the fragments is

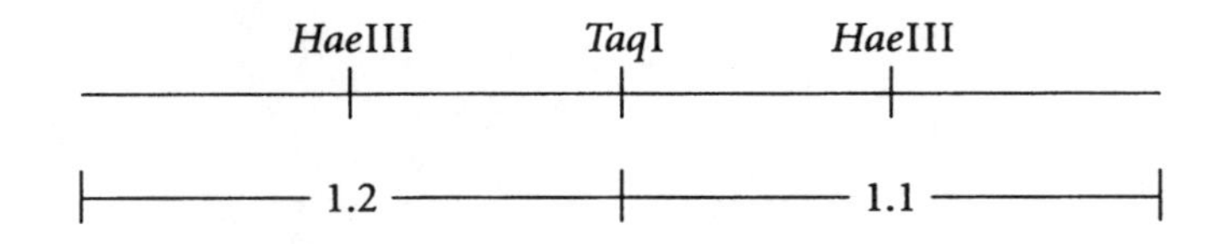

Only the 0.8 and 0.5 fragments contain the ends. Therefore, the complete map is as follows

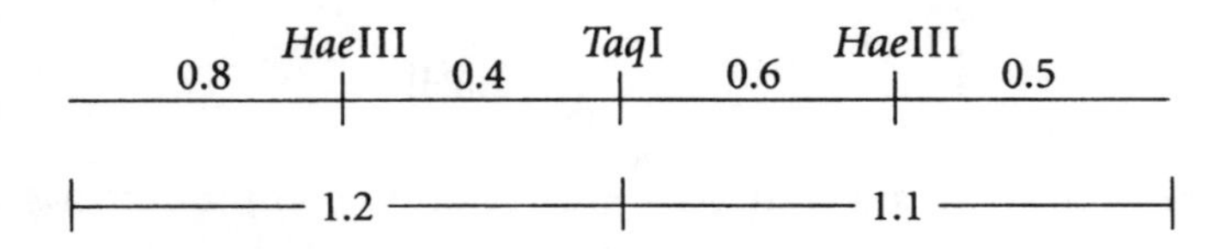

b. There are three bands, only two of which have labeled ends. The map is

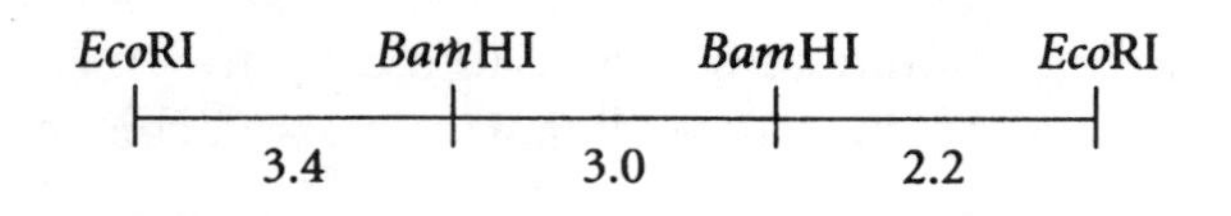

c. The 1.1 *Taq*I fragment should hybridize to the 2.2 fragment. However, without knowing the number of introns, it is possible that this *Taq*I fragment also contains sequences complementary to the 3.4 genomic fragment.

d. The 3.0 genomic fragment is very likely an intron. That is why the cDNA had no complementary sequences to this fragment.

17. a. If an individual is homozygous for an allele, there should be only one band on all the blots. While this does not prove homozygosity, it is consistent with that finding. On the other hand, two bands on any of these gels does show heterozygosity. Therefore, it is consistent that individuals 1, 3, 4, and 5 are homozygous.

b. Individual 3 makes no RNA. Therefore, the most likely conclusion is that the subject is homozygous for a mutation that blocks transcription.

Individual 4 makes a protein smaller than seen in unaffected people, but the mRNA is of the same size. This suggests that a chain termination (nonsense) mutation occurred.

Individual 5 makes a normal-length transcript but a slightly larger protein. One explanation is that a frameshift mutation occurred that eliminated the normal stop codon. Another explanation is a point mutation that altered the normal stop codon so that it coded for an amino acid.

18. a. Note that the genomic *Hin*dIII fragment is end-labeled before the *Bam* digest. This means that the 10 kb *Bam-Hin*dIII fragment is labeled only on its *Hin*dIII end. Only fragments extending from this end toward the *Bam* site are detectable in the partial digest with *Hpa*II (see Problem 5). The genomic map is

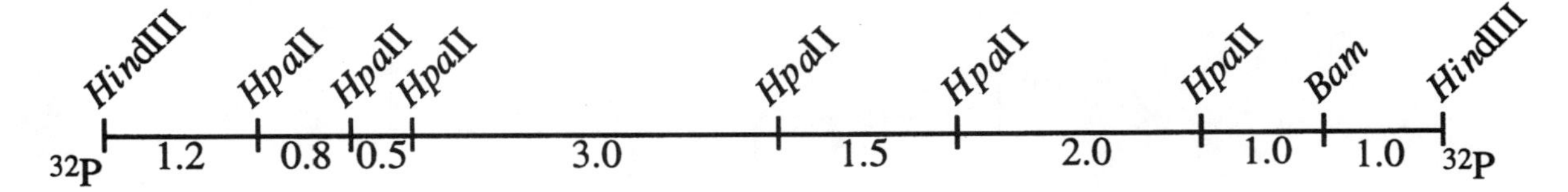

b. and c. In the following restriction map of the genomic DNA, the cDNA fragments are marked underneath by boxes and the "gaps" represent introns. There is a single *Hpa*II site within the cDNA, which is also marked. However, the exact ends and sizes of the cDNA fragments are not established. Fragment 2 contains sequences complementary to a 1-kb fragment, but there are two alternatives possible.

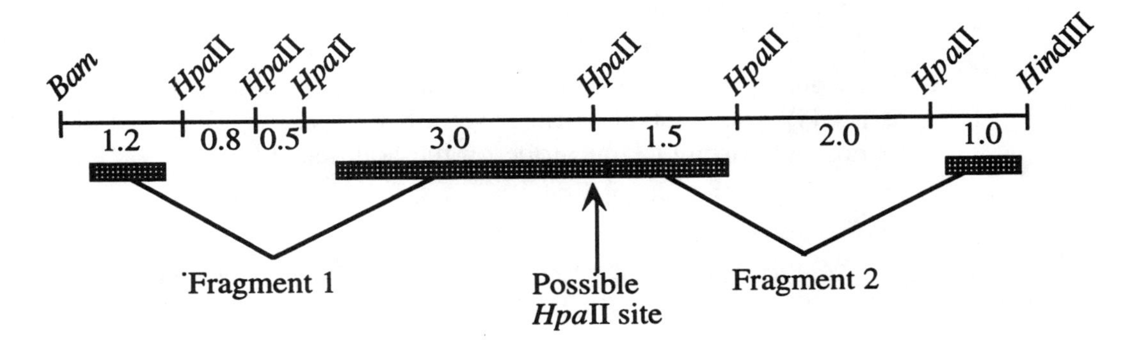

The genomic and cDNA maps differ dramatically because at least two introns have been removed from the genomic DNA to produce the cDNA.

19. a.

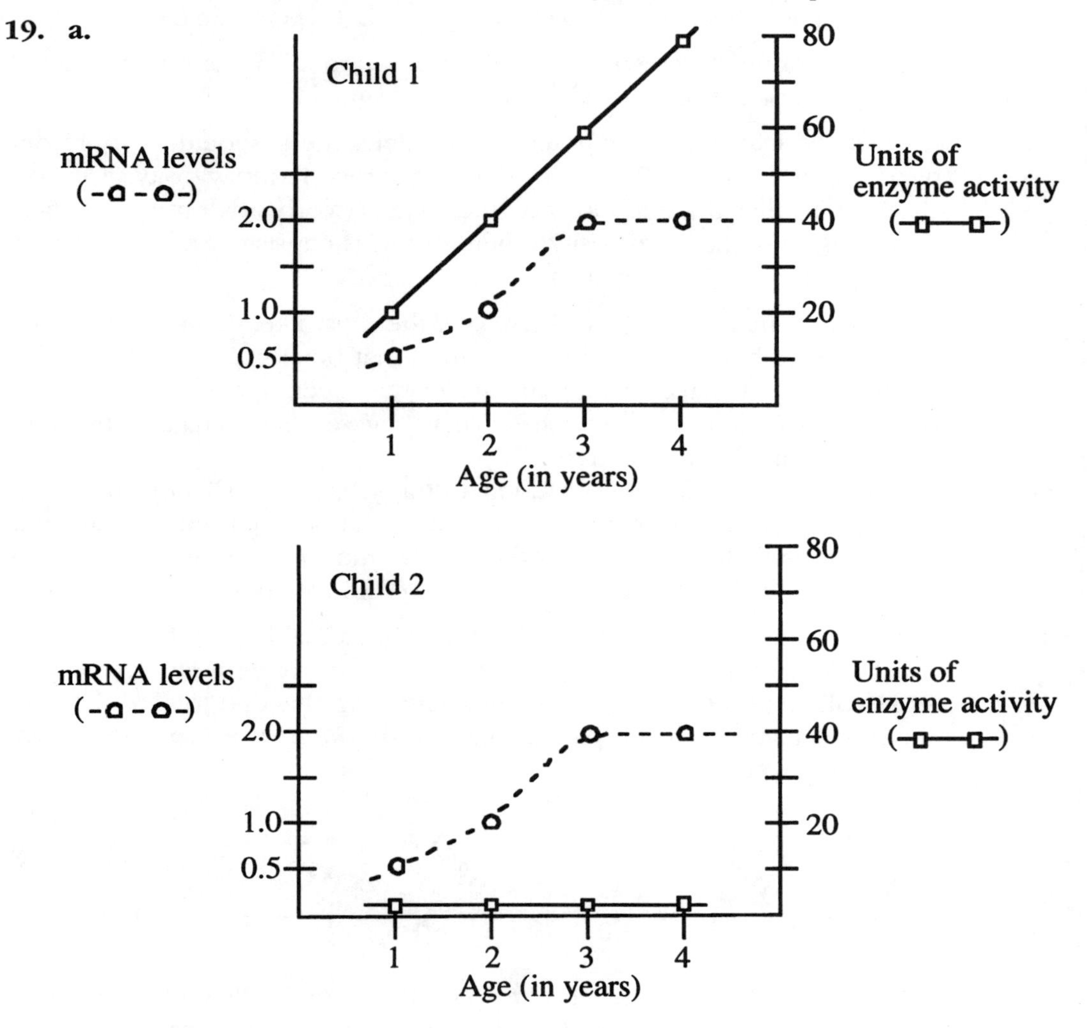

b. and c. Very low levels of active enzyme are caused by the introduction of an *Xho* site within the D gene. Because the size of the nonfunctional enzyme has not changed, the most likely event was a point mutation within the coding region of the gene that also created a new *Xho* site. It is likely that this point mutation also altered the active site, destroying enzyme activity.

d. Individual 1 would be defined as homozygous normal, while individual 2 is homozygous mutant. If either were heterozygous, there would be three bands hybridizing to the probe on the Southern blot.

20. a. The gel can be read from the bottom to the top in a 5′-to-3′ direction. The sequence is

5′ TTCGAAAGGTGACCCCTGGACCTTTAGA 3′

b. By complementarity, the template was

3´ AAGCTTTCCACTGGGGACCTGGAAATCT 5´

c. The double helix is

5´ TTCGAAAGGTGACCCCTGGACCTTTAGA 3´
3´ AAGCTTTCCACTGGGGACCTGGAAATCT 5´

d. Open reading frames have no stop codons. There are three frames for each strand, for a total of six possible reading frames. For the strand read from the gel, the transcript would be

5´ UC**UAA**AGGUCCAGGGGUCACCUUUCGAA 3´

And for the template strand

5´ UUCGAAAGG**UGA**CCCCUGGACCUU**UAG**A 3´

Stop codons are in bold and underlined. There are a total of four open reading frames of the six possible.

21. The region of DNA that encodes tyrosinase in "normal" mouse genomic DNA contains two *Eco*RI sites. Thus, after *Eco*RI digestion, three different-sized fragments hybridize to the cDNA clone. When genomic DNA from certain albino mice is subjected to similar analysis, there are no DNA fragments that contain complementary sequences to the same cDNA. This indicates that these mice lack the ability to produce tyrosinase because the DNA that encodes the enzyme must be deleted.

22. Conservatively, the amount of DNA necessary to encode this protein of 445 amino acids is $445 \times 3 = 1335$ base pairs. When compared with the actual amount of DNA used, 60 kb, the gene appears to be roughly 45 times larger than necessary. This "extra" DNA mostly represents the introns that must be correctly spliced out of the primary transcript during RNA processing for correct translation. (There are also comparatively very small amounts of both 5´ and 3´ untranslated regions of the final mRNA that are necessary for correct translation encoded by this 60-kb of DNA.)

13 Applications of Recombinant DNA Technology

1. Plant 1 shows the typical inheritance for a dominant gene that is heterozygous. Assuming kanamycin resistance is dominant to kanamycin sensitivity, the cross can be outlined as follows:

$$kan^R/kan^S \times kan^S/kan^S$$
$$\rightarrow$$
$$^1/_2\ kan^R/kan^S$$
$$^1/_2\ kan^S/kan^S$$

This would suggest that the gene of interest would be inserted once into the genome.

Plant 2 shows a 3:1 ratio in the progeny of the backcross. This suggests that there have been two unlinked insertions of the *kan*^{R} gene and presumably the gene of interest as well.

$$kan^{R1}/kan^{S1}\ ;\ kan^{R2}/kan^{S2} \times kan^{S1}/kan^{S1}\ ;\ kan^{S2}/kan^{S2}$$
$$\rightarrow$$
$$^1/_4\ \ kan^{R1}/kan^{S1}\ ;\ kan^{R2}/kan^{S2}$$
$$^1/_4\ \ kan^{R1}/kan^{S1}\ ;\ kan^{S2}/kan^{S2}$$
$$^1/_4\ \ kan^{S1}/kan^{S1}\ ;\ kan^{R2}/kan^{S2}$$
$$^1/_4\ \ kan^{S1}/kan^{S1}\ ;\ kan^{S2}/kan^{S2}$$

2. PFGE separates large DNA molecules (small chromosomes) by nature of their size. Unless the overall size of the DNA molecule (chromosome) has been changed, it will migrate in the same relative position.

a. The size of chromosome 1 has not been changed, so there will still be the same seven bands detected, one for each chromosome.

b. The same as (a).

c. Both chromosomes 1 and 7 have been altered. Compared with wild type, the largest and smallest chromosomes will disappear and two new intermediate bands will appear (unless they happen to comigrate with any of the other wild-type bands).

d. One band will be larger than expected and one will be smaller when compared with wild type.

e. The same as wild type, seven bands. The "extra" chromosome will comigrate with its homolog.

f. It is the number of different chromosomes, not the ploidy, that is observed on the gel. This will be the same as wild type (although *Neurospora* is typically haploid).

g. Compared with wild type, the largest band would disappear and be replaced by an even larger band.

3. One way to approach this question is to use the cloned DNA as a probe to see if the gene is transcribed in the nonphotosynthetic tissues. mRNA from various tissues would be isolated and separated on a gel, and a Northern blot performed with the clone as the probe. A band would be detected in only those tissues where the gene is transcribed.

4. There are four patterns of RFLPs possible: the strain 1 pattern; the strain 2 pattern; and the two recombinant patterns of strain 1 for probe A, strain 2 for probe B and strain 1 for probe B, strain 2 for probe A. When probed with both A and B, the following is expected.

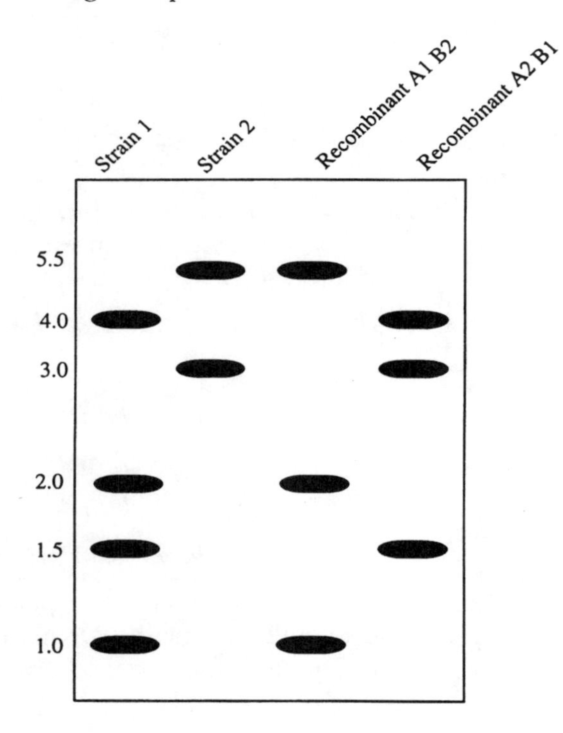

The pattern of bands tells you the "genotype" of each offspring. Although you are told that both RFLPs are on chromosome 5, you do not know if they are on the same or opposite sides of the centromere. The patterns of M_{II} segregation would help distinguish between these two possibilities. For illustration, assume that the two RFLPs are on opposite sides of and each 10 m.u. away from the centromere. For this case, 20% of the asci will show M_{II} segregation for the A RFLP, and 20% will show M_{II} segregation for the B RFLP. To analyze this, each spore would have to be grown separately, and from each colony, DNA would have to be isolated, digested with *Pst*I, and submitted to Southern blot testing. The patterns of bands observed would determine which RFLP marker each progeny inherited. Just considering the RFLP identified by clone A, the following asci would be observed:

A1	A2	A1	A1	A2	A2
A1	A2	A1	A1	A2	A2
A1	A2	A2	A2	A1	A1
A1	A2	A2	A2	A1	A1
A2	A1	A1	A2	A2	A1
A2	A1	A1	A2	A2	A1
A2	A1	A2	A1	A1	A2
A2	A1	A2	A1	A1	A2
40%	40%	5%	5%	5%	5%

The distance of the marker to the centromere would be calculated as $^1/_2$ of those asci showing M_{II} segregation. The same results would be observed for the B RFLP.

The map distance between the two RFLPs could be calculated using the formula

$$RF = {}^1/_2T + NPD$$

Tetratypes (T) will look like

A1B1	A1B2	A1B1	A1B2	A1B1	A1B1	A2B1	A2B1
A1B1	A1B2	A1B1	A1B2	A1B1	A1B1	A2B1	A2B1
A1B2	A1B1	A1B2	A1B1	A2B1	A2B1	A1B1	A1B1
A1B2	A1B1	A1B2	A1B1	A2B1	A2B1	A1B1	A1B1
A2B1	A2B2	A2B2	A2B1	A2B2	A1B2	A2B2	A1B2
A2B1	A2B2	A2B2	A2B1	A2B2	A1B2	A2B2	A1B2
A2B2	A2B1	A2B1	A2B2	A1B2	A2B2	A1B2	A2B2
A2B2	A2B1	A2B1	A2B2	A1B2	A2B2	A1B2	A2B2

and their "upside-down" versions, as well.

And nonparental ditypes (NPD) will look like

A1B2	A1B2	A1B2
A1B2	A1B2	A1B2
A1B2	A2B1	A2B1
A1B2	A2B1	A2B1
A2B1	A1B2	A2B1
A2B1	A1B2	A2B1
A2B1	A2B1	A1B2
A2B1	A2B1	A1B2

and their "upside-down" versions, as well.

5. Assuming that the DNA from this region is cloned, it could be used as a probe to detect this RFLP on Southern blots. DNA from individuals within this pedigree would be isolated (typically from blood samples containing white blood cells) and restricted with *Eco*RI, and Southern blots would be performed. Individuals with this mutant CF allele would have one band that would be larger (owing to the missing *Eco*RI site) when compared with wild type. Individuals that inherited this larger *Eco*RI fragment would, at minimum, be carriers for cystic fibrosis. In the specific case discussed in this problem, a woman that is heterozygous for this specific allele marries a man that is heterozygous for a different mutated CF allele. Just knowing that both are heterozygous, it is possible to predict that there is a 25% chance of their child's having CF. However, since the mother's allele is detectable on a Southern blot, it would be possible to test whether the fetus inherited this allele. DNA from the fetus (through either CVS or amniocentesis) could be isolated and tested for this specific *Eco*RI fragment. If the fetus did not inherit this allele, there would be a 0% chance of its having CF. On the other hand, if the fetus inherited this allele, there would be a 50% chance the child will have CF.

6. The typical procedure is to "knock out" the gene in question and then see if there is any observable phenotype. One methodology to do this is described in Figure 13-4 of the companion text. A recombinant vector carrying a selectable gene within the gene of interest is used to transform yeast cells. Grown under appropriate conditions, yeast that have incorporated the marker gene will be selected. Many of these will have the gene of interest disrupted by the selectable gene. The phenotype of these cells would then be assessed to determine gene function.

7. PFGE separates chromosomes by size. After electrophoresis, Southern blot the gel and probe with radioactive copies of the cloned gene. The clones will hybridize to the band that corresponds to the chromosome they were originally from.

8. The promoter and control regions of the plant gene of interest must be cloned and joined in the correct orientation with the glucuronidase gene. This places the reporter gene under the same transcriptional control as the gene of interest. Figure 11-15 in the companion text discusses the methodology used to create transgenic plants. Transform plant cells with the reporter gene construct, and, as discussed in the figure, grow into transgenic plants. The glucuronidase gene will now be expressed in the same developmental pattern as the gene of interest, and its expression can easily be monitored by bathing the plant in an X-Gluc solution and assaying for the blue reaction product.

9. **a.** The cross is B/b ; $\text{RFLP}^1/\text{RFLP}^2 \times b/b$; $\text{RFLP}^2/\text{RFLP}^2$, where B = bent-tail, b = wild type, RFLP^1 = 1.7-kb *Hin*dIII fragment, and RFLP^2 = 3.8-kb *Hin*dIII fragment.

The bent-tail progeny from this cross are

$$40\%\ B/b\ \text{RFLP}^2/\text{RFLP}^2$$

$$60\%\ B/b\ \text{RFLP}^1/\text{RFLP}^2$$

If the RFLP is unlinked to the bent-tail locus, the two markers should segregate independently. This is not what is observed. $RFLP^1$ appears to be linked to the B allele. If enough progeny were observed to have confidence in the observed ratios, the data suggest that the two markers are 40 m.u. apart (there is 40% recombination).

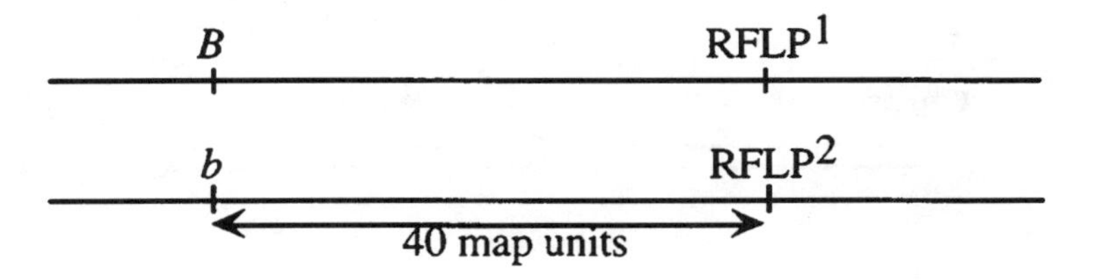

b. The wild-type progeny should be

60% "parental"	*b* $RFLP^2$/*b* $RFLP^2$
40% "recombinant"	*b* $RFLP^1$/*b* $RFLP^2$

10. a. The two strains of yeast were crossed (*A1 B1* × *A2 B2*) and the meiotic products were analyzed. If the two RFLP markers are on separate chromosomes, they should assort independently. The actual data are

Spore type	RFLPs	Frequency
1	*A1 B2*	15%
2	*A2 B1*	15%
3	*A1 B1*	35%
4	*A2 B2*	35%

The markers are not assorting randomly. Spore types 1 and 2 are recombinants and types 3 and 4 are parentals. The RFLPs are 30 m.u. apart.

b.

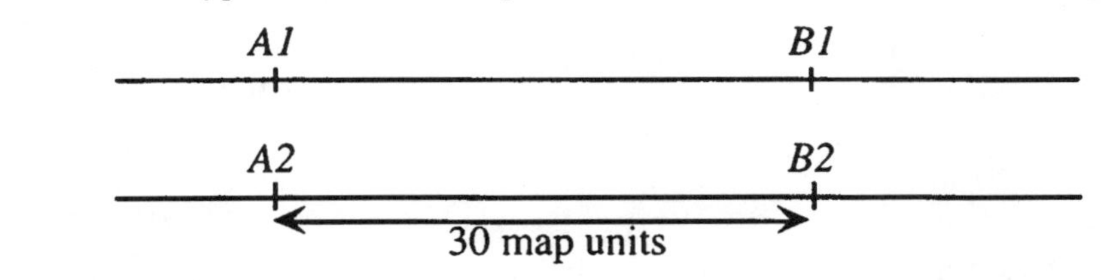

11 a. and b. During Ti plasmid transformation, the kanamycin gene will insert randomly into the plant chromosomes. Colony A, when selfed, has $^3/_4$ kanamycin-resistant progeny, and colony B, when selfed, has $^{15}/_{16}$ kanamycin-resistant progeny. This suggests that there was a single insertion into one chromosome in colony A and two independent insertions on separate chromosomes in colony B. This can be schematically represented by showing a single insertion within one of the pair of chromosomes "A" for colony A

Chromosomes "A"

T-DNA T-DNA

kanamycin^R

and two independent insertions into one of each of the pairs of chromosomes "B" and "C" for colony B

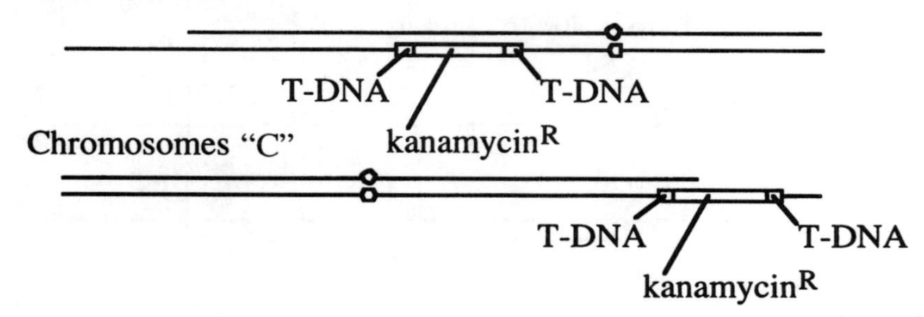

Genetically, this can be represented as

Colony A	kan^{RA}/kan^{SA}
Colony B	kan^{RB}/kan^{SB} ; kan^{RC}/kan^{SC}
When these are selfed	$kan^{RA}/kan^{SA} \times kan^{RA}/kan^{SA}$

$\downarrow$

$1/4$ kan^{RA}/kan^{RA}

$1/2$ kan^{RA}/kan^{SA}

$1/4$ kan^{SA}/kan^{SA}

kan^{RB}/kan^{SB} ; $kan^{RC}/kan^{SC} \times kan^{RB}/kan^{SB}$; kan^{RC}/kan^{SC}

$\downarrow$

$9/16$ $kan^{RB}/-$; $kan^{RC}/-$

$3/16$ $kan^{RB}/-$; kan^{SC}/kan^{SC}

$3/16$ kan^{SB}/kan^{SB} ; $kan^{RC}/-$

$1/16$ kan^{SB}/kan^{SB} ; kan^{Sv}/kan^{SC}

12. a. Yeast plasmids can exist free in the cytoplasm or can integrate into a chromosome; however, the patterns of inheritance will differ for the two states. Free plasmids are present in multiple copies and will be distributed to all progeny. This is observed in crosses with YP1 transformed cells: YP1 leu^+ $\times$ leu^- all progeny are leu^+ and all have vector DNA. Crosses with YP2 transformed cells show simple Mendelian inheritance and suggest that this plasmid has integrated into the yeast chromosome.

b. For YP1 transformed cells, the circular (and free) plasmid will be linearized by the single restriction cut, and when probed, a single band will be present on the Southern blot. Because the YP2 plasmid is integrated, a single cut within the plasmid will generate two fragments that contain plasmid DNA. However, the size of these fragments will depend on where in the genome other sites exist for this same restriction enzyme. This is schematically shown below:

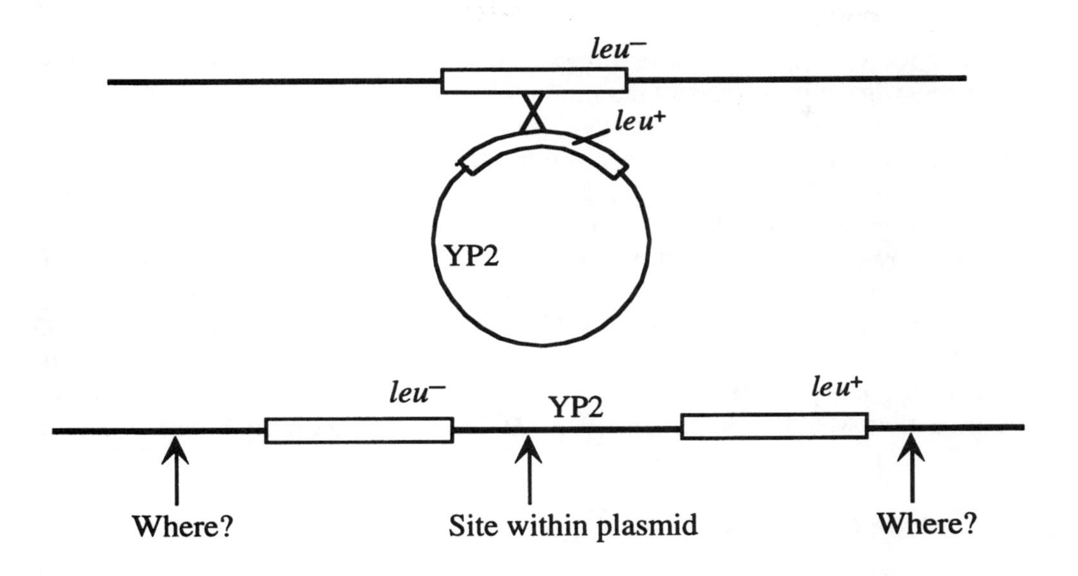

13. **a.** If the plasmid never integrates, the linear plasmid will be cut once by *Xba*I and two fragments will be generated that will both hybridize to the *Bgl*II probe. The autoradiogram will show two bands whose combined length will equal the full length of the plasmid.

b. If the plasmid integrates occasionally, most cells will still have free plasmids and these will be indicated by the two bands mentioned above. However, when the plasmid is integrated, two bands will still be generated, but their sizes will vary based on where other genomic *Xba*I sites are relative to the insertion point (see the figure for Problem 12b for similar logic). If integration is random, many other bands will be observed, but if it's at a specific site, only two other bands will be detected.

14. Assume the grandfather (I-1) is heterozygous for the dominant Huntington disease allele (which is normally much more likely for this rare disease). Assigning genotypes where possible, the following can be stated

I-1 H RFLP^A / h RFLP^A × h RFLP^B/h RFLP^B I-2

where H = Huntington disease allele, h = wild-type allele, RFLP^A = 3.7- and 1.2-kb fragments, and RFLP^B = 4.9-kb fragment.

All other members of this pedigree can also be assigned:

II-1 and II-4	? RFLP^A/h RFLP^B
II-2	h RFLP^B/h RFLP^B
II-3	h RFLP^A/h RFLP^A

where the ? indicates H or h, both being equally likely. It cannot be determined which chromosome II-1 and II-4 inherited from their father (I-1). Both have a 50% chance of being at risk for Huntington disease (i.e., inheriting the H RFLP^A chromosome).

For fetus III-1, RFLP^A was inherited from its father because RFLP^B must have been inherited from its mother.

For fetus III-2, $RFLP^B$ was inherited from its mother because $RFLP^A$ must have been inherited from its father.

Because the RFLP being analyzed in this question is 4 m.u. distant from the gene, recombination between the two must be considered. For fetus III-1, it is the probability of inheriting the *H* allele from its father that is relevant, and for fetus III-2 it is the probability of inheriting the *H* allele from its mother that is relevant. Both of these parents have a $^1/_2$ chance of having inherited this allele, and in both cases it would be linked to $RFLP^A$. Because fetus III-1 has inherited $RFLP^A$, there is $^1/_2 \times 96\%$ (the chance that $RFLP^A$ is still linked to *H*) = 48% chance that it has inherited Huntington disease. Fetus III-2 has inherited the $RFLP^B$ allele from its mother, so there is $^1/_2 \times 4\%$ (the chance that $RFLP^B$ is now linked to *H*) = 2% chance that it has inherited Huntington disease.

15. *Unpacking the Problem*

1. Hyphae in *Neurospora* are the threads of cells that grow out from the original ascospore. Therefore, *hyphal extension* refers to the pattern of these threads and the distance that they grow. Because this process is due to cell growth (and its control) and cell shape, anyone interested in a vast array of cell biological issues might be interested in the genes identified by such screens.

2. *Mutational dissection* is the attempt to identify all the genes and gene products involved in a particular process. In this experiment, the goal is to identify all the genes that can mutate to a small-colony phenotype by random insertion of unrelated DNA (in this case a bacterial plasmid with a selectable marker).

3. *Neurospora* is a haploid organism, and this is relevant to this problem. What might otherwise be a recessive mutation in a diploid organism (typical of gene knockouts) would instead be immediately expressed in a haploid one.

4. The source of the DNA is not relevant to the problem so long as it contains a selectable marker for the organism being transformed. The ease of growing, manipulating, and isolating bacterial plasmids makes them an attractive choice.

5. Transformation, as originally discovered by Griffith, is the uptake of DNA from one organism by another organism and its ultimate expression. In this situation, *Neurospora* has been pretreated in such a way as to cause the uptake of the bacterial plasmid. This is a well-used technique in molecular genetics as a way of introducing genes from virtually any source into the organisms under study.

6. Plant and fungal cells are generally prepared for transformation by removal of their cell walls. The cell membranes are then exposed to a high salt concentration and the exogenous DNA. Studies indicate that the DNA enters the cell in two ways: (1) phagocytosis and (2) localized temporary dissolving of the membrane by the high salt concentration.

7. With successful transformation, the exogenous DNA passes through the cytoplasm and enters the nucleus, where it becomes integrated into a host chromosome.
8. Integration and Transformation

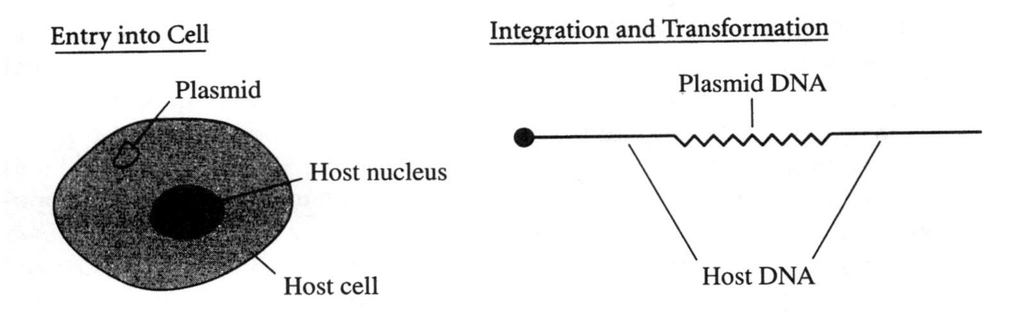

9. It is completely unnecessary to know what benomyl is. Its use simply allows for the selection of cells that received and integrated some exogenous DNA. Virtually any resistance marker could have been used. The choice of a resistance marker usually depends on what is easily available to the researcher, although questions of toxicity to humans may play a role in the choice.
10. Because hyphal extension occurs in colonies, not at the one-cell stage, the researcher must look for mutants that are expressed by a clone or colony. Therefore, he is looking for mutants that are "colonial." Mutations that produce an aberrant colony in size or shape are, by definition, involved with the extension of hyphae.
11. The "previous mutational analysis" could have been any random study. For example, in screens for specific auxotrophic mutants experimenters would have noticed this abnormal phenotype also appearing.
12.

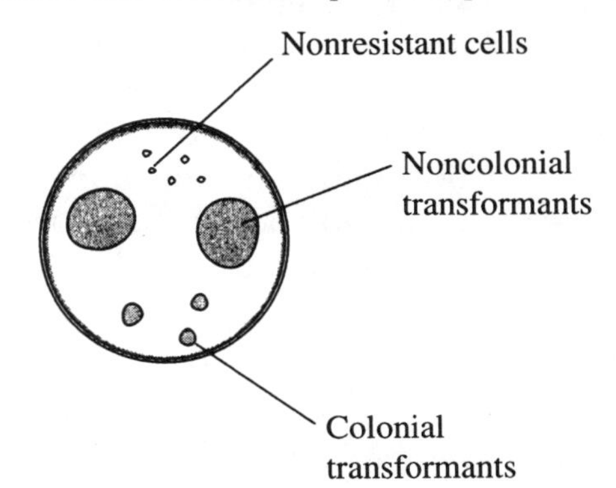

13. Tagging is a process to mutate and "mark" genes of interest by insertion of known DNA. In this case, the gene for benomyl resistance is being inserted randomly within the genome. Occasionally, insertion will occur within a gene involved with hyphal extension. These will cause the aberrant phenotype because the gene has been disrupted by the insertion of the selectable DNA. The disruption causes a knockout mutation within

the gene of interest and also supplies a "molecular handle" to later clone the DNA.

14. The orange-colored bread mold *Neurospora* is a multicellular haploid in which the cells are joined end to end to form hyphae. The hyphae grow through the substrate and also send up aerial branches that bud off haploid cells known as conidia (asexual spores). Conidia can detach and disperse to form new colonies, or alternatively they can act as paternal gametes and fuse with a maternal structure of a different individual. However, the different individual must be of the opposite mating type. In *Neurospora* there are two mating types, determined by the alleles *A* and *a*. A cross will succeed only if it is $A \times a$. An asexual spore from the opposite mating type fuses with a receptive hair, and a nucleus travels down the hair to pair with a maternal gamete that waits inside a specialized knot of hyphae. The *A* and *a* pair then undergo synchronous mitoses, finally fusing to form diploid meiocytes. Meiosis occurs, and in each meiocyte four haploid nuclei are produced, which represent the four products of meiosis. For an unknown reason, these four nuclei divide mitotically, resulting in eight nuclei, which develop into eight football-shaped sexual spores called ascospores. The ascospores are shot out of a flask-shaped fruiting body that has developed from the knot of hyphae that originally contained the maternal gametic cell. The ascospores can be isolated, each into a culture tube, where each ascospore will grow into a new culture by mitosis.

15. In order for the benomyl-resistance gene to be integrated within the host chromosome, it recombines with it. However, it is recombination between the *col* and ben^R genes that is interesting. It is either 0% (type 1), indicating that the insertion caused the small-colony phenotype, or 50% (type 2), showing the two events are unlinked.

16. Only two types are possible: integration into a "hyphal" gene (so the resistance and small-colony phenotype are linked) and ectopic integration and concurrent mutation of a gene causing the small-colony phenotype where the two are unrelated and also unlinked. Which type is more likely depends on mutation rates and the number of genes that can mutate to the small-colony phenotype (i.e., the ease of generating spontaneous *col* mutants) and on the rate of transformation and integration.

17. A probe in experiments such as this one is usually a sequence of DNA that can be used to identify a specific DNA sequence within a genome or colony. The probe is labeled in some way to indicate its presence. In this experiment, the probe was probably the bacterial plasmid (although it might have been the benomyl-resistance gene), most likely radioactively labeled. A genomic library from each *col* mutant would be screened with the probe to identify those clones that contain complementary sequences (and, with luck, some sequences of the gene of interest).

18. A probe specific to the bacterial plasmid could be made by growing bacteria with the plasmid. The plasmids could be isolated through cesium chloride centrifugation and then labeled.

Solution to the Problem

a. Type 1 isolates behave genetically as if the benomyl-resistance and small-colony phenotype are completely linked. The progeny are all parental in phenotype (either *col ben-R* or + *ben-S*). This would be the expected result if the insertion of the plasmid (and *ben-R* gene) caused the mutation that led to the *col* phenotype, which of course was the point of the experiment.

Type 2 isolates behave genetically as if the benomyl-resistance and small-colony phenotype are unlinked (i.e., the two markers are segregating randomly in the progeny). This is the expected result if the insertion of the *ben-R* gene was unrelated to the mutation that led to the *col* phenotype. In other words, the *col* mutation occurred randomly and separately from the insertion of the *ben-R* gene.

b. The type 1 isolates are the *col* genes that are mutated by the insertion of the plasmid, and therefore these are the genes that are tagged.

c. The type 1 isolates should be used to create genomic libraries. The libraries should be screened for clones that contain DNA adjacent to the insert by probing with known sequences from the plasmid. The identified clones will represent parts of the disrupted gene of interest. To recover the intact wild-type gene, a subclone of the disrupted gene sequence can be used to probe a wild-type genomic library.

d. All progeny that are benomyl-resistant will also contain DNA from the bacterial plasmid integrated into their chromosome or chromosomes. Thus, all *ben-R* strains from this experiment will have DNA that hybridizes to a probe specific for the plasmid.

14 GENOMICS

1. *Unpacking the Problem*

1. Two types of hybridizations that have already been discussed are hybridizations between strains of a species and hybridizations between species. A third type of hybridization is referred to in this problem: molecular hybridization. Molecular hybridization can involve either DNA-DNA hybridization or DNA-RNA hybridization. In both instances, it relies on the specificity of complementary pairing and can take place in solution, on a gel, on a filter, or on a slide. For example:

   5′—UACGGGAU—3′ RNA
   3′—ATGCCCTA—5′ DNA

2. In situ hybridization usually is conducted on a slide so that the stained chromosomes can be observed and the specific portion of a chromosome to which the probe hybridizes can be identified.

3. A YAC is a yeast artificial chromosome. It contains a yeast centromere, autonomous replication sequences (origins of replication), telomeres, and DNA that has been attached between them.

4. Chromosome bands are dark regions along the length of a chromosome that occur in a characteristic pattern for each chromosome within an organism. They can occur naturally, as with *Drosophila* polytene chromosomes,

or they can be induced by a number of chemical and physical agents, combined with staining to accentuate the bands and interbands.

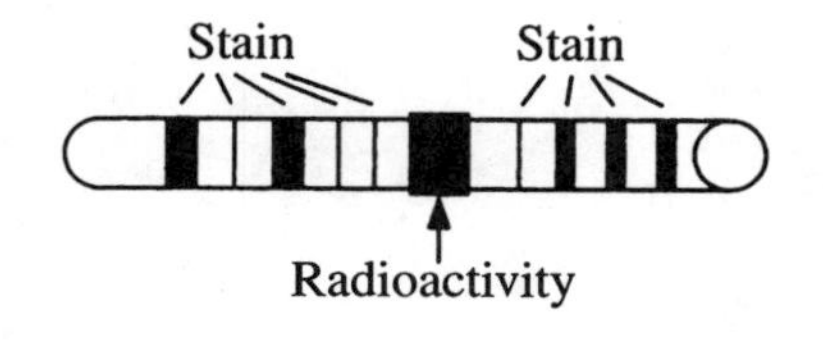

5. The five YACs could have been hybridized sequentially to the same chromosome preparation, which is, however, unlikely. Alternatively, the information could have been determined in five separate experiments. In either case, a YAC labeled with either radioactivity or fluorescence, and including the DNA of interest, was hybridized to a chromosome preparation. The chromosomes were properly treated to reveal the banding pattern, and the YACs were determined to hybridize to the same band.
6. A genomic fragment, by definition, contains a subportion of the genome being studied. In most instances, it actually contains a subportion of one chromosome. Five randomly chosen YACs would not be expected to contain the same genomic fragment or even fragments from the same chromosome. The fragments could have been produced by either physical (X-irradiation, shearing) or chemical (digestion, restriction) means, but it does not matter how they were produced.
7. A restriction enzyme is a naturally occurring bacterial enzyme that is capable of causing either single or double-stranded breaks in DNA at specific DNA sequences.
8. A long cutter is a restriction enzyme that produces very long fragments of DNA because the sequence it recognizes occurs infrequently within the genome.
9. The YACs were radioactively labeled so that their location after hybridization could be detected through autoradiography. To radioactively label is to attach an isotope that emits energy through decay. Commonly used radioactive labels are tritium (^3H) in place of hydrogen and ^{32}P in place of phosphorus.
10. An autoradiogram is a "self-picture" taken through radioactive decay from a labeled probe. When a gel or blot is used, the radioactive decay is captured by a piece of X-ray film. When in situ hybridization is performed on slides, the photographic emulsion coats the slide directly.
11. Free choice. Be sure you truly know the meaning of each term.
12. We are given a diagram of the composite autoradiographic results. The DNA from humans was isolated and subjected to digestion by a restriction enzyme that cuts very infrequently. Once the DNA was electrophoresed, it was Southern-blotted and then probed sequentially with radioactively labeled YACs, followed by sequential exposure to X-ray film. Between probings, the previous YAC hybrid was removed through denaturing of the DNA-DNA hybrid. Alternatively, five separate Southern blottings were done.

13. The haploid human genome is thought to contain approximately 3.3×10^6 kilobases of DNA.
14. Restriction digestion of human genomic DNA would be expected to produce hundreds of thousands of fragments.
15. The fragments produced by restriction of human genomic DNA would be expected to be mostly different.
16. When subjected to electrophoresis and then stained with a DNA stain, the digested human genome would produce a continuous "smear" of DNA, from very large fragments (in excess of tens of thousands of base pairs in length) to fragments that are very small (under a hundred bases in length).
17. In this question, only two distinct bands are produced, at most, in any one probing. The difference between what is seen with a DNA stain and what is seen with probing lies in the specificity of the agent being used. DNA stain will detect any DNA, while a DNA probe will detect only DNA that is complementary to the probe.
18. Number them from top to bottom, 1–3, across the gel. Thus, YACs A–C contain band 1, YACs C–D contain band 2, and YACs A and E contain band 3.
19. There are no restriction fragments on the autoradiogram. The fragments are on the filter (nitrocellulose, nylon) used to blot the gel. The radioactivity of the probes is captured by the X-ray film as it decays, producing an exposed region of film.
20. YACs B, D, and E hybridize to one fragment, and YACs A and C hybridize to two fragments.
21. A YAC can hybridize to two fragments if the YAC contains continuous DNA and there is a restriction site within that region. A YAC can also hybridize to two fragments if it contains discontinuous DNA from two locations in the genome that either are on different chromosomes (this is analogous to a translocation) or are separated by at least two restriction sites if they are on the same chromosome (this is analogous to a deletion). In this case, the former makes more sense. Because the YACs were selected for their binding to one specific chromosome band, it is unlikely that the YACs are composed of discontinuous DNA sequences. A YAC could hybridize to more than two fragments because the continuous DNA could contain many restriction sites or the discontinuous DNA could be composed of DNA from a number of regions in the genome.
22. Cytogeneticists use the term *band* to designate a region of a chromosome that is dark-staining. Molecular biologists use *band* to designate a region of dark appearing on an autoradiogram, which is produced by radioactive decay from a specific probe that reacted with a population of molecules localized by gel electrophoresis. In both cases, *band* refers to a localization.

Solution to the Problem

a. Note that fragments 1 and 3 occur together and fragments 1 and 2 occur together, but that fragments 2 and 3 do not occur together. This suggests that the sequence is 2 1 3 (or 3 1 2).

b. If the sequence of the fragments is 2 1 3, then the YACs can be shown in relation to these fragments. YAC A spans at least a portion of both 1 and 3. YAC B is within region 1. YAC C spans at least a portion of regions 1 and 2. YAC D is contained within region 2. YAC E is contained within region 3. A diagram of these results is shown below. In the diagram, there is no way to know the exact location of the ends of each YAC.

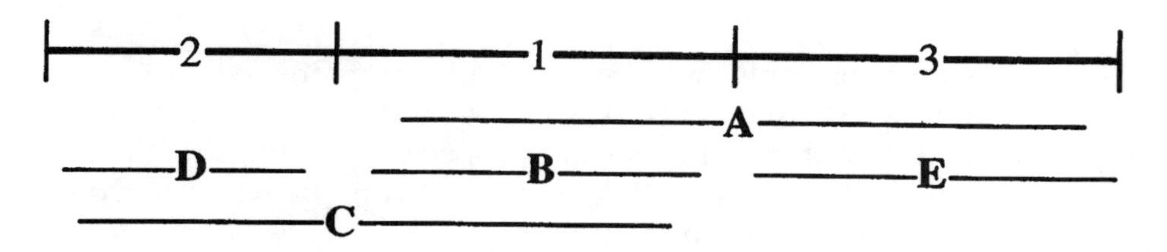

2. PFGE results in separation of the chromosomes, and the gel indicates the discreteness obtained by the separation. Because the *mata* probe does not hybridize to a band, there must not be a gene of similar sequence in *Neurospora.* On the other hand, the *leu2* probe hybridizes to a sequence contained within chromosome 4, and *ade3* to a sequence contained within chromosome 7.

3. The cross is

$$cys\text{-}1\ RFLP\text{-}1^O\ RFLP\text{-}2^O \times cys\text{-}1^+\ RFLP\text{-}1^M\ RFLP\text{-}2^M$$

Scoring the progeny, a parental type will have the genotype of either strain and, if the markers are all linked, be the most common. A recombinant type will have a mixed genotype and be less common. Clearly, the first two ascospore types are parental, with the remaining being recombinant.

a. The *cys-1* locus is in this region of chromosome 5. If it were not in this region, linkage to either of the RFLP loci would not be observed.

b. To calculate specific distances, you may need to review previous chapters. Here, it is assumed that you recall basic mapping strategies.

cys-1 to *RFLP-1* = $^{(2\,+\,3)}/_{100} \times 100\% = 5$ map units

cys-1 to *RFLP-2* = $^{(7\,+\,5)}/_{100} \times 100\% = 12$ map units

RFLP-1 to *RFLP-2* = $^{(2\,+\,3\,+\,7\,+\,5)}/_{100} \times 100\% = 17$ map units

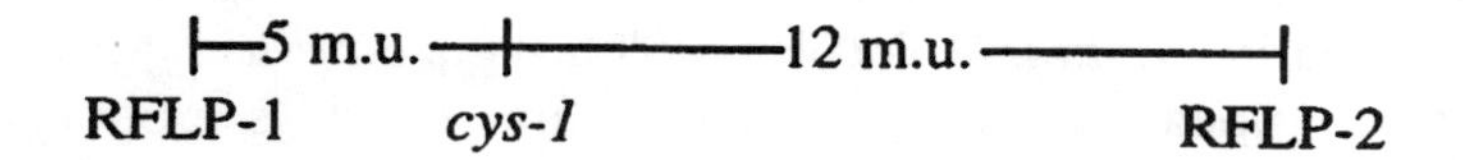

c. A number of strategies could be tried. Because this is an auxotrophic mutant, functional complementation can be attempted. Positional cloning or chromosome walking from the RFLPs is also a very common strategy.

4. **a. and b.** Compare each translocation gel to the wild-type gel and note the difference in bands that were obtained. Together, the translocations involve chromosomes 1, 2, and 4. The top band in all three gels is constant and must reflect chromosome 3, which is not involved in either translocation. Focusing on the three remaining bands in the wild-type gel, the last band is not altered in the 1;4 translocation but is altered in the 2;4 translocation. This means that it must reflect chromosome 2. The band second from the bottom in the wild-type gel is altered in the 1;4 translocation but is not altered in the 2;4 translocation and must reflect chromosome 1. The remaining band in the wild-type gel is altered in both the 1;4 and the 2;4 translocations, and it must reflect chromosome 4.

In the 1;4 lane, the two new bands in comparison with the wild-type lane are due to the 1;4 translocation, while in the 2;4 lane, they are due to the 2;4 translocation. In both translocation lanes, the P probe is associated with newly appearing bands, indicating that P is located on chromosome 4. This is confirmed in the wild-type lane.

5. Remember that a gene is one small region of a long strand of DNA and that a cloned gene will contain the entire sequence of the gene under normal circumstances. If there are two cuts within a gene, three fragments will be produced, all of which will interact with the probe, as was seen with enzyme 1. Cuts external to a gene will produce one fragment that will interact with the probe, which was seen with enzyme 2. One cut within a gene will produce two fragments that will interact with the probe, as was seen with enzyme 3.

6. **a.** To determine the physical map showing the STS order, simply list the STSs that are positive, using parentheses if the order is unknown, and align them with one another to form a consistent order.

YAC A:			1	4	3	
YAC B:		5	1			
YAC C:				4	3	7
YAC D:	(6 2)	5				
YAC E:					3	7

b. Once the sequence of STSs is known, the YACs can be aligned as follows, although precise details of overlapping and the locations of ends are unknown:

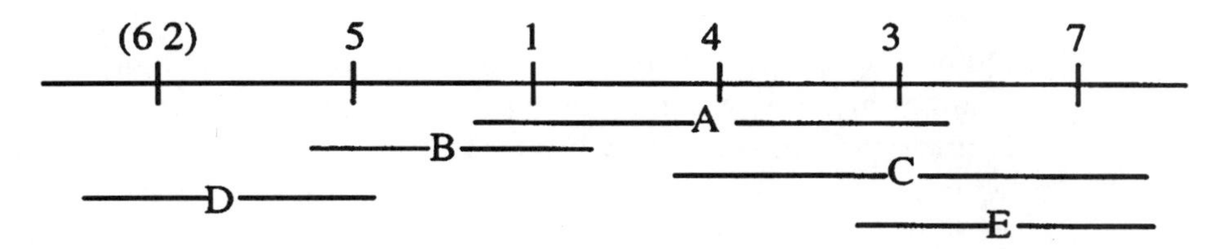

7. **a. and b.** There are four patterns that can be observed in the comparisons that can be made between these six markers: + +, – –, + –, and – +. The first two indicate concordance and the second two indicate a lack of concordance. Ideally, data would show either 100% concordance for the seven hybrids, indicating linkage, or 100% discordance for the seven hybrids, indicating a lack of linkage.

Because radiation hybrids involve chromosome breakage, two genes that are located very close together on the same chromosome may show some discordance despite the close linkage. Two genes that are located on different chromosomes may also show some concordance due to the chance that two separate fragments may become established within a single hybrid line. Therefore, the problem is how to distinguish between reduced concordance due to chromosome fragmentation and chance concordance due to two fragments from different chromosomes being in the same hybrid. Obviously, a statistical solution is needed, but there are not enough data in this problem for a statistical analysis.

Sort the data into three groups: 100% concordance, 100% discordance, and mixed (concordance/discordance). This follows below:

100% concordance	100% discordance	mixed
EF	None	A-B 2/5
		A-C 2/5
		A-D 6/1
		A-E 2/5
		A-F 2/5
		B-C 5/2
		B-D 1/6
		B-E 3/4
		B-F 3/4
		C-D 3/4
		C-E 3/4
		C-F 3/4
		D-E 3/4
		D-F 3/4

Markers E and F are most likely located on the same chromosome. Markers B and D may be located on different chromosomes.

In the absence of statistical analysis, with so few total hybrids, it is important to pay more attention to the + + patterns than the – – patterns simply because – – can arise either from linkage, with the specific chromosome missing in the hybrid, or from lack of linkage, with the two chromosomes (or fragments) lacking in the hybrid. Therefore, going back to the mixed category and focusing on those marker pairs that had a high degree, but not 100%, of concordance, one sees that the 6/1 pattern of A-D and the 5/2 pattern of B-C stand out. For the A-D pair, 3 of the 6 concordances are + +, whereas only 2 of the 5 concordances for B-C are + +. It is unclear from the data whether this is a significant difference, and significance cannot be determined in any fashion. Therefore, it would be important to collect more data before drawing further conclusions.

8. **a.** RAPDs are formed when regions of DNA are bracketed by two inverted copies of a "random" PCR primer sequence. Below, the primer is indicated by X's, and the amplified region appears in brackets. For convenience, the two amplified regions are shown on the same lengthy piece of DNA for strain 1.

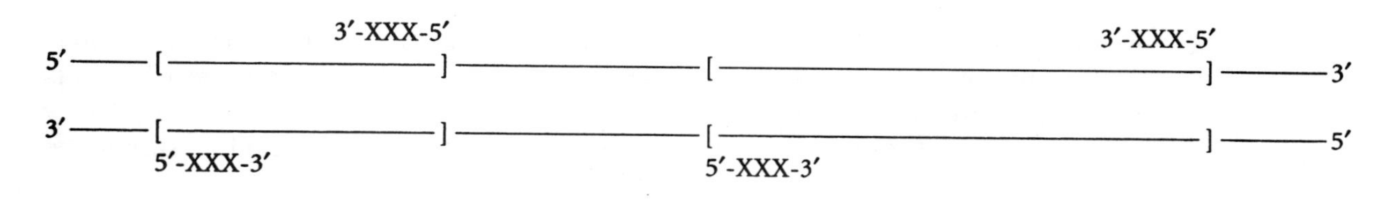

Strain 2 lacks one or two regions complementary to the primer.

b. Progeny 1 and 6 are identical with the strain 1 parent. Progeny 4 and 7 are identical with the strain 2 parent. Progeny 2 and 5 received the chromosome holding the upper band from the strain 1 parent and the chromosome holding the lower band from the strain 2 parent (resulting in no second band). Progeny 3 received the opposite: the chromosome holding the lower band from the strain 1 parent and the chromosome holding the upper band from the strain 2 parent (resulting in no second band). Therefore, bands 1 and 2 appear to be unlinked.

c. Recall that a nonparental ditype has two types only, both of which are recombinant. Therefore, the tetrad would be composed of two progeny like progeny 2 and two progeny like progeny 3.

9. **a.** Of the regions of overlap for cosmids C, D, and E, region 5 is the only region in common. Thus, gene x is localized to region 5.

b. The common region of cosmids E and F, or the location of gene y, is region 8.

c. Both probes are able to hybridize with cosmid E because the cosmid is long enough to contain part of genes x and y.

10. **a.** The following stylized schematic of a reciprocal translocation between chromosome 3 and 21 is arbitrarily chosen to show the salient details. Band 3.1 of the q arm of chromosome 3 is split by the translocations that are correlated to the *N* disease allele. Probe c hybridizes to the region of 3q3.1 that remains with chromosome 3, and probes a, b, and d hybridize to the region of 3q3.1 that is translocated in this case to chromosome 21.

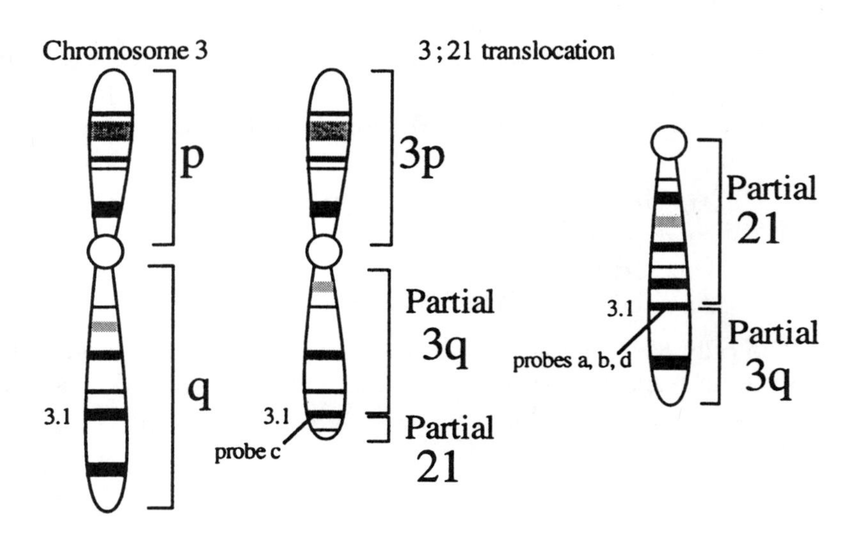

b. Because translocations of chromosome 3 that break band 3q3.1 are correlated to the disease, it is reasonable to assume that these rearrangements split the normal gene (*n*) in two, separating vital coding or regulatory regions. Therefore analysis and cloning of this specific region should be attempted.

In order to isolate and characterize the normal allele, chromosome walking from the known clones should be attempted in genomic libraries from individuals with the translocation and affected with the disease. Probe c is to one side of the breakpoint, while a, b, and d are on the other side. Also, translocation breakpoints serve as useful molecular landmarks because they are easily identified on Southerns as "split bands" when probed with cloned DNA spanning the breakpoint. Once the breakpoint has been identified and cloned, the appropriate subclones would be used to clone the normal allele from a "normal" genomic library. This would be in conjunction with the usual techniques to identify a gene: sequencing, open reading frame analysis, Northern blots, etc.

c. Once *n* is cloned, it can be used to clone the various alleles from individuals who have the disease but not a translocation. The various alleles could then be compared with *n* by sequence, regulation, etc.

11. a. DNA from each individual was obtained. It was restricted, electrophoresed, blotted, and then probed with the five probes. After each probing, an autoradiograph was produced.

b. First identify which chromosome came from the affected parent. This is easily determined by identifying which chromosome could not have come from the mother. For the first daughter, the chromosome with $2'$ was inherited from the father. Likewise, $2''$, $3''$, and $2''$ identify the paternal chromosome in the other children. In all cases, the chromosome drawn to the left is the one inherited from the mother.

Next compare the maternal chromosomes of affected offspring with unaffected offspring to determine which RFLP is most closely correlated to the disease. This analysis is based on the co-segregation of one of the RFLPs and the disease-causing gene. Notice that all these chromosomes show evidence of recombination. For example, when compared with the mother's chromosomes, it can be deduced that the maternally inherited chromosome of the unaffected daughter is the result of a double-crossover event.

Affected:	$1^{o}\ 2^{o}\ 3'\ 4^{o}\ 5^{o}$
Unaffected:	$1^{o}\ 2^{o}\ 3'\ 4'\ 5^{o}$
Unaffected:	$1'\ 2^{o}\ 3'\ 4'\ 5^{o}$
Affected:	$1'\ 2^{o}\ 3'\ 4^{o}\ 5'$

The only RFLP that correlates to the disease and therefore is likely closest to the disease allele is 4°. It is present in both affected children and absent in both unaffected children.

c. It appears that RFLP 4 is the closest marker to the gene and could be used for positional cloning by chromosome walking. However, with only four offspring, the genetic distance between the gene and this marker could be quite large. The number of markers for each human chromosome is already large and increasing almost daily. If possible, it makes sense to further analyze this family (and as many other families with the same trait that can be found) to see if the gene can be further localized before the arduous task of "walking" is attempted.

12. a., b., and c. Cystic fibrosis (CF) is a recessive, autosomally inherited disease. Both parents in this pedigree must be carriers because some of their children are affected. Because the problem states that the three probes used are very closely linked to the *CF* gene, recombination will be ignored.

The data from three probes are presented, but only probes 1 and 3 detect RFLPs in this pedigree and are therefore informative. Both probes detect either one or two bands depending on the allele present. Calling the one-band pattern allele *A* and the two-band pattern allele *B*, the individuals of the pedigree are

Father	$RFLP\text{-}1^B$ $RFLP\text{-}3^A$
Mother	$RFLP\text{-}1^A$ $RFLP\text{-}3^B$
Child 1 (II-1)	$RFLP\text{-}1^B$ $RFLP\text{-}3^A$ (does not have CF)
Child 2 (II-2)	$RFLP\text{-}1^B$ $RFLP\text{-}3^B$ (does have CF)
Child 3 (II-3)	$RFLP\text{-}1^B$ $RFLP\text{-}3^B$ (does have CF)
Child 4 (II-4)	$RFLP\text{-}1^A$ $RFLP\text{-}3^B$ (does not have CF)
Child 5 (II-5)	$RFLP\text{-}1^B$ $RFLP\text{-}3^A$ (does not have CF)
Child 6 (II-6)	$RFLP\text{-}1^B$ $RFLP\text{-}3^B$ (does have CF)
Child 7 (II-7)	$RFLP\text{-}1^A$ $RFLP\text{-}3^A$ (does not have CF)

The first step is to determine which RFLP alleles are linked to the disease-causing *CF* alleles. The pattern of inheritance suggests that $RFLP\text{-}1^B$ from the father and $RFLP\text{-}3^B$ from the mother are both linked to *CF* alleles because all children that are $RFLP\text{-}1^B$ $RFLP\text{-}3^B$ also have CF.

The oldest son (II-1) is a carrier because he has inherited a *CF* allele (linked to $RFLP\text{-}1^B$) from his father. Similarly, II-4 has inherited a *CF* allele from his mother, II-5 has inherited a *CF* allele from his father, and II-7 is homozygous normal.

15 Gene Mutation

1. All cells derived from the cell in which the reversion took place will now be w^+/w. Depending on when during development this took place, the petal will now be blue, either in part or in whole. Because the petal is part of the plant's soma, this reversion would not be inherited.

2. Grow spores in the absence of leucine and in the presence of an antibiotic that will kill only proliferating cells. Use filtration enrichment to obtain spores that were unable to grow because they were *leu*$^-$.

3. Starting with a yeast strain that is *pro-1*, plate the cells on medium lacking proline. Only those cells that are able to synthesize proline will form colonies. Most of these will be revertants; however, some will have second-site suppressors. (Treating the cells with a mutagen prior to plating them would significantly increase the yield.)

4. Streak the yeast on minimal medium plus arginine. When colonies appear, replica-plate them onto minimal medium. The absence of growth in minimal medium will identify the arginine requiring mutants.

5. Assume that you are working with 20 specific nutrients that were added to the minimal medium. Group the nutrients, with five to a group. (The choice of how many nutrients are included in each group is completely arbitrary.) Test each auxotroph against each group. When an auxotroph grows in one of the groups, test the auxotroph separately against each member of the group. A flow sheet would look something like the following:

Test 1	Growth		Test 2	Growth
group A (1 to 5)	–		11	–
group B (6 to 10)	–		12	–
group C (11 to 15)	+	→	13	+
group D (16 to 20)	–		14	–
			15	–

These results tell you that nutrient 13 was required for growth.

If a mutant cannot be identified to have a requirement for a specific nutrient, then it may be a double or multiple mutant. Possibly, it could require an unidentified component of the complete medium. Alternatively, many mutations that can prevent growth are not based on nutritional needs.

6. You need to apply the Poisson distribution (Chapter 6) to answer this problem. Mutants were not observed (zero class) on 37 plates out of 100. The formula is $e^{-\mu n}$ = number in zero class/total number. Or $e^{-\mu} \times 10^6 = 37/100$, and $\mu = -\ln(0.37) \times 10^{-6} = 1/10^{-6}$ per cell division.

7. There are many ways to test a chemical for mutagenicity. For example, the text discusses a detection system for recessive somatic mutations in mice (see Figure 15-14). Mice bred to be heterozygous for seven genes involved in coat color are exposed to a potential mutagen by injecting it into the uterus of their pregnant mother. Any somatic mutation from wild-type to mutant at one of the seven loci will result in a patch (or mutant sector) of differently colored fur. The number of mutant sectors later found on these chemically treated mice would be compared with the number found on genetically identical, but chemically untreated, mice (the control mice). (A proper control would expose the control mice to exactly the same experimental protocol except for the caffeine. This would include injection of whatever solvent was used into the uterus of their mother at the same developmental time as for the experimental mice.)

Much larger screens or selections could be done with fungi. For example, haploid *Neurospora* auxotrophic for the amino acid leucine could be exposed to caffeine and then plated onto minimal medium selecting for *leu*$^+$ colonies. Only those cells in which a reverse mutation (from *leu*$^-$ to *leu*$^+$) occurred would grow. Reversion rates of treated and untreated (control) cells would be compared to see if the caffeine was mutagenic. Alternatively, yeast cells could be exposed to caffeine and mutations in the gene *ade-3* could be scored and numbers compared between treated and untreated populations. (Mutations in this gene actually cause the yeast to be red instead of white, so large numbers of colonies can be screened rapidly.)

Although this question asks specifically for mutations in higher organisms, a rapid and widely used mutagen-detection system using bacteria was developed by Bruce Ames in the 1970s. Using a genetically modified bacterium (*Salmonella typhimurium*) that is auxotrophic for histidine and defective in DNA repair, the Ames test quickly ascertains the mutagenicity of various chemicals. Because the basic properties of DNA and mutation are the same in prokaryotes and eukaryotes, this test does have relevance. The Ames test has been further enhanced by using rat liver extracts to modify the tested chemicals to simulate human (and mammalian) metabolism. This is important because although the liver is responsible for most of the detoxification and metabolism of ingested chemicals (hence the connection between alcohol and liver disease!), some chemicals are modified in ways that actually make them toxic or mutagenic.

8. Stain pollen grains, which are haploid, from a homozygous *Wx* parent with iodine. Look for red pollen grains, indicating mutations to *wx*, under a microscope.

9. The most straightforward explanation is that a mutation from wild type to black occurred in the germ line of the male wild-type mouse. Thus, he was a gonadal mosaic of wild-type and black germ cells.

10. An X-linked disorder cannot be passed from father to son. Because the gene for hemophilia must have come from the mother, the nuclear power plant cannot be held responsible.

It is possible that the achondroplastic gene mutation was caused by exposure to radiation.

11. The mutation rate needs to be corrected for achondroplastic parents and put on a "per gamete" basis. Mutation rate = $(10 - 2)/[2 \times (94{,}075 - 2)] = 4.25 \times 10^{-5}$.

For this problem, you do not have to worry about revertants because you are asked only for the net mutation frequency.

12. The commission was looking for induced recessive X-linked lethal mutations, which would show up as a shift in the sex ratio. A shift in the sex ratio is the first indication that a population has sustained lethal genetic damage. Other recessive mutations might have occurred, of course, but they would not be homozygous and therefore would go undetected. All dominant mutations would be immediately visible, unless they were lethal. If they were lethal, there would be lowered fertility, an increase in detected abortions, or both, but the sex ratio would not shift as dramatically.

13. **a.** reddish all over

b. reddish all over

c. many small, red spots

d. a few large, red spots

e. like part c, but with fewer reddish patches

f. like part d, but with fewer reddish patches

g. some large spots and many small spots

14. ***Unpacking the Problem***

1.

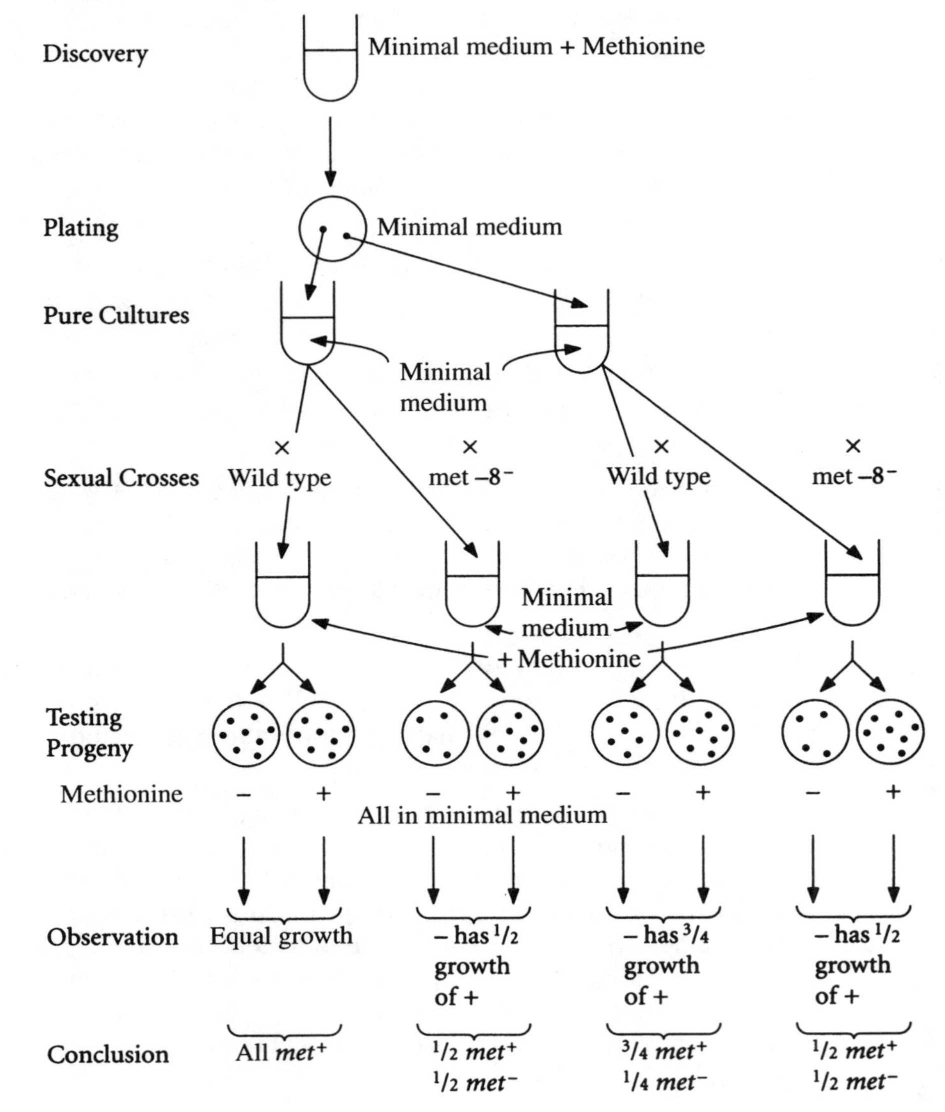

2. **Haploid** refers to possessing a single genome.

   Auxotrophic means that an organism requires dietary provision of some substance that normally is not required by members of its species.

   Methionine is an amino acid.

   Asexual spores are a mode of propagation used by some species in which the spores are derived from an organism without a genetic contribution from another organism. Of necessity, the spores have the same number of chromosomes as the organism from which they are derived.

   Prototrophic means that an organism does not have any special dietary requirements beyond those normal for the species.

   A **colony** is a collection of cells or organisms all derived mitotically, from a single cell or organism, and all possessing the same genotype.

   A **mutation** is the process that generates alternative forms of genes, and it results in an inherited difference between parent and progeny.

3. The "8" in *met-8* refers to the eighth locus found that leads to a methionine requirement. It is unnecessary to know the specifics of the mutation in order to work the problem.

4. The following crosses were made in this problem:

   Cross 1: prototroph 1 × wild type
   Cross 2: prototroph 1 × backcross to *met-8*
   Cross 3: prototroph 2 × wild type
   Cross 4: prototroph 2 × backcross to *met-8*

5. Use *met-8** to indicate the prototroph derived from the *met-8* strain.

   Cross 1: *met-8** 1 × *met-8*$^+$
   Cross 2: *met-8** 1 × *met-8*
   Cross 3: *met-8** 2 × *met-8*$^+$
   Cross 4: *met-8** 2 × *met-8*

6. In this organism, asexual spores give rise to an organism that is capable of forming sexual spores following a mating. Therefore, the original mutation occurred in somatic tissue that subsequently gave rise to germinal tissue.

7. Because the trait being selected is the ability to grow in the absence of methionine, a reversion is being studied.

8. Only two revertants were observed because reversion occurs at a much lower frequency than forward mutation.

9. The millions of asexual spores did not grow because they required methionine and the medium used did not contain methionine.

10. A low percentage of the millions of spores that did not grow would be expected to have other mutations that rendered them incapable of growth. In addition, a low percentage would be expected to have chromosome abnormalities that would lead to death.

11. The wild type used in this experiment was prototrophic, by definition; that is, *wild type* refers to the norm for a species, which means "prototrophic."

12. It is highly unlikely that visual inspection could distinguish between wild type and prototrophic revertants.

13. One way to select for a *met-8* mutation is to grow a large number of spores on a medium that lacks methionine. Filtration will separate those spores capable of growth from those incapable of growth. Once spores have been isolated that are incapable of growth in a medium lacking methionine, they can be tested for a methionine mutation by plating them on medium containing methionine. If they are capable of growth on this second medium, they are methionine auxotrophs.

14. The starting auxotrophic spores were haploid. Both mitotic crossing-over and haploidization require diploids. Therefore, it is unlikely that either process is involved with producing the observed results.

15. Cross 1: *met-8* × wild type → 1 *met-8*:1 wild type

    Cross 2: *met-8* × *met-8* → all *met-8*

    Cross 3: wild type × wild type → all wild type

16. While the analysis could have been conducted using tetrad analysis, it is more likely that random selection of progeny was used.

17.

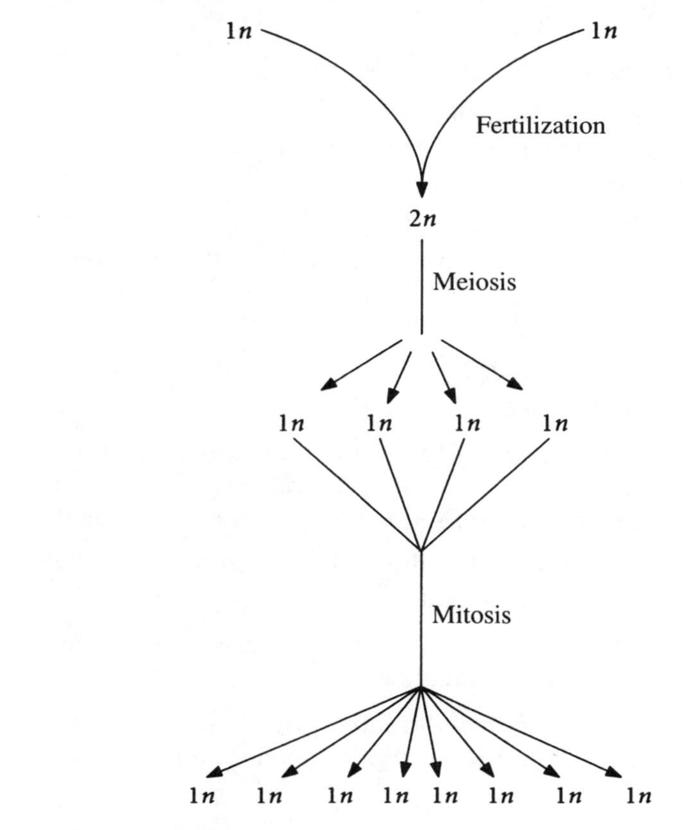

18. If a 3:1 ratio is obtained in haploids, then two genes must be segregating.

Solution to the Problem

a. and b. The pattern of growth for prototroph 1 suggests that it is a reversion of the original mutation. When crossed with wild type, a reversion would be expected to produce all *met*$^+$ progeny, and when backcrossed, it would be expected to produce a 1:1 ratio. Let the reversion be symbolized by *met-8**. The crosses are

P	*met-8** × *met-8*$^+$	P	*met-8** × *met-8*
	↓		↓
F_1	$^1/_2$ *met-8** prototrophic	F_1	$^1/_2$ *met-8** prototrophic
	$^1/_2$ *met-8*$^+$ prototrophic		$^1/_2$ *met-8* auxotrophic

The pattern of growth for prototroph 2 suggests that a suppressor at another, unlinked locus is responsible for its prototrophic growth. Let *met-s* symbolize this suppressor. Then the crosses are

P	*met-8* ; *met-s* × *met-8*$^+$; *met-s*$^+$		P	*met-8*; *met-s* × *met-8*; *met-s*$^+$	
	↓			↓	
F_1	$^1/_4$ *met-8*; *met-s*	prototrophic	F_1	$^1/_4$ *met-8*; *met-s*	prototrophic
	$^1/_4$ *met-8*$^+$; *met-s*$^+$	prototrophic		$^1/_4$ *met-8*; *met-s*$^+$	auxotrophic
	$^1/_4$ *met-8*$^+$; *met-s*	prototrophic		$^1/_4$ *met-8*; *met-s*	prototrophic
	$^1/_4$ *met-8*; *met-s*$^+$	auxotrophic		$^1/_4$ *met-8*; *met-s*$^+$	auxotrophic

15. The most likely explanations for the five observed colony types are

a. White colonies: the cells are parental in phenotype; that is, the cells of most colonies are *ad*$^+$/*ad*.

b. (1) Diploid red, round cells: the *ad*$^+$ gene was inactivated by mutation, deleted, or mutated to the *ad* allele.

(2) Diploid red, elongated cells: the chromosome carrying round and white was broken and these two genes were lost (a deletion). Mitotic crossing-over is another alternative.

c. (1) No growth originally but grew in a complete medium: mutation of *met*$^+$, *nic*$^+$, or both.

(2) No growth originally, nor in a complete medium: loss of some function that was essential for viability. Alternatively, a dominant lethal mutation occurred.

16. The mutants can be categorized as follows:

Mutant 1: an auxotrophic mutant
Mutant 2: a non-nutritional, temperature-sensitive mutant
Mutant 3: a leaky, auxotrophic mutant
Mutant 4: a leaky, non-nutritional, temperature-sensitive mutant
Mutant 5: a non-nutritional, temperature-sensitive, auxotrophic mutant

17. The basic cross is *trp-x* · *trp-y*$^+$ × *trp-x*$^+$ · *trp-y*. Trp$^+$ progeny represent half the recombinants of such a cross.

a. A is 100%(135 × 2)/1000, or 27 map units from B.
A is 100%(179 × 2)/1000, or 35.8 map units from C.
B is 100%(50 × 2)/1000, or 10 map units from C.

The map is

b. There are three classes of Trp^- spores from cross 2: $A^- C^+$, $A^+ C^-$, and $A^- C^-$.

Cross 4-: 2-1 × B: Because the number of Trp^+ recombinants indicate that 2-1 is the same distance as A from B, the 2-1 spore is type $A^- C^+$ and the cross is $A^- B^+ C^+ \times A^+ B^- C^+$.

Cross 5: 2-2 × B: Because the number of Trp^+ recombinants is low, most likely a double crossover was required indicating that the 2-2 spore is $A^- C^-$ and the cross is $A^- B^+ C^- \times A^+ B^- C^+$. Ignoring interference, the double crossover should occur (27%)(10%) = 2.7% of the time, which fits nicely with the 13 Trp^+ recombinants scored.

Cross 6: 2-3 × B: The number of Trp^+ recombinants indicate that 2-3 is the same distance as C from B, suggesting that the 2-3 spore is type $A^+ C^-$ and the cross is $A^+ B^+ C^- \times A^+ B^- C^+$.

Cross 7: 2-4 × B: The same as 2-1.

Cross 8: 2-5 × B: The same as 2-3.

18. a. Intercrossing mutant strains that all share a common recessive phenotype is the basis of the complementation test. This test is designed to identify the number of different genes that can mutate to a particular phenotype. In this problem, if the progeny of a given cross still express the twitcher phenotype, the mutations fail to complement and are considered alleles of the same gene; if the progeny are wild type, the mutations complement and the two strains carry mutant alleles of separate genes.

b. From the data,

1 and 5	fail to complement	gene A
2, 6, 8, and 10	fail to complement	gene B
3 and 4	fail to complement	gene C
7, 11, and 12	fail to complement	gene D
9	complements all others	gene E

There are five complementation groups (genes) identified by these data.

c. mutant 1: $a^1/a^1 \cdot b^+/b^+ \cdot c^+/c^+ \cdot d^+/d^+ \cdot e^+/e^+$ (although only the mutant alleles are usually listed)

mutant 2: $a^+/a^+ \cdot b^2/b^2 \cdot c^+/c^+ \cdot d^+/d^+ \cdot e^+/e^+$

mutant 5: $a^5/a^5 \cdot b^+/b^+ \cdot c^+/c^+ \cdot d^+/d^+ \cdot e^+/e^+$

1/5 hybrid: $a^1/a^5 \cdot b^+/b^+ \cdot c^+/c^+ \cdot d^+/d^+ \cdot e^+/e^+$ phenotype: twitcher 1 and 5 are both mutant for gene A

(the relevant cross: $a^+/a^+ \cdot b^2/b^2 \times a^5/a^5 \cdot b^+/b^+$ gives)

2/5 hybrid: $a^+/a^5 \cdot b^+/b^2 \cdot c^+/c^+ \cdot d^+/d^+ \cdot e^+/e^+$ phenotype: wild type 2 and 5 are mutant for different genes

19. a. If the nonsense mutation is homozygous, then complete translation will occur only with T^S. The traditional definition of a recessive allele is that it is seen in the phenotype when it is the only type of allele present. By this definition, T^+ is recessive to the dominant T^S.

b. In a trihybrid cross, the white phenotype will occur only when one or both of the enzymes is homozygous for a nonsense mutation and the wild-type form of T is also homozygous. The frequencies and phenotypes seen are

27	$A/– \; ; \; B/– \; ; \; T^S/–$	purple
9	$a^n/a^n \; ; \; B/– \; ; \; T^S/–$	purple
9	$A/– \; ; \; b^n/b^n \; ; \; T^S/–$	purple
9	$A/– \; ; \; B/– \; ; \; T^+/T^+$	purple
3	$a^n/a^n \; ; \; b^n/b^n \; ; \; T^S/–$	purple
3	$a^n/a^n \; ; \; B/– \; ; \; T^+/T^+$	white
3	$A/– \; ; \; b^n/b^n \; ; \; T^+/T^+$	white
1	$a^n/a^n \; ; \; b^n/b^n \; ; \; T^+/T^+$	white

or 57 purple:7 white

20. Let the new mutation be designated m. Cross the homozygote with different lines of tomatoes, each of which has mutations on different chromosomes. That is, if you let the known mutations be designated as a, you will be crossing $a^+/a^+ \cdot m/m$ with $a/a \cdot m^+/m^+$. All progeny are $a^+/a \cdot m^+/m$. Next, let these F_1 plants self and score the segregation of a and m. If independent assortment is observed, the two genes are not linked. If there is not independent assortment, the genes are linked. As an example of a specific cross, assume that the new mutation is on chromosome 1. You could cross a plant that is mottled, dwarf, peach, oblate, normal (nonwooly), necrotic, compound, beaked, and multiloculed to the new mutation in a background that is otherwise wild type. The new mutation will show independent assortment with all genes except those within 50 map units of it.

As for the very practical questions asked in these problems, such as how much field space is needed, they cannot be answered from the text. You can imagine, however, that a rather large field space would be required, and at least two growing seasons would be needed to be confident that the new mutation had been mapped correctly.

21. The cross is $arg^r \times arg^+$, where arg^r is the revertant. However, it might also be $arg^- \cdot su \times arg^+ \cdot su^+$, where su^+ has no effect on the arg gene and su is a suppressor of arg^-.

a. If the revertant is a precise reversal of the original change that produced the mutant allele, 100% of the progeny would be arginine independent.

b. If a suppressor mutation on a different chromosome is involved, then the cross is $arg^- \; ; \; su \times arg^+ \; ; \; su^+$. Independent assortment would lead to the following:

1 $arg^- \; ; \; su$	arginine independent
1 $arg^+ \; ; \; su^+$	arginine independent
1 $arg^- \; ; \; su^+$	arginine dependent
1 $arg^+ \; ; \; su$	arginine independent

c. If a suppressor mutation 10 map units from the arg locus occurred, then the cross is $arg^- \; su \times arg^+ \; su^+$ and the diploid intermediate is $arg^- \; su \, /arg^+$

su^+. The two parental types would occur 90% of the time, and the two recombinant types would occur 10% of the time. The progeny would be

45%	arg^- su	arginine independent
45%	arg^+ su^+	arginine independent
5%	arg^- su^+	arginine dependent
5%	arg^+ su	arginine independent

22. a. The results show that the new mutants have the following epistatic relationships: $w^- > p^- > y^- > b^- > o^-$ representing a possible biochemical pathway:

$$\text{white} \xrightarrow{w} \text{pink} \xrightarrow{p} \text{yellow} \xrightarrow{y} \text{beige} \xrightarrow{b} \text{orange} \xrightarrow{o}$$

b. All heterokaryon pairs would have a wild-type phenotype because of complementation.

c. The cross is $o^- p^- \times o^+ p^+$. Remember that haploid fungi immediately enter meiosis after gamete fusion, producing haploid progeny. Because the two genes are 16 map units apart, the recombinants will total 16% of the progeny and the parentals will be 84%. The genotypes and phenotypes of the progeny expected would be

42% $o^- p^-$	pink
42% $o^+ p^+$	wild type
8% $o^- p^+$	orange
8% $o^+ p^-$	pink

23. The cross is $trp^r \times trp^+$, where trp^r is the revertant. However, it might also be $trp^- \cdot su \times trp^+ \cdot su^+$, where su^+ has no effect on the *trp* gene and *su* is a suppressor of trp^-.

a. If the revertant is a precise reversal of the original change that produced the mutant allele, 100% of the progeny would be tryptophan independent.

b. If a suppressor mutation on a different chromosome is involved, then the cross is trp^- ; $su \times trp^+$; su^+. Independent assortment would lead to the following:

1 trp^- ; su	tryptophan independent
1 trp^+ ; su^+	tryptophan independent
1 trp^- ; su^+	tryptophan dependent
1 trp^+ ; su	tryptophan independent

c. If a suppressor mutation 24 map units from the *trp* locus occurred, then the cross is trp^- su $\times$ trp^+ su^+ and the diploid intermediate is trp^- su /trp^+ su^+. The two parental types would occur 76% of the time, and the two recombinant types would occur 24% of the time. The progeny would be

38%	trp^- su	tryptophan independent
38%	trp^+ su^+	tryptophan independent
12%	trp^- su^+	tryptophan dependent
12%	trp^+ su	tryptophan independent

d. A common reason that a mutant is nonreverting is because it is the result of a deletion. However, because only "one" revertant was isolated from mutant B, it is possible that not enough cells were looked at to find an A revertant.

16 Mechanisms of Gene Mutation

1. In higher organisms, the sequence CpG is often methylated to give 5-methyl-CpG. Spontaneous deamination of the 5-methylcytosine generates thymine and thus causes C $\rightarrow$ T transitions. Tandemly repeated sequences can lead to mistakes during replication. In the DNA sequence below the sequence CATC (underlined) is repeated three times. Through "slipped mispairing," a four-base repeat can be added or deleted during replication, leading to possible frameshift mutations. Finally, repeated but separated sequences (marked by boxes) can lead to spontaneous deletions or duplications either through slipped mispairing or by misalignment during recombination.

```
AAGGCTAGCTTTAGGAGATCCCG
ATCTCAAAGCTATCTAGCTTTAGG
TATATAGATCTATGCTCTCTGATC
TAGCATCCCTAGCATCATATCGGG
ATCCTACGAATCTTTGATCGGTAT
CGGGATACGTATGAAGGCTAGCC
TCATCCATCCATCCAAGCTTAATA
TCGATCGGATCCTGGAATTGGATT
CCCAGAGATCTTTTTAGCTAGCTC
CCGCCTAGCTTTCGGAGCTTAATC
CTAATGAGCAACCACCGGTATATA
GCCAATACAAGCCGGATTCGGGA
TCCTTAGGA[blacked out]TCGA
GAAATCGGATCGGGATCTTATCGT
C[blacked out]CCTAGTTCCAAT
CTTTATCGGATCGGAAGGCTAT
```

2. The Streisinger model proposed that frameshifts arise when loops in single-stranded regions are stabilized by slipped mispairing of repeated sequences. In the *lac* gene of *E. coli*, a four-base-pair sequence is repeated three times in tandem, and this is the site of a hot spot.

The sequence is 5′-CTGG CTGG CTGG-3′. During replication the DNA must become single stranded in short stretches for replication to occur. As the new strand is synthesized, transient disruptions of the hydrogen bonds holding the new and old strands together may be stabilized by the incorrect base-pairing of bases that are now out of register by the length of the repeat, or in this case, a total of four bases. Depending on which strand, new or template, loops out with respect to the other, there will be an addition or deletion of four bases, as diagrammed below:

```
                  T G
                 C   G
5′-C T G G            C T G G-3′    →  DNA synthesis
3′-G A C C ————————   G A C C  G A C C -5′
```

In this diagram, the upper strand looped out as replication was occurring. The loop is stabilized by base pairing one repeat out of register. As replication continues at the 3′ end, an additional copy of CTGG will be synthesized, leading to an addition of four bases. This will result in a frameshift mutation.

3. In Problem 2, had the lower strand (the template) looped, the result would have been a deletion in the newly synthesized upper strand.

Misalignment of homologous chromosomes during recombination results in a duplication in one strand and a corresponding deletion in the other.

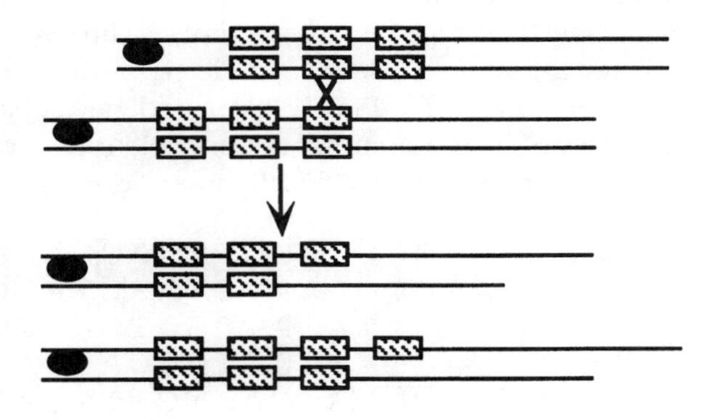

Recombination between two homologous repeats in a looped DNA molecule can lead to deletion.

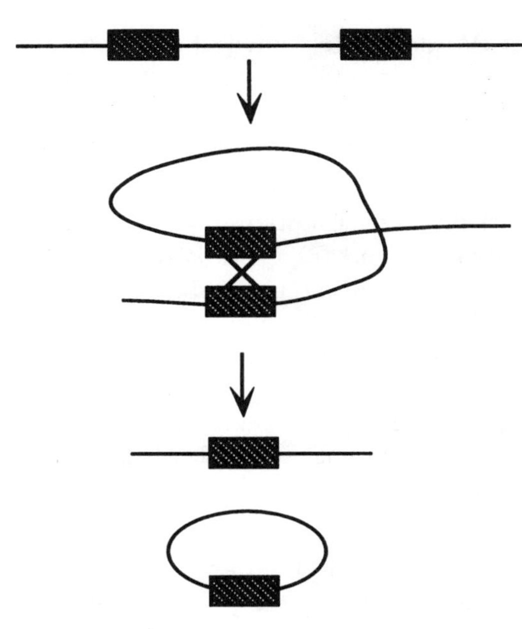

All these mechanisms are supported by DNA sequencing results.

4. **a.** Sequence (4) would have a hotspot for frameshift mutations. As marked below, the sequence CCCBA is tandemly repeated four times. This could lead to the deletion or addition of this five-base repeat during replication.

ACCAEACBBBCEEABAECBA**CCCBACCCBACCCBACCCBA**CCCEBCAACBAEABCEBAEC

b. Sequence (6) would have a hotspot for deletions. It contains the repeated (and separated) sequence marked below.

BCEAAEECABBEBCECCAECAABCBBCEECBACAECCEC**BCEAAEECABBEBC**CAE

5. **a.** Assuming replication is otherwise similar to Earth-based systems, short tandemly repeated sequences can lead to additions or deletions of the repeat. An example of such a sequence might be:

MPPJQLJKLMPQ**QPJMQPJMQPJMQPJM**JMLL

b. Based on the principle above, tandemly repeated sequences should be avoided. Repeated sequences that are apart, rather than in tandem, can lead to deletion of the intervening sequence, so these should also be avoided.

6. Depurination results in the loss of adenine or guanine from the DNA. Because the resulting apurinic site cannot specify a complementary base, replication is blocked. Under certain conditions, replication proceeds with a near random insertion of a base opposite the apurinic site. In three-fourths of these insertions, a mutation will result.

Deamination of cytosine yields uracil. If left unrepaired, the uracil will be paired with adenine during replication, ultimately resulting in a transition mutation.

8-OxodG (8-oxo-7-hydrodeoxyguanosine) can pair with adenine, resulting in a transversion.

7. 5-BU is an analog of thymine. It undergoes tautomeric shifts at a higher frequency than does thymine and therefore is more likely to pair with G than thymine is during replication. At the next replication this will lead to a GC pair rather than the original AT pair. On the other hand, 5-BU can also be incorporated into DNA by mispairing with guanine. In this case it will convert a GC pair to an AT pair.

EMS is an alkylating agent that produces *O*-6-ethylguanine. This alkylated guanine will mispair with thymine, which leads from a GC pair to an AT pair at the next replication.

8. O^{-6}-methyl G leads predominantly to high levels of GC $\rightarrow$ AT transitions. 8-oxodG gives rise predominantly to high levels of G $\rightarrow$ T transversions. Finally, C-C photodimers will most often cause C $\rightarrow$ T transitions, but some transversions are also possible.

9. An AP site is an apurinic or apyrimidinic site. AP endonucleases introduce chain breaks by cleaving the phosphodiester bonds at the AP sites. Some exonuclease activity follows, so that a number of bases are removed. The resulting gap is filled by DNA pol I and then sealed by DNA ligase.

General excision repair is used to remove damaged DNA, including photodimers. It cleaves the phosphodiester backbone on either side of the damage, removing 12 or 13 nucleotides in bacteria and 27 to 29 nucleotides in eukaryotes. The resulting gap is filled by DNA pol I and then sealed by DNA ligase.

Photodimers can be repaired by photolyase (found in bacteria and lower eukaryotes), which binds to and splits the dimer in the presence of certain wavelengths of light. In bacteria the damaged DNA can also be bypassed by the SOS system, which enables DNA polymerase to "fill in" random bases where it encounters photodimers on the template strand. Finally, although photodimers on the template strand cause DNA pol III to stall, replication can restart downstream from the dimer, leaving a region of single-stranded DNA. The single-stranded DNA will attract single-stranded-binding protein and another protein, RecA, which can affect recombinational repair using DNA from the sister chromatid to patch the gap.

10. Mismatch repair occurs if a mismatched nucleotide is inserted during replication. The new, incorrect base is removed and the proper base is inserted. In *E. coli*, the enzymes involved can distinguish between new and old strands because the old strand is methylated.

Recombination repair occurs if lesions such as AP sites and UV photodimers block replication (there is a gap in the new complementary strand). Recombination fills this gap with the corresponding segment from the sister DNA molecule, which is normal in both strands. This produces one DNA molecule with a gap across from a correct strand, which can then be filled by complementation, and one with a photodimer across from a correct strand.

11. Leaky mutants are mutants with an altered protein product that retains a low level of function. Enzyme activity may, for instance, be reduced rather than abolished by a mutation.

12. The new "wild-type" isolate contains an allele for a gene that increases the spontaneous reversion rate of *ad-3*. This gene appears to be unlinked to *ad-3* because independent assortment is observed. Call this gene rev. The cross becomes

ad-3 ; *rev*$^+$ (Strain A) ∞ *ad-3*$^+$; *rev* (wild-type isolate)

Of the *ad-3* progeny, half would be *rev*$^+$ (low reversion rate) and half would be *rev* (high reversion rate).

Further crosses to map *rev* could be performed as well as testing the allele and gene specificity of the high reversion rate to see if it is the result of increased spontaneous mutation rates (gene non-specific) or some other process (*ad-3* specific).

13. **a.** Because 5´-UAA-3´ does not contain G or C, a transition to a GC pair in the DNA cannot result in 5´-UAA-3´. 5´-UGA-3´ and 5´-UAG-3´ have the DNA antisense-strand sequence of 3´-ACT-5´ and 3´-ATC-5´, respectively. A transition to either of these stop codons occurs from the nonmutant 3´-ATT-5´. However, a DNA sequence of 3´-ATT-5´ results in an RNA sequence of 5´-UAA-3´, itself a stop codon.

b. Yes. An example is 5´-UGG-3´, which codes for Trp, to 5´-UAG-3´.

c. No. In the three stop codons the only base that can be acted upon is G (in UAG, for instance). Replacing the G with an A would result in 5´-UAA-3´, a stop codon.

14. **a. and b.** Mutant 1: most likely a deletion. It could be caused by radiation.

Mutant 2: because proflavin causes either additions or deletions of bases and because spontaneous mutation can result in additions or deletions, the most probable cause was a frameshift mutation by an intercalating agent.

Mutant 3: 5-BU causes transitions, which means that the original mutation was most likely a transition. Because HA causes GC-to-AT transitions and HA cannot revert it, the original must have been a GC-to-AT transition. It could have been caused by base analogs.

Mutant 4: the chemical agents cause transitions or frameshift mutations. Because there is spontaneous reversion only, the original mutation must have been a transversion. X-irradiation or oxidizing agents could have caused the original mutation.

Mutant 5: HA causes transitions from GC to AT, as does 5-BU. The original mutation was most likely an AT-to-GC transition, which could be caused by base analogs.

c. The suggestion is a second-site reversion linked to the original mutant by 20 map units and therefore most likely in a second gene. Note that auxotrophs equal half the total number of recombinants.

15. a. A lack of revertants suggests either a deletion or an inversion within the gene.

b. To understand these data, recall that half the progeny should come from the wild-type parent.

Prototroph A: because 100% of the progeny are prototrophic, a reversion at the original mutant site may have occurred.

Prototroph B: half the progeny are parental prototrophs, and the remaining prototrophs, 28%, are the result of the new mutation. Notice that 28% is approximately equal to the 22% auxotrophs. The suggestion is that an unlinked suppressor mutation occurred, yielding independent assortment with the *nic* mutant.

Prototroph C: there are 496 "revertant" prototrophs (the other 500 are parental prototrophs) and 4 auxotrophs. This suggests that a suppressor mutation occurred in a site very close [$100\%(4 \times 2)/1000 = 0.8$ m.u.] to the original mutation.

16. a. To select for a nerve mutation that blocks flying, place *Drosophila* at the bottom of a cage and place a poisoned food source at the top of the cage.

b. Make antibodies against flagellar protein and expose mutagenized cultures to the antibodies.

c. Do filtration through membranes with variously sized pores.

d. Screen visually.

e. Go to a large shopping mall and set up a rotating polarized disk. Ask the passersby to look through the disk for a free evaluation of their vision and their need for sunglasses. People with normal vision will see light with a constant intensity through the disk. Those with polarized vision will see alternating dark and light.

f. Set up a Y tube (a tube with a fork giving the choice of two pathways) and observe whether the flies or unicellular algae move to the light or the dark pathway.

g. Set up replica cultures and expose one of the two plates to low doses of UV.

17. a. uracil DNA glycosylase

b. MutM glycosylase

c. general excision repair

d. methyl-directed mismatch repair

e. photolyase or general excision repair

f. endonuclease

g. alkyl transferase

17 Chromosome Mutation I: Chromosome Structure

1. **a.** Cytologically, deletions lead to shorter chromosomes with missing bands (if banded) and an unpaired loop during meiotic pairing when heterozygous. Genetically, deletions are usually lethal when homozygous, do not revert, and when heterozygous, lower recombinational frequencies and can result in "pseudodominance" (the expression of recessive alleles on one homolog that are deleted on the other). Occasionally, heterozygous deletions express an abnormal (mutant) phenotype.

   b. Cytologically, duplications lead to longer chromosomes and, depending on the type, unique pairing structures during meiosis when heterozygous. These may be simple unpaired loops or more complicated twisted loop structures. Genetically, duplications can lead to asymmetric pairing and unequal crossing-over events during meiosis, and duplications of some regions can produce specific mutant phenotypes.

   c. Cytologically, inversions can be detected by banding, and when heterozygous, they show the typical twisted "inversion" loop during homologous pairing. Pericentric inversions can result in a change in the p:q ratio. Genetically, no viable crossover products are seen from recombination within the inversion when heterozygous, and as a result, flanking genes show a decrease in RF.

   d. Cytologically, reciprocal translocations may be detected by banding, or they may drastically change the size of the involved chromosomes as well as the positions of their centromeres. Genetically, they establish new linkage relationships. When heterozygous, they show the typical cross structure during meiotic pairing and cause a diagnostic 50% reduction of viable gamete production, leading to semisterility.

2. **a.** The products of crossing-over within the inversion will be inviable when the inversion is heterozygous. This paracentric inversion spans 25% of the region between the two loci and therefore will reduce the observed recombination between these genes by a similar percentage (i.e., 9%). The observed RF will be 27%.

b. When the inversion is homozygous, the products of crossing-over within the inversion will be viable, so the observed RF will be 36%.

3. The cross is

P $\quad A/- \cdot B/- \cdot C/- \cdot D/- \cdot E/- \cdot F/- \times a\ b\ c\ d\ e\ f/a\ b\ c\ d\ e\ f$

$F_1 \quad ^1/_2\ A\ B\ C\ D\ E\ F/a\ b\ c\ d\ e\ f$

$\quad ^1/_2\ A\ B\ C\ ?\ ?\ F/a\ b\ c\ d\ e\ f$

Because all progeny flies are A B C F, the wild type must have been homozygous for these genes. However, to explain the appearance of d e progeny, it is possible that the wild-type fly was heterozygous for a deletion spanning genes *D* and *E.* With this explanation, the second class of F_1 progeny would be hemizygous for *d* and *e*, or *A B C* – – *F/a b c d e f.*

An alternative explanation is that the wild-type fly was heterozygous *D E/d e* and that a heterozygous inversion spanned these two genes, preventing recombination.

4.

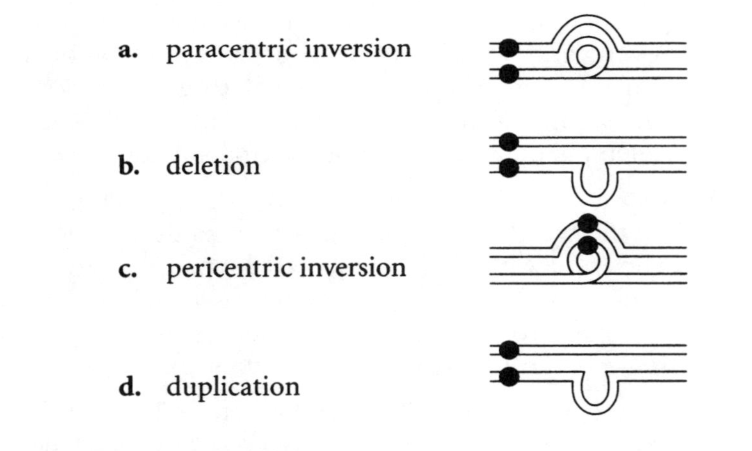

5. From each of the statements concerning the rare cells, you should be able to draw the following conclusions:

Statement	Conclusion
require leucine	leu^+ lost
do not mate	one mating type lost
will not grow at 37°C	un^+ lost
cross only with *a* type	*a* lost
only nucleus 1 alleles recovered	loss of nucleus 2 alleles

Because his^+, $ad\text{-}3^+$, nic^+, and met^+ function are required by the heterokaryon, these alleles must have been retained. Therefore, the most reasonable explanation is that a deletion occurred in the left arm of the large chromosome in nucleus 2 and that leu^+, mating type *a*, and un^+ were lost.

6. The following represents the crosses that are described in this problem:

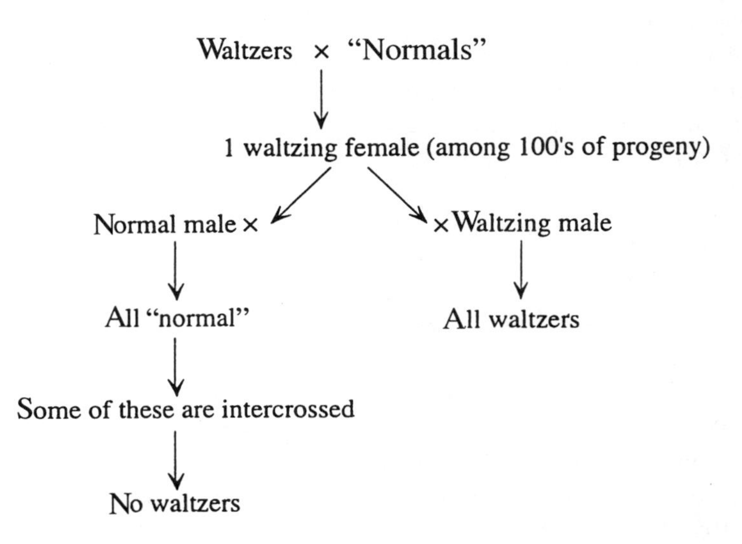

The single waltzing female that arose from a cross between waltzers and normals is expressing a recessive gene. It is possible that this represents a new "waltzer" mutation that was inherited from one of the "normal" mice, but given the cytological evidence (the presence of a shortened chromosome), it is more likely that this exceptional female inherited a deletion of the wild-type allele, which allowed expression of the mutant recessive phenotype.

When this exceptional female was mated to a waltzing male, all the progeny were waltzers; when mated to a normal male, all the progeny were normal. When some of these normal offspring were intercrossed, there were no progeny that were waltzers. If a "new" recessive waltzer allele had been inherited, all these "normal" progeny would have been w^+/w. Any intercross should have therefore produced 25% waltzers. On the other hand, if a deletion had occurred, half the progeny would be w^+/w and half would be $w^+/w^{deletion}$. If $w^+/w^{deletion}$ are intercrossed, 25% of the progeny would not develop (the homozygous deletion would likely be lethal), and no waltzers would be observed. This is consistent with the data.

7. The colonies that would not revert most likely had a deletion within the *ad-3B* gene. These colonies could grow only with adenine supplementation.

8. This problem uses a known set of overlapping deletions to order a set of mutants. This is called **deletion mapping** and is based on the expression of the recessive mutant phenotype when heterozygous with a deletion of the corresponding allele on the other homolog. For example, mutants *a*, *b*, and *c* are all expressed when heterozygous with Del1. Thus it can be assumed that these genes are deleted in Del1. When these results are compared with the crosses with Del2 and it is discovered that these progeny are b^+, the location of gene *b* is mapped to the region deleted in Del1 that is not deleted in Del2. This logic can be applied in the following way:

Compare deletions 1 and 2: this places allele *b* more to the left than alleles *a* and *c*. The order is *b* (*a*, *c*), where the parentheses indicate that the order is unknown.

Compare deletions 2 and 3: this places allele *e* more to the right than (*a*, *c*). The order is *b* (*a*, *c*) *e*.

Compare deletions 3 and 4: allele *a* is more to the left than *c* and *e*, and *d* is more to the right than *e*. The order is *b a c e d*.

Compare deletions 4 and 5: allele *f* is more to the right than *d*. The order is *b a c e d f*.

Allele	Band
b	1
a	2
c	3
e	4
d	5
f	6

9. **a. and b.** When a deletion is crossed with a recessive point mutation in the same gene, the recessive point mutation is expressed. When the deletion and the point mutation are in different genes, wild type is observed.

Mutant	Defect
1	deletion of at least part of genes *h* and *i*
2	deletion of at least part of genes *k* and *l*
3	deletion of at least part of gene *m*
4	deletion of at least part of genes *k*, *l*, and *m*
5	deletion not within the *h* through *m* genes, or a recessive point mutation

10. **a.** Eighteen map units were either deleted or inverted. A large inversion would result in semisterility, whereas a large deletion might be lethal. Thus, an inversion is more likely.

b.

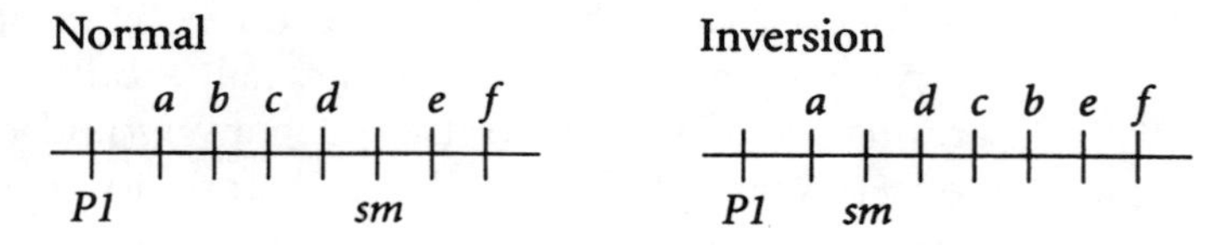

Alternatively, *sm* could be located external to the *e* locus on both the normal and the inversion chromosomes.

c. The semisterility is the result of crossing-over in the inverted region. All products of crossing-over would have either duplications or deletions.

11. The testcross is *P B Q*/*p B q* × *p b q*/*p b q*.

a. To obtain normal eye shape, the bar allele must be deleted. Actually, bar-eye is the result of a tandem duplication of a normal allele rather than a variant allele that results in bar-eye. Thus, to delete the extra copy of the gene, synapsis must occur out of register between the two chromosomes:

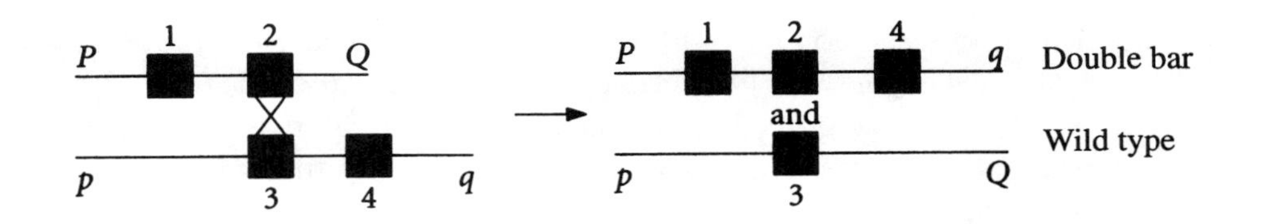

b. In the diagram above, the flanking markers are $P\,q$ and $p\,Q$ after crossing-over. If gene 1 had paired with gene 4, then the wild-type phenotype would have been associated with $P\,q$, and the double bar-eye would have been with $p\,Q$.

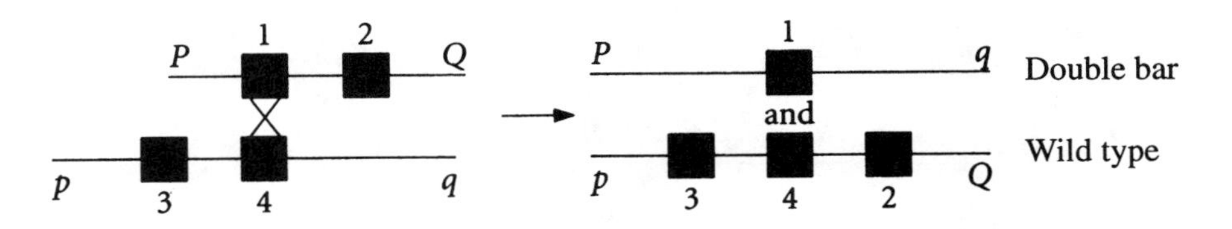

12. a. Classical dominance assumes that one copy of the dominant allele is sufficient for the dominant phenotype. In this case, the female progeny have one copy of the dominant allele, yet they do not have the dominant phenotype. To explain these results, it must be assumed that both the dominant and the variant alleles have a product and that the phenotype is the result of the ratio of one product to the other.

b. The female parents are v^+v^-/v^+v^-. The ratio of the two alleles is 1:1, which results in the wild-type phenotype. The male parents are v^-/Y. No dominant alleles are present, so the males are vermilion eyed.

c. The male progeny are v^+v^-/Y, and the ratio of the two alleles is 1:1, yielding a wild-type phenotype identical with that of their female parents. The female progeny are v^+v^-/v^-. The ratio of v^+ to vermilion is 1:2. Thus, they have the vermilion phenotype.

13. The data suggest that one or both breakpoints of the inversion are located within an essential gene, causing a recessive lethal mutation.

14. a. Single crossovers between a gene and its centromere lead to a tetratype (second-division segregation). Thus a total of 20% of the asci should show second-division segregation, and 80% will show first-division segregation. The following are representative asci:

$un3^+$ $ad3^+$	$un3^+$ $ad3^+$	$un3^+$ $ad3^+$	$un3^+$ $ad3$	$un3^+$ $ad3$
$un3^+$ $ad3^+$	$un3^+$ $ad3^+$	$un3^+$ $ad3^+$	$un3^+$ $ad3$	$un3^+$ $ad3$
$un3^+$ $ad3^+$	$un3^+$ $ad3$	$un3^+$ $ad3$	$un3^+$ $ad3^+$	$un3^+$ $ad3^+$
$un3^+$ $ad3^+$	$un3^+$ $ad3$	$un3^+$ $ad3$	$un3^+$ $ad3^+$	$un3^+$ $ad3^+$
$un3$ $ad3$	$un3$ $ad3^+$	$un3$ $ad3$	$un3$ $ad3^+$	$un3$ $ad3$
$un3$ $ad3$	$un3$ $ad3^+$	$un3$ $ad3$	$un3$ $ad3^+$	$un3$ $ad3$
$un3$ $ad3$	$un3$ $ad3$	$un3$ $ad3^+$	$un3$ $ad3$	$un3$ $ad3^+$
$un3$ $ad3$	$un3$ $ad3$	$un3$ $ad3^+$	$un3$ $ad3$	$un3$ $ad3^+$
80%	5%	5%	5%	5%

In all cases, the "upside-down" version would be equally likely.

b. The aborted spores could result from a crossing-over event within an inversion of the wild type compared with the standard strain. Crossing-over within heterozygous inversions leads to unbalanced chromosomes and nonviable spores. This could be tested by using the wild type from Hawaii in mapping experiments of other markers on chromosome 1 in crosses with the standard strain and looking for altered map distances.

15. a.

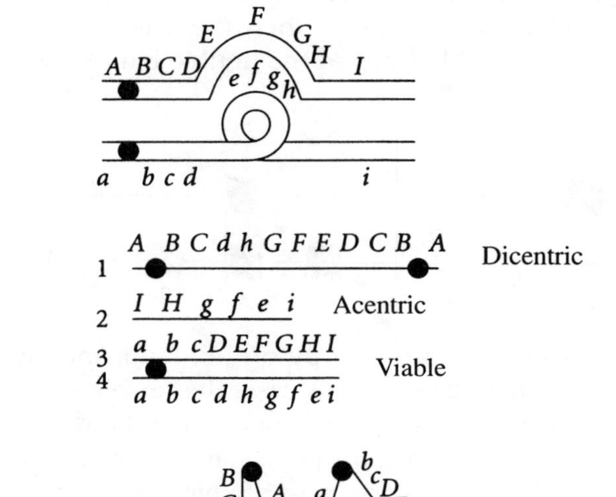

b.

c.

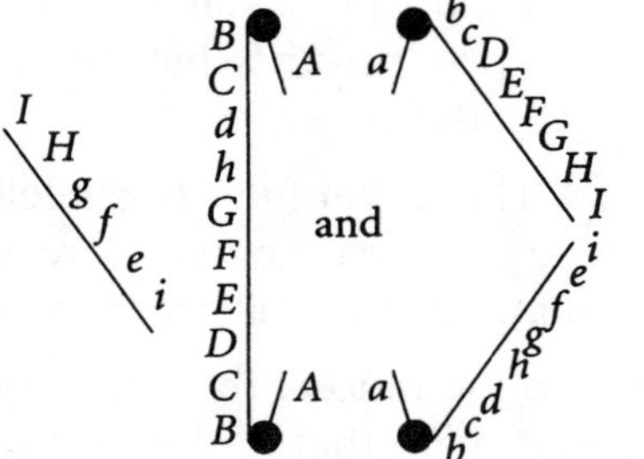

d. The chromosomes numbered 3 and 4 will give rise to viable progeny. The genotypes of those progeny will be *A B C D E F G H I/a b c D E F G H I* and *A B C D E F G H I/a b c d h g f e i*.

16. a. This is a paracentric inversion.

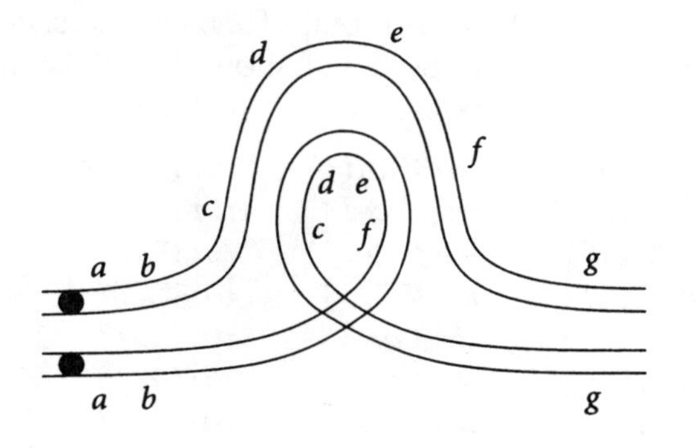

b.

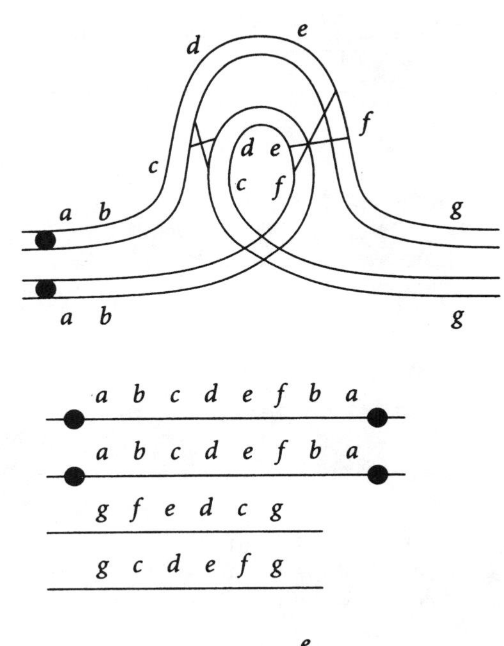

c.

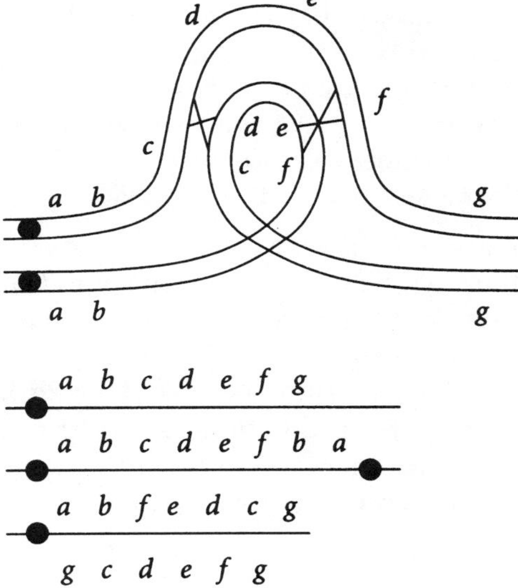

d.

e. The strand not involved in the three-strand double crossover will be normal.

17. a. The Sumatra chromosome contains a pericentric inversion when compared with the Borneo chromosome.

b.

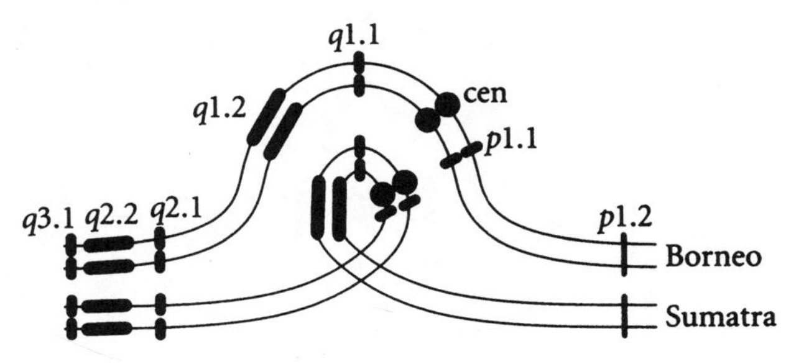

c.

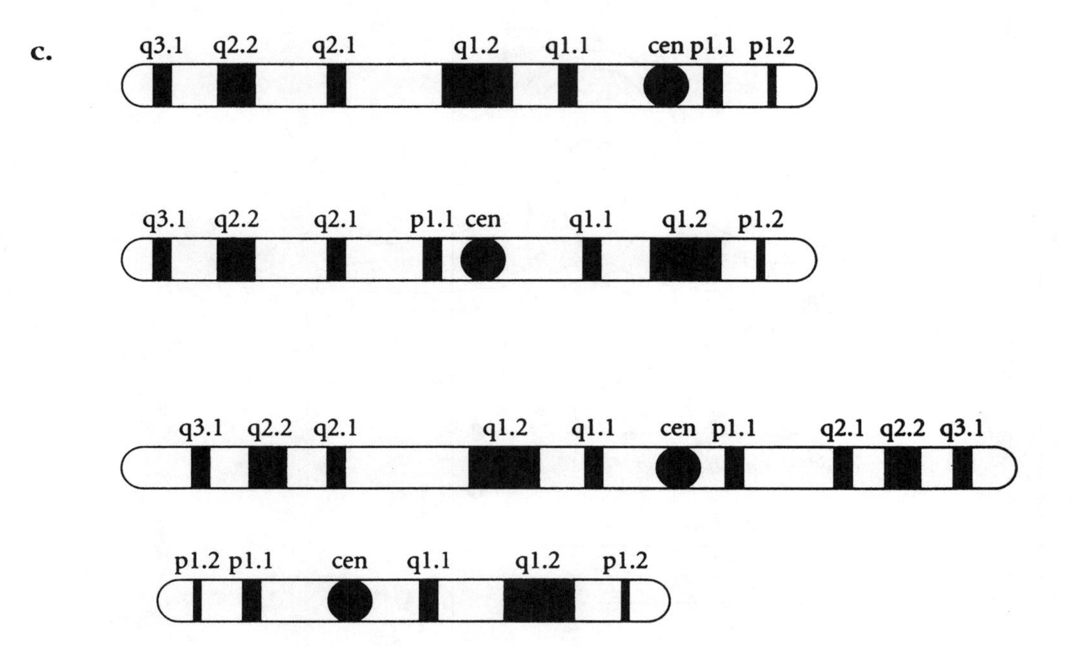

d. Recall that all single crossovers within the inverted region will lead to four meiotic products: two that will be viable, nonrecombinant (parental) types and two that will be extremely unbalanced (most likely nonviable), recombinant types. In other words, if 30% of the meioses have a crossover in this region, 15% of the gametes will not lead to viable progeny. That means that 85% of the gametes should produce viable progeny.

18. a. Two crosses show 28 map units between the loci for body color and eye shape in a testcross of the F_1: California × California and Chile × Chile. The third type of cross, California × Chile, leads to only 4 map units between the two genes when the hybrid is testcrossed. This indicates that the genetic distance has decreased by 24 map units, or 100%(24/28) = 85.7%. A deletion cannot be used to explain this finding, nor can a translocation. Most likely the two lines are inverted with respect to each other for 85.7% of the distance between the two genes.

b.

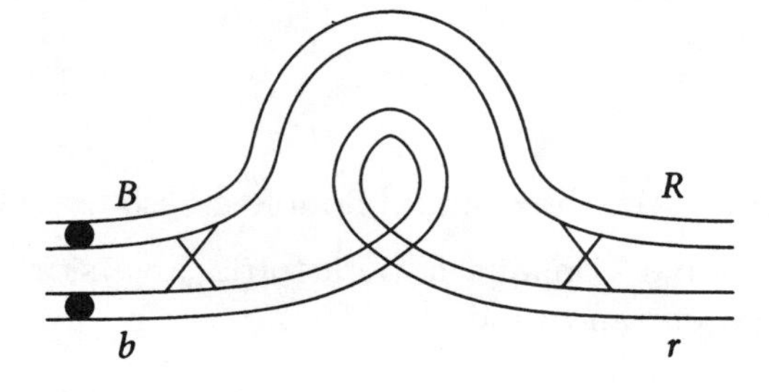

A single crossover in either region would result in 4% crossing-over between *B* and *R*. The products are

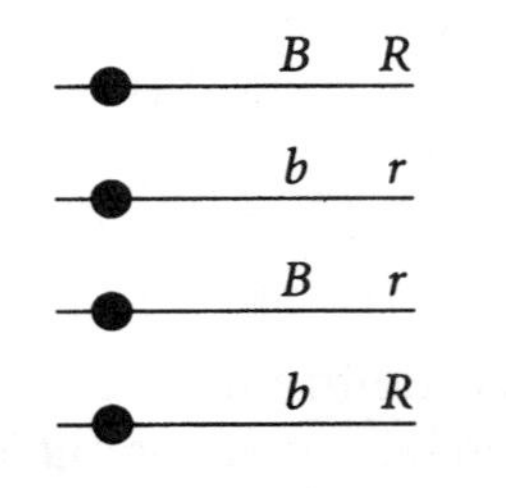

19. a. To construct the maps, look at the data three genes at a time. For example, the Okanagan sequence for *a*, *b*, and *c* is

a to *b*: 12 m.u.
a to *c*: 14 m.u.
b to *c*: 2 m.u.

The best fit is the gene order *a b c*.

b. Diagram the heterozygote during homologous pairing. Crossing-over can occur between *c* and *f* and between *d* and *b*, as indicated in the following figure.

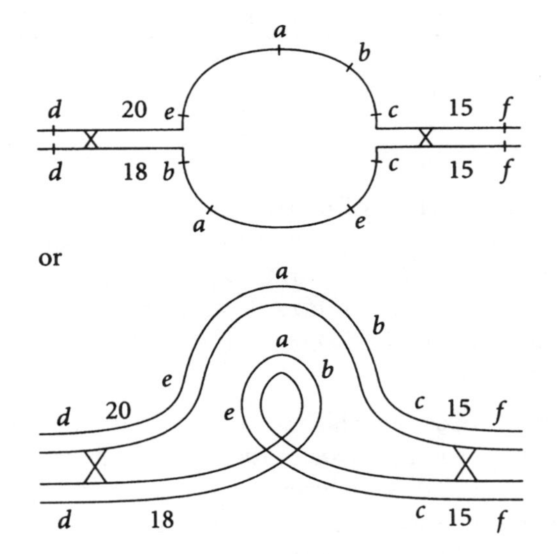

The only recombination events that will lead to viable progeny occur in two regions: (1) the 18 map units between *d* and the breakpoint and (2) the 15 map units between *c* and *f*. Therefore, the recombination frequency between *d* and any gene in the inverted region is 18%. The *c*–*f* recombination frequency is obviously 15%. Finally, the *d*–*f* recombination frequency is 18% + 15%, or 33%.

	a	*b*	*c*	*d*	*e*	*f*
a	0	0	0	18	0	15
b		0	0	18	0	15
c			0	18	0	15
d				0	18	33
e					0	15
f						0

20. a. The aberrant plant is semisterile, which suggests an inversion. Because the *d–f* and *y–p* frequencies of recombination in the aberrant plant are normal, the inversion must involve *b* through *x*.

b. To obtain recombinant progeny when an inversion is involved, either a double crossover occurred within the inverted region or single crossovers occurred between *f* and the inversion, which occurred someplace between *f* and *y*.

21. The cross is

P *c bz wx sh d/c bz wx sh d* × *C Bz Wx Sh D/C Bz Wx Sh D*

F_1 *C Bz Wx Sh D/c bz wx sh d*

backcross *C Bz Wx Sh D/c bz wx sh d* × *c bz wx sh d/c bz wx sh d*

a. The total number of progeny is 1000. Classify the progeny as to where a crossover occurred for each type. Then, total the number of crossovers between each pair of genes. Calculate the observed map units.

Region	#CO	M.U. observed	M.U. expected
C–Bz	103	10.3	12
Bz–Wx	13	1.3	8
Wx–Sh	13	1.3	10
A–D	186	18.6	20

Notice that a reduction of map units, or crossing-over, is seen in two intervals. Results like this are suggestive of an inversion. The inversion most likely involves the *Bz, Wx,* and *Sh* genes.

Further notice that all those instances in which crossing-over occurred in the proposed inverted region involved a double crossover. This is the expected pattern.

b. A number of possible classes are missing: four single-crossover classes resulting from crossing-over in the inverted region, eight double-crossover classes involving the inverted region and the noninverted region, and triple crossovers and higher. The 10 classes detected were the only classes that were viable. They involved a single crossover outside the inverted region or a double crossover within the inverted region.

c. Class 1: parental; increased owing to nonviability of some crossovers

Class 2: parental; increased owing to nonviability of some crossovers

Class 3: crossing-over between *C* and *Bz;* approximately expected frequency

Class 4: crossing-over between *C* and *Bz*; approximately expected frequency

Class 5: crossing-over between *Sh* and *D*; approximately expected frequency

Class 6: crossing-over between *A* and *D*; approximately expected frequency

Class 7: double crossover between *C* and *Bz* and between *Sh* and *D*; approximately expected frequency

Class 8: double crossover between *C* and *Bz* and between *A* and *D*; approximately expected frequency

Class 9: double crossover between *Bz* and *Wx* and between *Wx* and *Sh*; approximately expected frequency

Class 10: double crossover between *Bz* and *Wx* and between *Wx* and *Sh*; approximately expected frequency

d. Cytological verification could be obtained by looking at chromosomes during meiotic pairing. Genetic verification could be achieved by mapping these genes in the wild-type strain and observing their altered relationships.

22. a. and b. The F_1 females are $y\ cv\ v\ f\ B^+\ car/y^+\ cv^+\ v^+\ f^+\ B^+\ car^+$. These are crossed with *y cv v f B car*/Y males.

Class 1: parental

Class 2: parental

Class 3: DCO *y–cv* and *B–car*

Class 4: reciprocal of class 3

Class 5: DCO *cv–v* and *v–f*

Class 6: reciprocal of class 5

Class 7: DCO *cv–v* and *f–car*

Class 8: reciprocal of class 7

Class 9: DCO *v–cv* and *v–f*

Class 10: reciprocal of class 9

Class 11: This class is identical with the male parent's X chromosome and could not have come from the female parent. Thus, the male sperm must have donated it to the offspring. In *Drosophila*, sex is determined by the ratio of X chromosomes to the number of sets of autosomes. The ratio in males is 1X:2A, where A stands for the autosomes contributed by one parent (the ratio in females is 2X:2A). Thus, this class of males must have arisen from the union of an X-bearing sperm with an egg that was the product of nondisjunction for X and contained only autosomes.

Because all nonparental progeny are double crossovers, there must be an inversion spanning the genes that were studied.

c. Class 11 should have only one sex chromosome, which could be checked cytologically. Also, XO (O is used to denote the absence of another sex chromosome) males are sterile because the production of functional sperm requires genes found on the Y chromosome.

23. The inversion results in no viable crossover products from heterozygous females. When *Cu pr/Cu pr* females are crossed with irradiated wild-type males, all female progeny will be heterozygous for the inversion and for any possible recessive lethal mutation induced by irradiation. They will have curled wings and wild-type eyes ($Cu\ pr/Cu^+\ pr^+\ lethal?$). Each female will carry a different mutation (indicated by "*?*"), if any were induced. Cross the females individually with homozygous *Cu pr* males to generate groups of flies with the same mutation. Then for each, cross the normal-eyed progeny among themselves ($Cu\ pr/Cu^+\ pr^+\ ?$). This results in

1/4	$Cu^+\ pr^+\ ?/Cu^+\ pr^+\ ?$	normal wings and eyes
1/2	$Cu\ pr/Cu^+\ pr^+\ ?$	curled wings and normal eyes
1/4	*Cu pr/Cu pr*	curled wings and purple eyes

If a lethal mutation had been induced, the class with normal wings and eyes would be missing.

24.

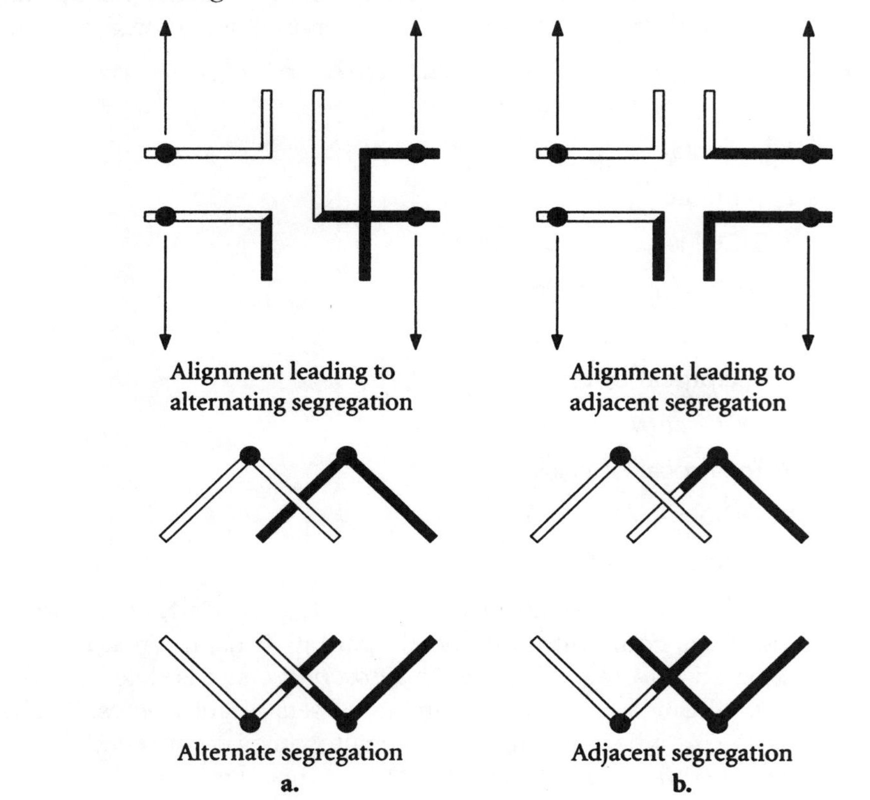

25. If the *a* and *b* genes are on separate chromosomes, independent assortment should occur, giving equal frequencies of $a\ b$, $a^+\ b$, $a\ b^+$, and $a^+\ b^+$. This was not observed; instead, the two genes are behaving as if they were linked, with 10 m.u. between them. This behavior is indicative of a reciprocal translocation in one of the parents, most likely the wild type from nature.

At meiosis prior to progeny formation, the chromosomes would look like the following figure (centromere not included because its position has no effect on the results):

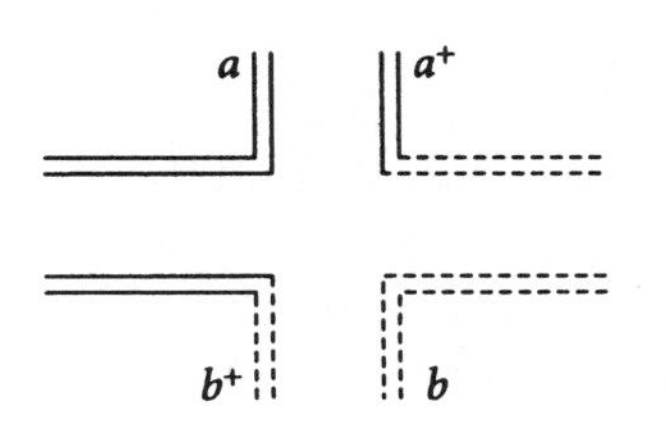

Only alternate segregation avoids duplications and deletions for many genes. Therefore, the majority of the progeny would be parental *a b* and a^+ b^+. The *a* b^+ and a^+ *b* progeny would result from crossing-over between either gene and the translocation breakpoints.

26. Notice that the males have the male parent phenotype and the females have the female parent phenotype. This suggests a translocation of the Cy chromosome to the Y chromosome:

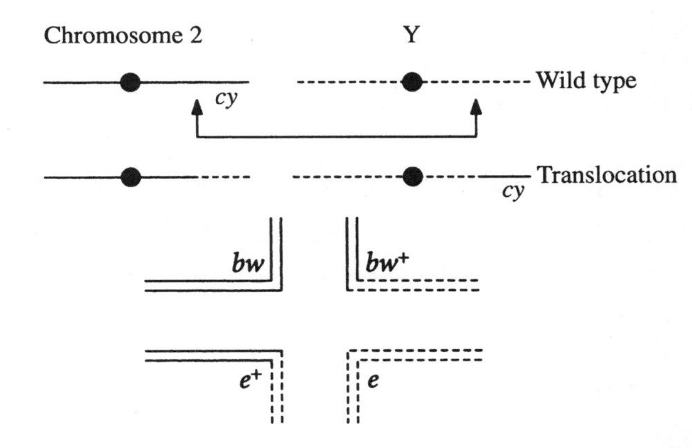

27. a.

b. The surviving offspring would result from alternate segregation and would be either *bw e* (brown eye, ebony body) or bw^+ e^+ (wild type), in a 1:1 ratio.

28. The size of the insertional translocation will determine whether the translocated region or the rest of the region will dominate homologous pairing. It is assumed that the insertion is quite small in relation to the rest of both chromosomes and that meiotic pairing would behave as if one chromosome (the one that received the translocation) had an insertion compared with its wild-type homolog and the other would have a deletion compared with its wild-type homolog. Because these would segregate independently, PD = NPD and the frequency of T would be dependent on the centromere-to-insertion and centromere-to-deletion distances. Representative genotypes and phenotypes of these asci would be (N = normal homolog, I = insertion, D = deletion)

PD	NPD	T
I D wild type	I N_2 duplication	I D wild type
I D wild type	I N_2 duplication	I D wild type
I D wild type	I N_2 duplication	I N_2 duplication
I D wild type	I N_2 duplication	I N_2 duplication
N_1 N_2 wild type	N_1 D inviable	N_1 D inviable
N_1 N_2 wild type	N_1 D inviable	N_1 D inviable
N_1 N_2 wild type	N_1 D inviable	N_1 N_2 wild type
N_1 N_2 wild type	N_1 D inviable	N_1 N_2 wild type

29. ***Unpacking the Problem***

1. A "gene for tassel length" means that there is a gene with at least two alleles (*T* and *t*) that controls the length of the tassel. A "gene for rust resistance" means that there is a gene that determines whether the corn plant is resistant to a rust infection or not (*R* and *r*).

2. The precise meaning of the allelic symbols for the two genes is irrelevant to solving the problem because what is being investigated is the distance between the two genes.

3. A locus is the specific position occupied by a gene on a chromosome. It is implied that gene loci are the same on both homologous chromosomes. The gene pair can consist of identical or different alleles.

4. Evidence that the two genes are normally on separate chromosomes would have come from previous experiments showing that the two genes independently assort during meiosis.

5. Routine crosses could consist of F_1 crosses, F_2 crosses, backcrosses, and testcrosses.

6. The genotype *T/t* ; *R/r* is a double heterozygote, or dihybrid, or F_1 genotype.

7. The pollen parent is the "male" parent that contributes to the pollen tube nucleus, the endosperm nucleus, and the progeny.

8. Testcrosses are crosses that involve a genotypically unknown and a homozygous recessive organism. They are used to reveal the complete genotype of the unknown organism and to study recombination during meiosis.

9. The breeder was expecting to observe 1 *T/t* ; *R/r*:1 *T/t* ; *r/r*:1 *t/t* ; *R/r*:1 *t/t* ; *r/r*.

10. Instead of a 1:1:1:1 ratio indicating independent assortment, the testcross indicated that the two genes were linked, with a genetic distance of 100%(3 + 5)/210 = 3.8 map units.

11. The equality and predominance of the first two classes indicate that the parentals were *T R*/*t r*.

12. The equality and lack of predominance of the second two classes indicate that they represent recombinants.

13. The gametes leading to this observation were

    46.7% *T R*   1.4% *T r*
    49.5% *t r*   2.4% *t R*

14. 46.7% *T R*
    49.5% *t r*

15. 1.4% *T r*
    2.4% *t R*

16. *Tr* and *t R*

17. *T* and *R* are linked, as are *t* and *r*.
18. Two genes on separate chromosomes can become linked through a translocation.
19. One parent of the hybrid plant contained a translocation that linked the *T* and *R* alleles and the *t* and *r* alleles.
20. A corn cob is a structure that holds on its surface the progeny of the next generation.
21.

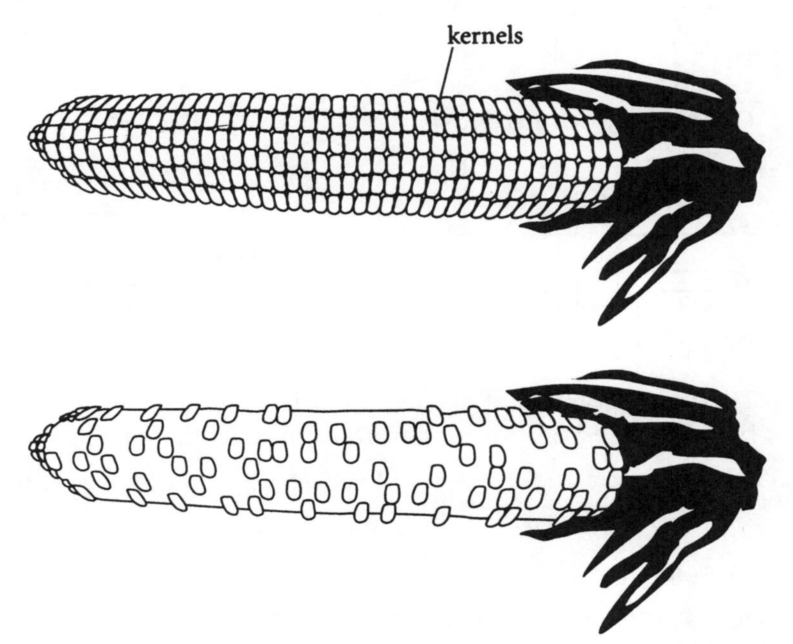

22.

23. A kernel is one progeny on a corn cob.
24. Absence of half the kernels, or 50% aborted progeny (semisterility), could result from the random segregation of one normal with one translocated chromosome (T1 + N2 and T2 + N1) during meioses in a parent that is heterozygous for a reciprocal translocation.
25. Approximately 50% of the progeny died. It was the "female" that was heterozygous for the translocation.

Solution to the Problem

a. The progeny are not in the 1:1:1:1 ratio expected for independent assortment; instead, the data indicate close linkage. And, half the progeny did not develop, indicating semisterility.

b. These observations are best explained by a translocation that brought the two loci close together.

c. Parents: *T R/t r* × *t/t* ; *r/r*

Progeny:		
	98	*T R/t* ; *r*
	104	*t r/t* ; *r*
	3	*T r/t* ; *r*
	5	*t R/t* ; *r*

d. Assume a translocation heterozygote in coupling. If pairing is as diagrammed below, then you would observe the following:

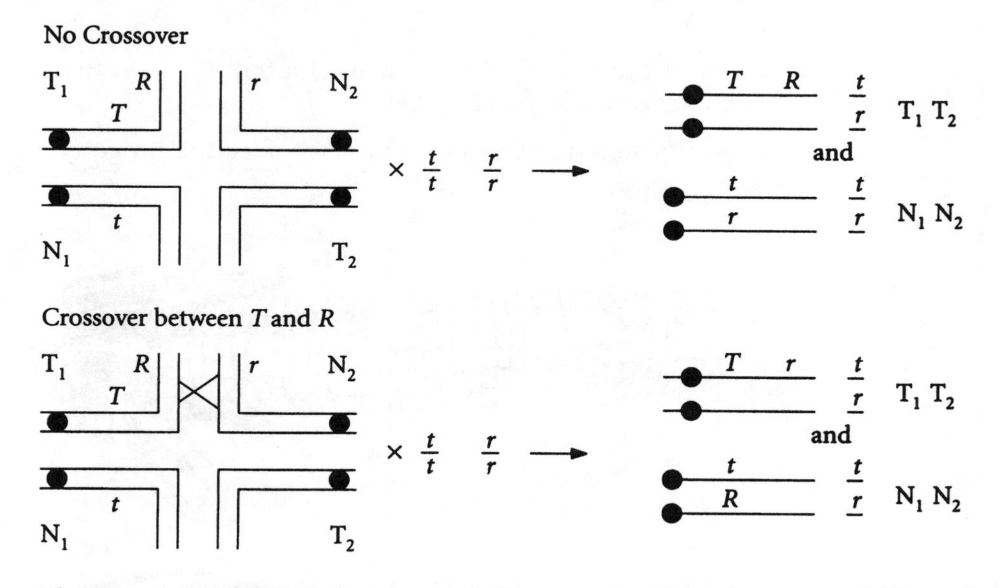

e. The two recombinant classes result from a recombination event followed by proper segregation of chromosomes, as diagrammed above.

30. a. The cross was *nic-2 leu-3* × nic-2^+ *leu-*3^+. The expected results were

$^1/_4$ *nic-2 leu-3* $^1/_4$ *nic-*2^+ *leu-3*

$^1/_4$ *nic-*2^+ *leu-*3^+ $^1/_4$ *nic-2 leu-*3^+

b. Only parentals were obtained. This could be observed if a translocation located both genes on the same chromosome. Alignment at metaphase I could be as follows, which would lead to abortion in the case of duplication and deletion progeny:

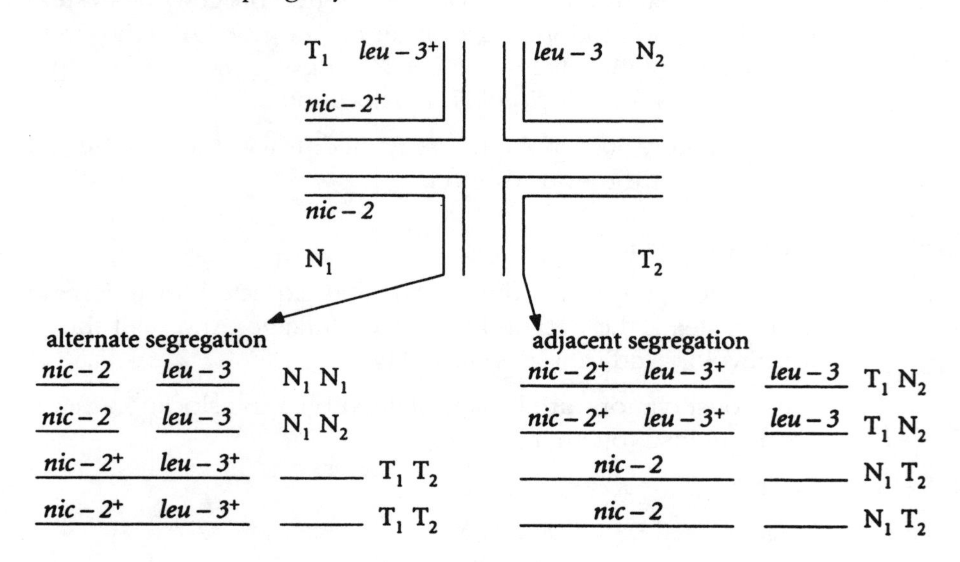

31. a. The homozygous wild-type parental could have been homozygous for a translocation. The translocation could result in the two genes' being on two chromosomes or on one. The latter is assumed below:

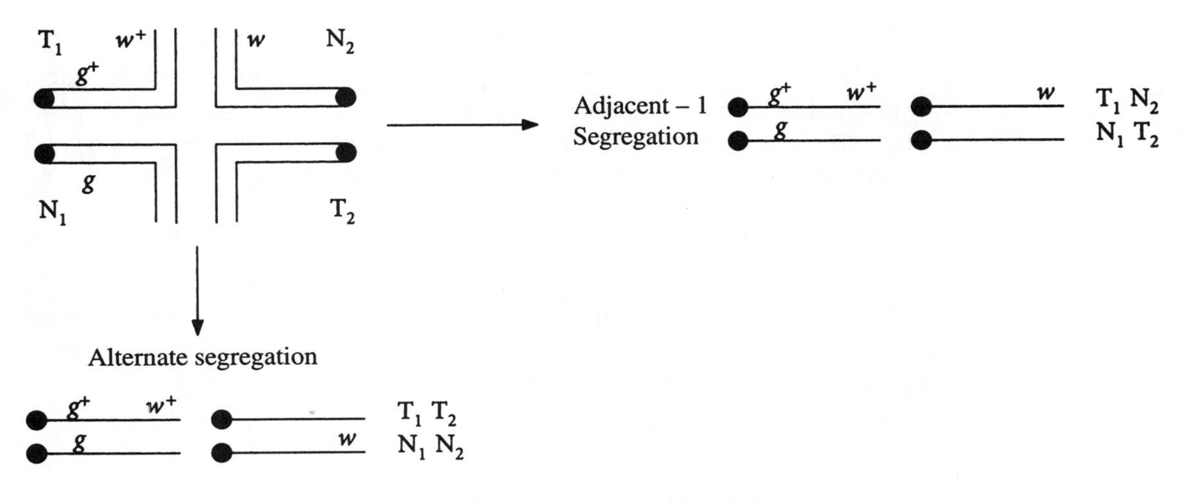

b. In order to solve this problem, list the gametes and their frequencies, and put this information into a Punnett square. Assume that unbalanced gametes are not viable, that adjacent-2 segregation does not occur, and that alternate and adjacent-1 segregation occur at equal frequencies. The gamete composition, but not the phenotypic outcomes, differ with the choice of whether the two genes are on two or one chromosome.

	1/4 g^+w^+ T_1T_2	1/4 gw N_1N_2	1/4 g^+w^+w T_1N_2	1/4 g N_1T_2
1/4 g^+w^+ T_1T_2	1/16 ++	1/16 ++	abort	abort
1/4 gw N_1N_2	1/16 ++	1/16 gw	abort	abort
1/4 g^+w^+w T_1N_2	abort	abort	abort	1/16 ++
1/4 g N_1T_2	abort	abort	1/16 ++	abort

The viable progeny are

$^5/_{16}$ $g^+ w^+$

$^1/_{16}$ $g\ w$

or 5 wild type to 1 green, waxy.

Of the wild type, four are semisterile and would have 50% abortion, and one is homozygous for the translocation and would produce 100% normal progeny. The green, waxy would produce 100% normal progeny. This assumes that abnormal gametes are capable of compensating for each other.

32. The cross was

P	X^{e+}/Y (irradiated) × X^e/X^e
F_1	most X^e/Y yellow males two ? gray males

a. Gray male 1 was crossed with a yellow female, yielding yellow females and gray males, which is reversed sex linkage. If the e^+ allele was translocated to the Y chromosome, the gray male would be X^e/Y^{e+}, or gray. When crossed with yellow females, the results would be

X^e/Y^{e+}	gray males
X^e/X^e	yellow females

b. Gray male 2 was crossed with a yellow female, yielding gray and yellow males and females in equal proportions. If the e^+ allele was translocated to an autosome, the progeny would be as below, where "A" indicates autosome:

P	A^{e+}/A ; X^e/Y × A/A X^e/X^e	
F_1	A^{e+}/A ; X^e/X^e	gray female
	A^{e+}/A ; X^e/Y	gray male
	A/A ; X^e/X^e	yellow female
	A/A ; X^e/Y	yellow male

33. Cross 1: independent assortment of the two genes (expected for genes on separate chromosomes).

Cross 2: the two genes now appear to be linked (the observed RF is 1%); also, half the progeny are inviable. These data suggest a reciprocal translocation occurred and both genes are very close to the breakpoints.

Cross 3: the viable spores are of two types: half contain the normal (nontranslocated chromosomes) and half contain the translocated chromosomes.

34. The F_1 is heterozygous for both the translocation and *P*. It is semisterile. A crossover that occurs within the region from the breakpoint to the centromere will be viable. Among the viable crossovers, two-thirds will maintain the link between the *P* allele and semisterility and one-third will disrupt that linkage. Therefore, the frequency of crossingover that needs to be considered is 30 m.u. – 20 m.u. = 10 m.u. When the F_1 is backcrossed to the *p/p* parent, which does not contain the translocation, the following progeny are obtained, where * denotes the translocation chromosome.

Parentals	45%	*P*/p*	purple, semisterile
	45%	*p/p*	green, fully fertile
Recombinants	5%	*P/p*	purple, fully fertile
	5%	*p*/p*	green, semisterile

a. green, semisterile = 5%

b. green, fully fertile = 45%

c. purple, semisterile = 45%

d. purple, fully fertile = 5%

e. To solve this problem, recognize that the gametes from the F_1 will occur in the same frequency as the progeny in the backcross. Green, fully fertile plants can arise from the union of two nontranslocated gametes, *p*, or from the union of two gametes bearing the translocation, p^*. Therefore, the probability of obtaining green, fully fertile plants from an F_1 selfing is

$$p(p/p) + p(p^*/p^*) = (0.45)(0.45) + (0.05)(0.05) = 0.2050$$

A third possibility does exist, although it is rather unlikely. If there is fusion of two unbalanced gametes in which the unbalanced genetic components are complementary, then a balanced embryo would result.

35. The original plant had two reciprocal translocation chromosomes that brought genes *P* and *S* very close together. Because of the close linkage, a ratio suggesting a monohybrid cross, instead of a dihybrid cross, was observed, both with selfing and with a testcross. All gametes are fertile because of homozygosity.

original plant: *P S/p s*
tester: *p s/p s*

F_1 progeny: heterozygous for the translocation:

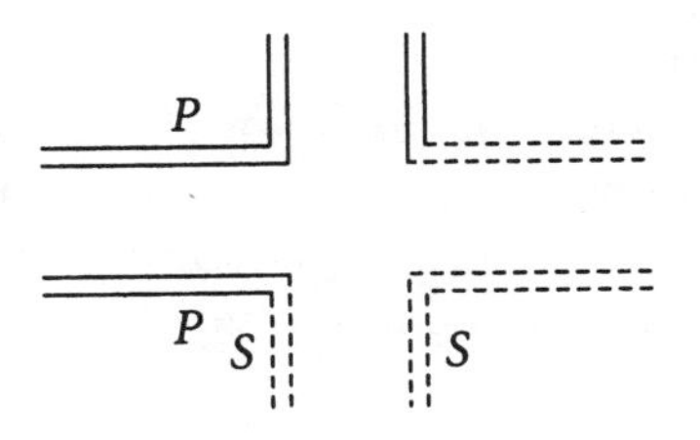

The easiest way to test this is to look at the chromosomes of heterozygotes during meiosis I.

36. The breakpoint can be treated as a gene with two "alleles," one for normal fertility and one for semisterility. The problem thus becomes a two-point cross.

Parentals	764	semisterile *Pr*
	727	normal *pr*
Recombinants	145	semisterile *pr*
	186	normal *Pr*
	1822	

100%(145 + 186)/1822 = 18.17 m.u.

37.

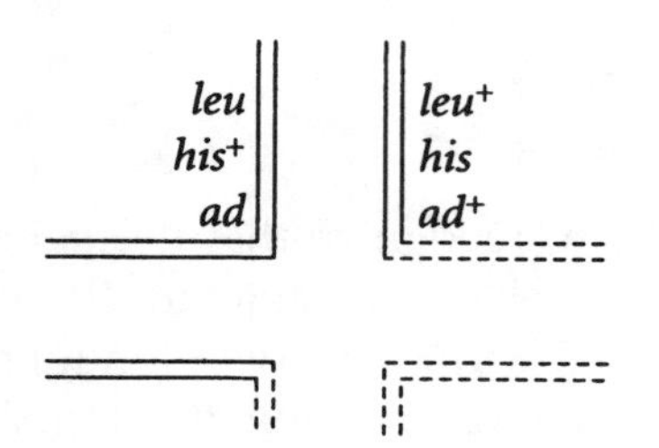

Because the short arm carries no essential genes, adjacent-1 segregation will yield progeny that are viable. Select for leu^+, his^+, and ad^+ by omitting their products from the medium.

38. The percent degeneration seen in the progeny of the exceptional rat is roughly 50% larger than that seen in the progeny from the normal male. Semisterility is an important diagnostic for translocation heterozygotes. This could be verified by cytological observation of the meiotic cells from the exceptional male.

39. **a.** Heterozygous reciprocal translocations lead to duplications and deletions. Therefore, in asci in which crossing-over occurs within the translocated region, each crossing-over event would lead to two white ascospores that abort and two viable dark (with duplications) ascospores. In asci in which no crossing-over occurs within the translocated region, the ascospores would be normal color. Alternate segregation is assumed.

b. Heterozygous pericentric inversions lead to duplications and deletions. Therefore, in asci in which crossing-over occurs within the pericentric inversion, each crossing-over event would lead to two white ascospores that abort and two viable but dark ascospores. In asci in which no crossing-over occurs within the pericentric inversion, the ascospores would be normal color.

c. Heterozygous paracentric inversions result in an acentric fragment that has lost some genetic material (deletion) and a dicentric chromosome that has gained some genetic material (duplication) if crossing-over occurs within the paracentric inversion. Therefore, in asci in which crossing-over occurs within the paracentric inversion, each crossing-over event would lead to two white ascospores that abort and two viable dark ascospores. In asci in which no crossing-over occurs within the paracentric inversion, the ascospores would be normal color.

40. Species B is probably the "parent" species. A paracentric inversion in this species would give rise to species D. Species E could then occur by a translocation of *z x y* to *k l m*. Next, species A could result from a translocation of *a b c* to *d e f*. Finally, species C could result from a pericentric inversion of *b c d e*.

41. The assumptions are that half the gametes from a single heterozygous translocation are nonviable and that the two parents have the same chromosomes involved in translocations.

The progeny of crosses between parents with translocations is referring to the F_1 progeny, in which the parental generation was heterozygous. Designate the parents as follows:

A: (T1 T2 N1 N2)

B: (T1 T2 N1 N2)

The gametes will be

A:			B:		
	$^1/_4$	(T1 T2)		$^1/_4$	(TI T2)
	$^1/_4$	(N1 N2)		$^1/_4$	(N1 N2)
	*$^1/_4$	(T1 N2)		*$^1/_4$	(T1 N2)
	*$^1/_4$	(T2 N1)		*$^1/_4$	(T2 N1)

where * signifies an unbalanced gamete.

Fertilization between balanced gametes will occur $4(^1/_4)(^1/_4)$, or $^1/_4$, of the time. An additional $^1/_8$ of the fertilizations will lead to a balanced genome even though both gametes were unbalanced, for example, (T2 N1) × (T1 N2). Therefore, $^3/_8$ of the progeny are viable and $^{10}/_{16}$ are nonviable.

18 Chromosome Mutation II: Changes in Chromosome Number

1. Klinefelter syndrome — XXY male
   Down syndrome — trisomy 21
   Turner syndrome — XO female

2. Create a hybrid by crossing the two plants and then double the chromosomes with a treatment that disrupts mitosis such as colchicine treatment. Alternatively, diploid somatic cells from the two plants could be fused and then grown into plants through various culture techniques.

3. **a.** 3, 3, 3, 3, 3 3, 3 3, 0, 0

   b. 7, 7, 7, 7, 8, 8, 6, 6

4. **a.** If a 6x were crossed with a 4x, the result would be 5x.

   b. Cross *A/A* with *a/a/a/a* to obtain *A/a/a*.

   c. The easiest way is to expose the *A/a** plant cells to colchicine for one cell division. This will result in a doubling of chromosomes to yield *A/A/a*/a**.

   d. Cross 6x (*a/a/a/a/a/a*) with 2x (*A/A*) to obtain *A/a/a/a*.

   e. In culture, expose haploid plants cells to the herbicide and select for resistant colonies. Treat cells that grow with colchicine to obtain diploid cells.

5. b

6. To solve this problem, first recognize that *B* can pair with *B* one-third of the time (leaving *b* to pair with *b*) and that *B* can pair with *b* two-thirds of the time. If *B* pairs with *B*, the resulting tetrad is *B/b*, *B/b*, *B/b*, *B/b*. If *B* pairs with *b*,

the resulting tetrad can be of two equally frequent types, depending on how the pairs align with respect to each other. One type is *B/B*, *B/B*, *b/b*, *b/b*; the second is *B/b*, *B/b*, *B/b*, *B/b*. Summing, one-third of the tetrads would be *B/B*, *B/B*, *b/b*, *b/b* and two-thirds would be *B/b*, *B/b*, *B/b*.

7. If there is no pairing between chromosomes from the same parent, then all pairs are *A/a*. Because the pairs can align with respect to each other, as diagrammed below,

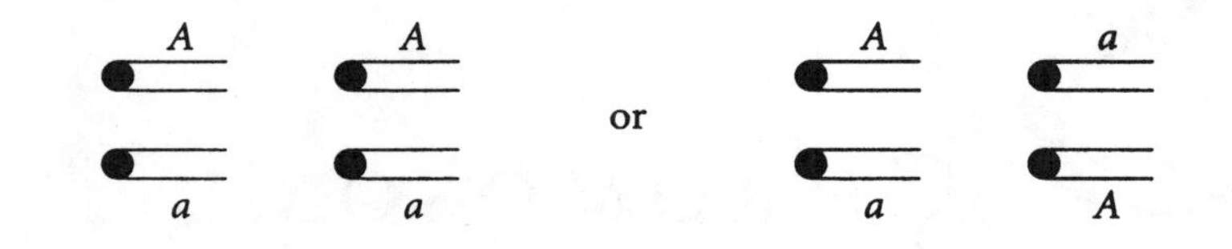

the gametes are 1 *A/A*:2 *A/a*:1 *a/a*. With selfing, the progeny are

	$^1/_4$ *A/A*	$^1/_2$ *A/a*	$^1/_4$ *a/a*
$^1/_4$ *A/A*	$^1/_{16}$ *A/A/A/A*	$^1/_8$ *A/A/A/a*	$^1/_{16}$ *A/A/a/a*
$^1/_2$ *A/a*	$^1/_8$ *A/A/A/a*	$^1/_4$ *A/A/a/a*	$^1/_8$ *A/a/a/a*
$^1/_4$ *a/a*	$^1/_{16}$ *A/A/a/a*	$^1/_8$ *A/a/a/a*	$^1/_{16}$ *a/a/a/a*

or

$^1/_{16}$	*A/A/A/A*
$^1/_4$	*A/A/A/a*
$^3/_8$	*A/A/a/a*
$^1/_4$	*A/a/a/a*
$^1/_{16}$	*a/a/a/a*

If there is no pairing between chromosomes from different parents, then all pairs are *B/B* and *b/b* and all gametes are *B/b*. Thus, 100% of the progeny are *B/B/b/b*.

8. **a.** Let *T. turgidum* be designated by X, *T. monococcum* by Y, and *T. aestivum* by Z. Then note that bivalents contain two chromosomes while univalents contain one. You can write the following equations:

$$X + Y = 21$$
$$Z + Y = 28$$
$$Z + X = 35$$

Next, solve two of the three equations in terms of X:

$$X + Y = 21 \text{ becomes } Y = 21 - X$$

and

$$Z + X = 35 \text{ becomes } Z = 35 - X$$

Now, substitute these values for Y and Z into the third equation and solve it for X:

$Z + Y = 28$ becomes $(35 - X) + (21 - X) = 28$
$56 - 2X = 28$
$28 = 2X$
$14 = X$

Finally, substitute the value of X into the other two equations:

$X + Y = 21$ becomes $14 + Y = 21$
$Y = 21 - 14 = 7$

and

$Z + X = 35$ becomes $Z + 14 = 35$
$Z = 35 - 14 = 21$

Because of the original definitions of X, Y, and Z, the following holds: *T. turgidum* gametes have 14 chromosomes, *T. monococcum* gametes have 7 chromosomes, and *T. aestivum* gametes have 21 chromosomes that are contributed to the hybrids. Therefore, the $2n$ value for *T. turgidum* is 28 chromosomes, for *T. monococcum* is 14 chromosomes, and for *T. aestivum* is 42 chromosomes.

b. and c. Let the 14 chromosomes in *T. monococcum* be designated by A/A, indicating it is diploid. Because 7 bivalents occurred in its hybridization with *T. turgidum*, the hybrid can be designated as A/A + B. Because *T. turgidum* contributed half its chromosomes, it must be A/A + B/B, indicating it is an allopolyploid.

When *T. turgidum* (A/A + B/B) was crossed with *T. aestivum*, the hybrid had 14 bivalents and 7 univalents. *T. turgidum* had to contribute A + B. Therefore, *T. aestivum* also had to contribute A + B to achieve 14 bivalents. It also contributed 7 univalents. In order to identify these 7 chromosomes, look at the cross between *T. aestivum* (A/A + B/B + ?/?) and *T. monococcum* (A/A). The hybrids produced 7 bivalents, which have to be A/A and 14 univalents, which have to be B and something else, which can be designated as C. Therefore, *T. aestivum* can be designated A/A + B/B + C/C, which is an allopolyploid.

To check these conclusions, substitute them back into the original crosses:

A/A + B/B × A/A → A/A + B
A/A + B/B + C/C × AA → A/A + B + C
A/A + B/B + C/C × A/A + B/B → A/A + B/B + C

9. Consider the following table, in which "L" and "S" stand for 13 large and 13 small chromosomes, respectively:

Hybrid	Chromosomes
G. hirsutum × *G. thurberi*	S, S, L
G. hirsutum × *G. herbaceum*	S, L, L
G. thurberi × *G. herbaceum*	S, L

Each parent in the cross must contribute half its chromosomes to the hybrid offspring. It is known that *G. hirsutum* has twice as many chromosomes as the other two species. Furthermore, its chromosomes are composed of chromosomes donated by the other two species. Therefore, the genome of *G. hirsutum* must consist of one large and one small set of chromosomes. Once this is realized, the rest of the problem essentially solves itself. In the first hybrid, the genome of *G. thurberi* must consist of one set of small chromosomes. In the second hybrid, the genome of *G. herbaceum* must consist of one set of large chromosomes. The third hybrid confirms the conclusions reached from the first two hybrids.

The original parents must have had the following chromosome constitution:

G. hirsutum	26 large, 26 small
G. thurberi	26 small
G. herbaceum	26 large

G. hirsutum is a polyploid derivative of a cross between the two Old World species. This could easily be checked by looking at the chromosomes.

10. a. For a moment, pretend that you can distinguish all the alleles of each gene; for simplicity's sake, for example, number them 1 (*F*), 2 (*F*), 3 (*f*), and 4 (*f*). Through random pairing during meiosis, the combinations possible are 1-2, 1-3, 1-4, 2-3, 2-4, and 3-4. In other words, for each gene there are six combinations. Changing the numbers into letters, the gametes for *F* would be $^1/_6$ *F/F*, $^4/_6$ *F/f*, and $^1/_6$ *f/f*.

When two genes are considered, the gametes are

1/6 *FF*	1/6 *GG*	= 1/36	*FF GG*
	4/6 *Gg*	= 4/36	*FF Gg*
	1/6 *gg*	= 1/36	*FF gg*
4/6 *Ff*	1/6 *GG*	= 4/36	*Ff GG*
	4/6 *Gg*	= 16/36	*Ff Gg*
	1/6 *gg*	= 4/36	*Ff gg*
1/6 *ff*	1/6 *GG*	= 1/36	*ff GG*
	4/6 *Gg*	= 4/36	*ff Gg*
	1/6 *gg*	= 1/36	*ff gg*

b. The cross is *F/F/f/f* ; *G/G/g/g* × *F/F/f/f* ; *G/G/g/g*. To calculate the frequency of progeny that would be *F/F/F/f* ; *G/G/g/g*, consider each gene separately. The combination *F/F/F/f* can be achieved in two ways:

$$p(F/F/F/f) = [p(F/F) \times p(F/f)] + [p(F/f) \times p(F/F)] = (^1/_6 \times {}^4/_6) + (^4/_6 \times {}^1/_6) = {}^2/_9$$

The combination *G/G/g/g* can be achieved in three ways:

$$p(G/Gg/g) = [p(G/G) \times p(g/g)] + [p(g/g) \times p(G/G)] + [p(G/g) \times p(G/g)] = (^1/_6 \times {}^1/_6) + (^1/_6 \times {}^1/_6) + (^4/_6 \times {}^4/_6) = {}^1/_{36} + {}^1/_{36} + {}^{16}/_{36} = {}^1/_2$$

Therefore

$p(F/F/F/f\ ;\ G/G/g/g) = {}^{2}/_{9} \times {}^{1}/_{2} = {}^{1}/_{9}$

and

$p(f/f/f/f\ ;\ g/g/g/g) = p(f/f/f/f) \times p(g/g/g/g) =$

$({}^{1}/_{6} \times {}^{1}/_{6}) \times ({}^{1}/_{6} \times {}^{1}/_{6}) = {}^{1}/_{1296}$

11. b, f, h, and sometimes c.

12. One of the parents of the woman with Turner syndrome (XO) must have had an allele for colorblindness, an X-linked recessive disorder. Because her father has normal vision, she could not have obtained her sole X from him. Therefore, nondisjunction occurred in her father. A sperm lacking a sex chromosome fertilized an egg carrying the colorblindness allele. The nondisjunction event could have occurred during either meiotic division.

If the colorblind patient had Klinefelter syndrome (XXY), then both X's must carry the allele for colorblindness. Therefore, nondisjunction had to occur in the mother. Remember that during meiosis I, given no crossover between the gene and the centromere, allelic alternatives separate from each other. During meiosis II, identical alleles on sister chromatids separate. Therefore, the nondisjunctive event had to occur during meiosis II because both alleles are identical.

13. **a.** If most individuals were female, this suggests that the normal allele has been lost (by nondisjunction or deletion) or is nonfunctional (through X-inactivation) in the colorblind eye.

b. If most of the individuals were male, this suggests that the male might have two or more cell lines (he is a mosaic, $X^{normal}Y/X^{cb}Y$) or that he has two X chromosomes (he has Klinefelter syndrome) and that the same processes as in females could be occurring.

14. If the fluorescent spot indicates a Y chromosome, then two spots are indicative of nondisjunction. Presumably, exposure to dibromochloropropane increases the rate of nondisjunction. This could be tested in several ways. The most straightforward would be to expose male animals to the chemical, look for an increase in double-spotted sperm over those not exposed, and also examine testicular cells to observe the rate of nondisjunction with and without exposure. Alternatively, specific crosses could be set up that would reveal nondisjunction upon exposure to the chemical. If X linkage in fruit flies is used for the assay, white-eyed females exposed to the chemical should have a higher rate of nondisjunction than do unexposed females. When crosses to red-eyed males are done, nondisjunction of the X chromosome would result in white-eyed females and red-eyed males.

15. **a. and b.** Nystagmus appears to be an X-linked recessive disorder. That means that the individual with Turner's syndrome had to have obtained her sole X from her mother. She did not obtain a sex chromosome from her father, which indicates that nondisjunction occurred in him. The nondisjunction could have occurred at either M_I or M_{II}.

16. One possibility is that the mean age of mothers giving birth dropped significantly. Because the older mother is at higher risk for nondisjunction, this would result in the observation. Hospital records could be used to check the mother's age when giving birth in 1952 and 1972. Another possibility is an increase of amniocentesis amongst older mothers followed by induced abortion of trisomy-21 fetuses. Because pregnant women 35 and older routinely undergo amniocentesis, the rate for this population may have fallen whereas the rate for the younger population remained unchanged. This also could be checked through hospital records.

17. **a.** Loss of one X in the developing fetus after the two-cell stage.

b. Nondisjunction leading to Klinefelter syndrome (XXY), followed by a nondisjunctive event in one cell for the Y chromosome after the two-cell stage, resulting in XX and XXYY.

c. Nondisjunction of the X at the one-cell stage.

d. Fused XX and XY zygotes (from the separate fertilizations either of two eggs or of an egg and a polar body by one X-bearing and one Y-bearing sperm).

e. Nondisjunction of the X at the two-cell stage or later.

18. Remember that nondisjunction at meiosis I leads to the retention of both chromosomes in one cell, while at meiosis II it leads to the retention of both sister chromatids in one cell.

(1) trisomy 21; nondisjunction at meiosis II in the female

(2) trisomy 21; nondisjunction at meiosis I in the female

(3) normal; normal meiosis in both parents

(4) trisomy 21; nondisjunction at meiosis II in the male

(5) normal; nondisjunction in the female (meiosis I) and the male (either meiotic division)

(6) XYY; nondisjunction for the sex chromosomes in the male meiosis II

(7) monosomy 21-trisomy 21 in a male zygote; occurrence of mitotic nondisjunction for the 21^c chromosome fairly early in development

(8) sexual mosaic; fused XX and XY zygotes or, as in Problem 17d, fused fertilized egg and fertilized polar body

19. Type a: The extra chromosome must be from the mother. Because the chromosomes are identical, nondisjunction had to have occurred at M_{II}.

Type b: The extra chromosome must be from the mother. Because the chromosomes are not identical, nondisjunction had to have occurred at M_I.

Type c: The mother correctly contributed one chromosome, but the father did not contribute any chromosome 4. Therefore, nondisjunction occurred in the male during either meiotic division.

Type d: One cell line lacks a maternal contribution while the other has a double maternal contribution. Because the two lines are complementary, the best explanation is that nondisjunction occurred in the developing embryo during mitosis.

Type e: Each cell line is normal, indicating that nondisjunction did not occur. The best explanation is that the second polar body was fertilized and was subsequently fused with the developing embryo. Alternatively, a pair of twins fused.

20. *Unpacking the Problem*

1. *Homozygous* means that an organism has two identical alleles.

   A *mutation* is any deviation from wild type.

   An *allele* is one particular form of a gene.

   Closely linked means that two genes are physically very close to each other on the same chromosome.

   Recessive refers to a type of allele that is expressed only when it is the sole type of allele for that gene found in an individual.

   Wild type is the most frequent type found in a laboratory population or in a population in the "wild."

   Crossing-over refers to the physical exchange of alleles between homologous chromosomes.

   Nondisjunction is the failure of separation of either homologous chromosomes or sister chromatids in the two meiotic divisions.

   A *testcross* is a cross to a homozygous recessive organism for the trait or traits being studied.

   Phenotype is the appearance of an organism.

   Genotype is the genetic constitution of an organism.

2. No, the genes in question are on an autosome, specifically, number 4.

3. The most common lab species, *Drosophila melanogaster*, has 8 chromosomes.

4.

P	$b\ e^+/b\ e^+ \times b^+\ e/b^+\ e$	(cross 1)
F_1	$b\ e^+/b^+\ e \times b\ e/b\ e$	(cross 2)
Progeny	$b\ e^+/b\ e$	expected parental
	$b^+\ e/b\ e$	expected parental
	$b^+\ e^+/b\ e$	unexpected recombinant ("very closely linked," so rare)
	$b\ e/b\ e$	unexpected recombinant ("very closely linked," so rare)
	rare wild type $\times$ $b\ e/b\ e$	(cross 3)
Progeny	$1/6$	wild type
	$1/6$	bent, eyeless
	$1/3$	bent
	$1/3$	eyeless

5. $b\ e^+$ and $b^+\ e$

6. $b\ e^+/b^+\ e$

7. It is not at all surprising that the F_1 are wild type. This means that both mutations are recessive and complement (are in different genes).

8. *b e/b e* → gametes: *b e*

9. The two common gametes are $b\ e^+$ and $b^+\ e$. The two rare gametes are $b^+\ e^+$ and *b e*.

10. Normal

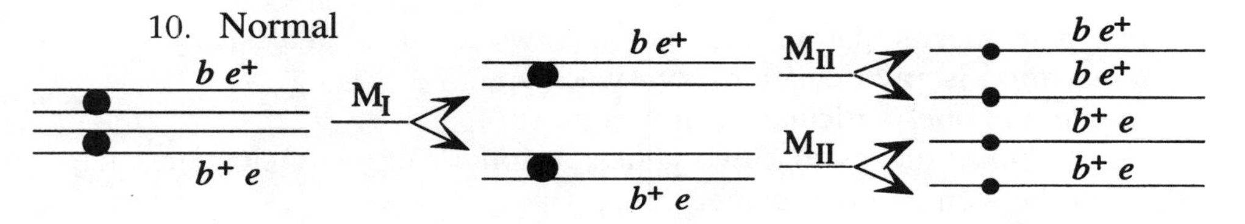

First-Division Nondisjunction: all gametes are aneuploid

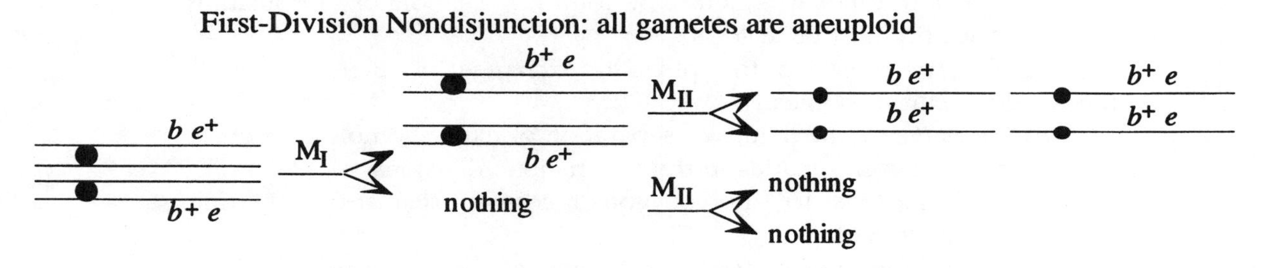

Second-Division Nondisjunction: half the gametes are aneuploid

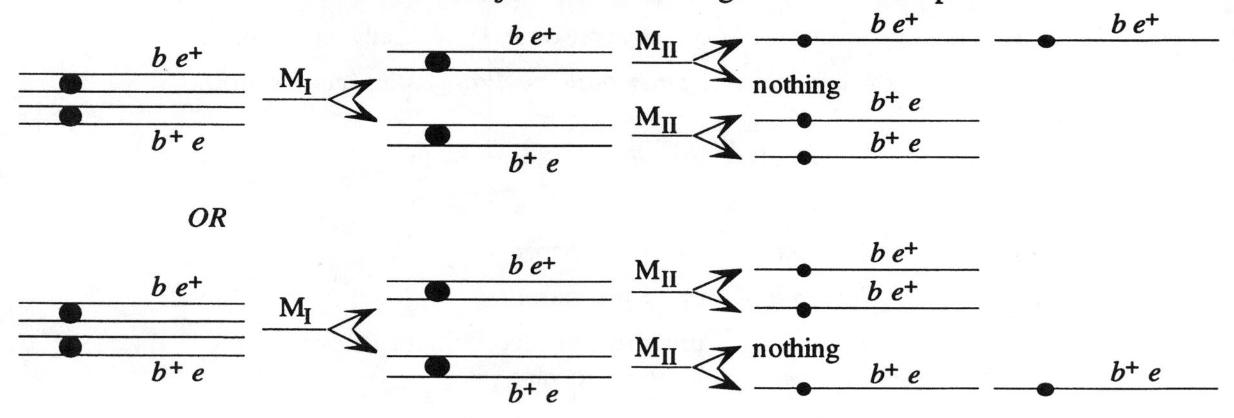

11. This is answered in 10, above.

12. Viable progeny may be able to arise from aneuploidic gametes because chromosome 4 is very small and, percentage-wise, contributes little to the genome. The progeny would be monosomic and trisomic.

13. Listed below are the gametes from 9 and 10 above, the contribution of the male parent, and the phenotype of the progeny.

Female gamete	Male gamete	Phenotype
$b^+ e$	$b\ e$	eyeless
$b\ e^+$	$b\ e$	bent
$b^+ e^+$	$b\ e$	wild type
$b\ e$	$b\ e$	bent, eyeless
$b^+ e/b\ e^+$	$b\ e$	wild type
—	$b\ e$	bent, eyeless
$b\ e^+/b\ e^+$	$b\ e$	bent
$b^+ e/b^+ e$	$b\ e$	eyeless

14. The ratio points to meiosis of a trisomic.

15. Research with artificial chromosomes has indicated that extremely small chromosomes segregate improperly at higher rates than longer chromosomes. It is suspected that the chromatids from homologous chromosomes need to intertwine in order to remain together until the onset of anaphase. Very short chromosomes are thought to have some difficulty in doing this and therefore have a higher rate of nondisjunction. In this instance, which deals with natural chromosomes as opposed to artificial chromosomes, very small chromosomes would be expected to have very little genetic material in them, and therefore their loss or gain may not be of too much importance during development.

16. rare wild type × $e\ b/e\ b$ (cross 3)

If the rare wild type is from recombination, then the cross becomes

$b^+ e^+/b\ e \times b\ e/b\ e$

Progeny	$b^+ e^+/b\ e$	parental: wild type
	$b\ e/b\ e$	parental: bent, eyeless
	$b^+ e/b\ e$	rare recombinant: eyeless
	$b\ e^+/b\ e$	rare recombinant: bent

If the rare wild type is from nondisjunction, then the cross becomes

$b\ e^+/b^+ e/b\ e \times b\ e/b\ e$

Progeny	$b\ e^+/b^+ e/b\ e$	wild type
	$b\ e^+/b\ e/b\ e$	bent
	$b^+ e/b\ e/b\ e$	eyeless
	$b\ e^+/b\ e$	bent
	$b^+ e/b\ e$	eyeless
	$b\ e/b\ e$	bent, eyeless

Solution to the Problem

Cross 1: P $b\ e^+/b\ e^+ \times b^+ e/b^+ e$

F_1 $b^+ e/b\ e^+$

Cross 2: P X/X ; $b^+ e/b\ e^+$ × X/Y ; $b\ e/b\ e$

F_1 expect 1 $b\ e^+/b\ e$:1 $b^+ e/b\ e$, X/X and X/Y

one rare observed X/X ; $b^+ e^+$

a. The common progeny are $b^+ e/b\ e$ and $b\ e^+/b\ e$.

b. The rare female could have come from crossing-over, which would have resulted in a gamete that was $b^+ e^+$. The rare female also could have come from nondisjunction that gave a gamete that was $b\ e^+/b^+ e$. Such a gamete might give rise to viable progeny.

c. If the female had been wild type ($b^+ e^+/b\ e$) as a result of crossing-over, her progeny would have been as follows:

Parental:	$b^+ e^+/b\ e$	wild type (common)
	$b\ e/b\ e$	bent, eyeless (common)
Recombinant:	$b\ e^+/b\ e$	bent (rare)
	$b^+ e/b\ e$	eyeless (rare)

These expected results are very far from what was observed, so the rare female was not the result of recombination.

If the female had been the product of nondisjunction ($b\ e^+/b^+ e/b\ e$), her progeny when crossed to $b\ e/b\ e$ would be as follows:

$^1/_6$	$b^+ e/b\ e$	eyeless
$^1/_6$	$b\ e^+/b\ e/b\ e$	bent
$^1/_6$	$b^+ e/b\ e/b\ e$	eyeless
$^1/_6$	$b\ e^+/b\ e$	bent
$^1/_6$	$b\ e/b\ e$	bent, eyeless
$^1/_6$	b $e^+/b^+ e/b\ e$	wild type

Overall, 2 bent:2 eyeless:1 bent eyeless:1 wild type

These results are in accord with the observed results, indicating that the female was a product of nondisjunction.

21. a. The cross is $P/P/p \times p/p$.

The gametes from the trisomic parent will occur in the following proportions:

$^1/_6$	p
$^2/_6$	P
$^1/_6$	P/P
$^2/_6$	P/p

Only gametes that are p can give rise to potato leaves, because potato is recessive. Therefore, the ratio of normal to potato will be 5:1.

b. If the gene is not on chromosome 6, there should be a 1:1 ratio of normal to potato.

22. The generalized cross is $A/A/A \times a/a$, from which $A/A/a$ progeny were selected. These progeny were crossed with a/a individuals, yielding the results given. Assume for a moment that each allele can be distinguished from the other, and let 1 = A, 2 = A and 3 = a. The gametic combinations possible are

1-2 (*A/A*) and 3 (*a*)
1-3 (*A/a*) and 2 (*A*)
2-3 (*A/a*) and 1 (*A*)

Because only diploid progeny were examined in the cross with *a/a*, the progeny ratio should be 2 wild type:1 mutant if the gene is on the trisomic chromosome. With this in mind, the table indicates that *y* is on chromosome 1, *cot* is on chromosome 7, and *h* is on chromosome 10. Genes *d* and *c* do not map to any of these chromosomes.

23. Radiation could have caused point mutations or induced recombination, but nondisjunction is the more likely explanation.

24. P $a\ b^+\ c\ d^+\ e \times a^+\ b\ c^+\ d\ e^+$

Selection is for the $a^+\ b^+\ c^+\ d^+\ e^+$ phenotype.

Because this rare colony gave rise to both parental types among asexual (haploid) spores, the best explanation is that the rare colony initially contained both marked chromosomes owing to nondisjunction. That is, it was disomic. Subsequent mitotic nondisjunction yielded the two parental types, possibly because the disomic was unstable.

25. Before attempting these problems, draw the chromosomes. The cross is

a. The aborted spores arise from nondisjunction. Nondisjunction at meiosis I would produce 4 n+1 (*pan1*$^+$ *leu2*/*pan1 leu2*$^+$, viable):4 n–1 (nonviable).

b. If, at meiosis II, the chromosome carrying *pan1* experienced nondisjunction, the progeny would be 2 *pan1 leu2*$^+$ (n+1):2 n–1:4 *pan1*$^+$ *leu2* (n).

c. Consider the following crossover (2-4):

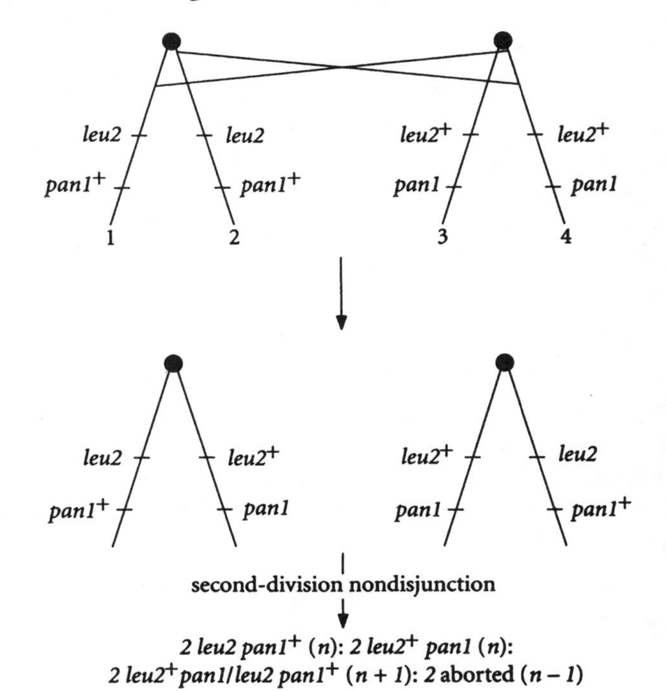

The four chromatids would be (*leu2 pan1*$^+$, *leu2*$^+$ *pan1*) and (*leu2*$^+$ *pan1*, *leu2 pan1*$^+$). If the second chromosome experienced nondisjunction, it would give rise to 2 fully prototrophic:2 aborted spores. The first chromosome would give rise to 2 leu-requiring:2 pan-requiring spores.

26. The two chromosomes are

If one of the centromeres becomes functionally duplex before meiosis I, the homologous chromosomes will separate randomly. This will result in one daughter cell having one chromosome and the second daughter cell having one chromosome and a second chromosome with two chromatids. Meiosis II will lead to a nullisomic (white) and a monosomic (buff) from the first daughter cell and a disomic (black) and a monosomic (buff) from the second daughter cell.

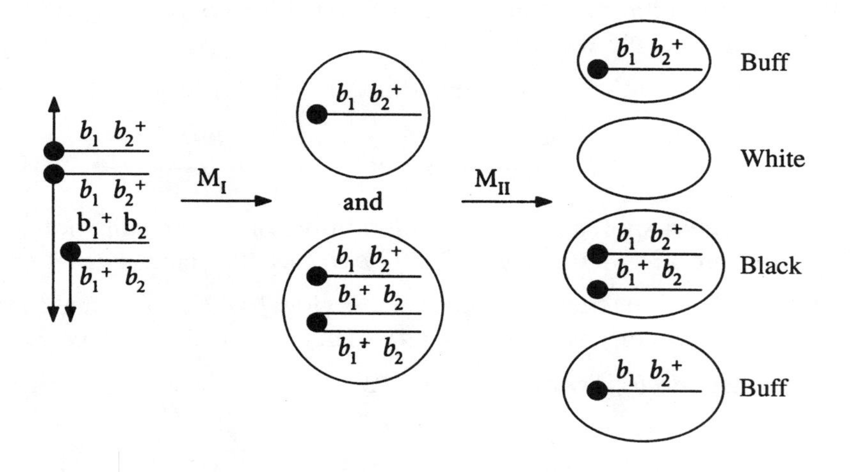

If both centromeres divide prematurely, each daughter cell will get two chromosomes and all the ascospores will be black.

If nondisjunction occurred at meiosis I, the result would be half black (disomic) and half white (nullisomic) spores. Nondisjunction at meiosis II for one of the cells would result in two buff spores (normal meiosis II) and one white (nullisomic) and one buff (two copies of the same gene).

Normal meiosis will yield all buff spores because no crossing-over occurs between the two genes.

27. Aneuploidy is the result of nondisjunction. Therefore, any system that will detect nondisjunction will work. One of the easiest follows. Using *Drosophila*, cross white-eyed females with red-eyed males and look for white-eyed females and red-eyed males in the progeny (the reverse of what is expected). Compare populations unexposed to any environmental pollutants with those exposed to different suspect agents.

28. a. Before considering gametic viability, simply list the gametes and their frequency in both parents.

$^2/_6$ P
$^2/_6$ P/p
$^1/_6$ P/P
$^1/_6$ p

Next, consider whether the gametes are functional in each sex:

Female	Functional	Relative proportion	Final frequency
$^2/_6$ P	100%	4	$^4/_9$
$^2/_6$ P/p	50%	2	$^2/_9$
$^1/_6$ P/P	50%	1	$^1/_9$
$^1/_6$ p	100%	2	$^2/_9$

Female	Functional	Relative proportion	Final frequency
$^2/_6$ P	100%	2	$^2/_3$
$^2/_6$ P/p	0%	0	0
$^1/_6$ P/P	0%	0	0
$^1/_6$ p	100%	1	$^1/_3$

Now, a table can be constructed for the frequency of functional gametes from each sex:

		Female			
		$^4/_9$ P	$^2/_9$ P/p	$^1/_9$ P/P	$^2/_9$ p
Male	$^2/_3$ P	$^8/_{27}$ P/P	$^4/_{27}$ $P/P/p$	$^2/_{27}$ $P/P/P$	$^4/_{27}$ P/p
	$^1/_3$ p	$^4/_{27}$ P/p	$^2/_{27}$ $P/p/p$	$^1/_{27}$ $P/P/p$	$^2/_{27}$ p/p

Summing, the ratio of purple to white is 25:2.

The same method was used to solve the remaining parts of this problem. The results are given below.

b. 17 purple:10 white

c. 4 purple:5 white

29. a. *B. campestris* was crossed with *B. napus*, and the hybrid had 29 chromosomes consisting of 10 bivalents; and 9 univalents. *B. napus* had to have contributed a total of 19 chromosomes to the hybrid. Therefore, *B. campestris* had to have contributed 10 chromosomes. The $2n$ number of *B. campestris* is 20.

When *B. nigra* was crossed with *B. napus*, *B. nigra* had to have contributed 8 chromosomes to the hybrid. The $2n$ number of *B. nigra* is 16.

B. oleracea had to have contributed 9 chromosomes to the hybrid formed with *B. juncea*. The $2n$ number in *B. oleracea* is 18.

b. First list the haploid and diploid number for each species:

Species	Haploid	Diploid
B. nigra	8	16
B. oleracea	9	18
B. campestris	10	20
B. carinata	17	34
B. juncea	18	36
B. napus	19	38

Now, recall that a bivalent in a hybrid indicates that the chromosomes are essentially identical. Therefore, the more bivalents formed in a hybrid, the closer the two parent species. Three crosses result in no bivalents, suggesting that the parents of each set of hybrids are not closely related:

Cross	Haploid number
B. juncea × *B. oleracea*	18 vs. 9
B. carinata × *B. campestris*	17 vs. 10
B. napus × *B. nigra*	19 vs. 8

Three additional crosses resulted in bivalents, suggesting a closer relationship among the parents:

Cross	Haploid number	Bivalents	Univalents
B. juncea × *B. nigra*	18 vs. 8	8	10
B. napus × *B. campestris*	19 vs. 10	10	9
B. carinata × *B. oleracea*	17 vs. 9	9	8

Note that in each cross the number of bivalents is equal to the haploid number of one species. This suggests that the species with the larger haploid number is a hybrid composed of the second species and some other species. In each case, the haploid number of the unknown species is the number of univalents. Therefore, the following relationships can be deduced:

> *B. juncea* is an amphidiploid formed by the cross of *B. nigra* and *B. campestris.*
>
> *B. napus* is an amphidiploid formed by the cross of *B. campestris* and *B. oleracea.*
>
> *B. carinata* is an amphidiploid formed by the cross of *B. nigra* and *B. oleracea.*

These conclusions are in accord with the three crosses that did not yield bivalents:

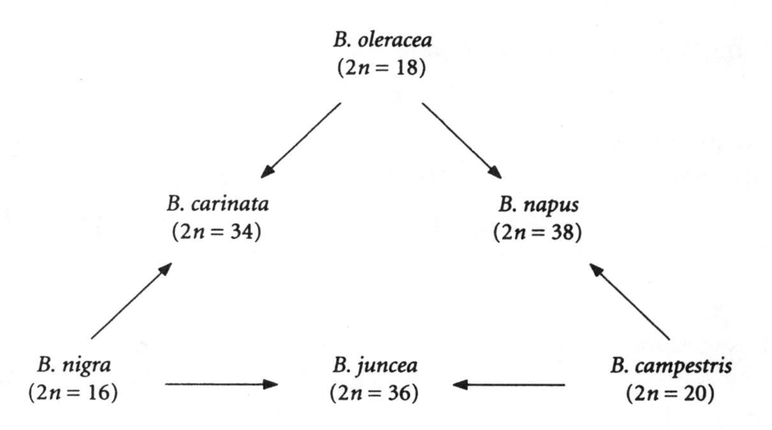

30. Recall that ascospores are haploid. The normal genotype associated with the phenotype of each spore is given below:

1	2	3
$b^+ f^+$	$b\, f^+$	$b^+ f$
$b^+ f^+$	$b\, f^+$	$b^+ f$
$b^+ f^+$	abort	$b^+ f^+$
$b^+ f^+$	abort	$b^+ f^+$
abort	$b^+ f$	$b\, f$
abort	$b^+ f$	$b\, f$
abort	$b^+ f$	$b\, f^+$
abort	$b^+ f$	$b\, f^+$

a. For the first ascus, the most reasonable explanation is that nondisjunction occurred at the first meiotic division. Second-division nondisjunction or chromosome loss are two explanations of the second ascus. Crossing-over best explains the third ascus.

b.

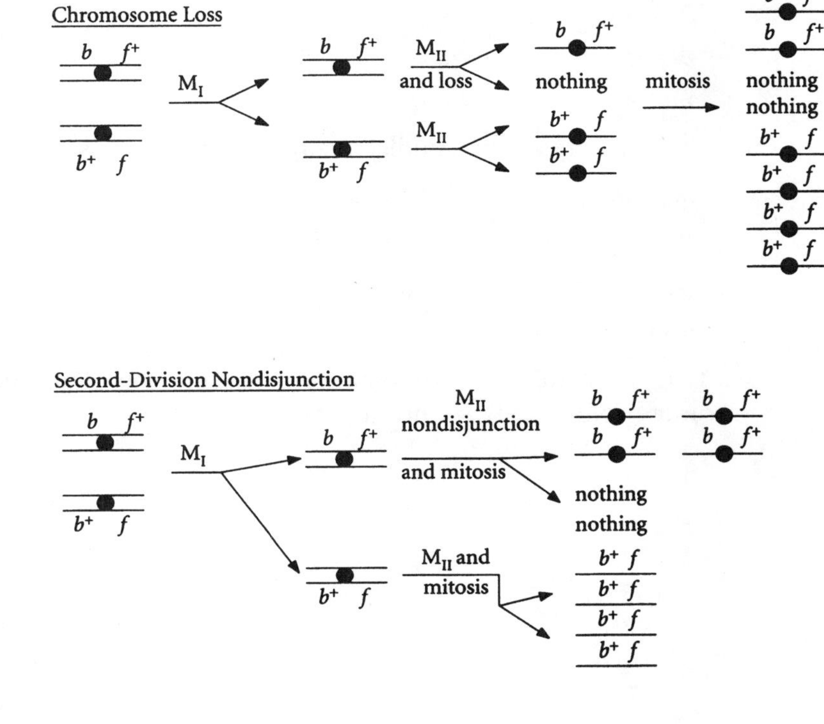

31. a. Each mutant is crossed with wild type, or

$$m \times m^+$$

The resulting tetrads (octads) show 1:1 segregation indicating that each mutant is the result of a mutation in a single gene.

b. The results from crossing the two mutant strains indicate either both strains are mutant for the same gene,

$$m_1 \times m_2$$

or they are mutant in different but closely linked genes

$$m_1\, m_2^+ \times m_1^+ m_2$$

c. and d. Because phenotypically black offspring can result from nondisjunction (notice that in case C and case D, black appears in conjunction with aborted spores), it is likely that mutant 1 and mutant 2 are mutant in different but closely linked genes. The cross is therefore

$$m_1\, m_2^+ \times m_1^+ m_2$$

Case A is an NPD tetrad and would be the result of a four-strand double cross-over.

$m_1^+\, m_2^+$	black
$m_1^+\, m_2^+$	black
$m_1\, m_2$	fawn
$m_1\, m_2$	fawn

Case B is a tetratype and would be the result of a single cross-over between one of the genes and the centromere.

$m_1^+\, m_2^+$	black
$m_1^+\, m_2$	fawn
$m_1\, m_2^+$	fawn
$m_1\, m_2$	fawn

Case C is a the result of nondisjunction during meiosis I.

$m_1^+\, m_2;\ m_1\, m_2^+$	black
$m_1^+\, m_2;\ m_1\, m_2^+$	black
no chromosome	abort
no chromosome	abort

Case D is the result of recombination between one of the genes and the centromere followed by nondisjunction during meiosis II. For example

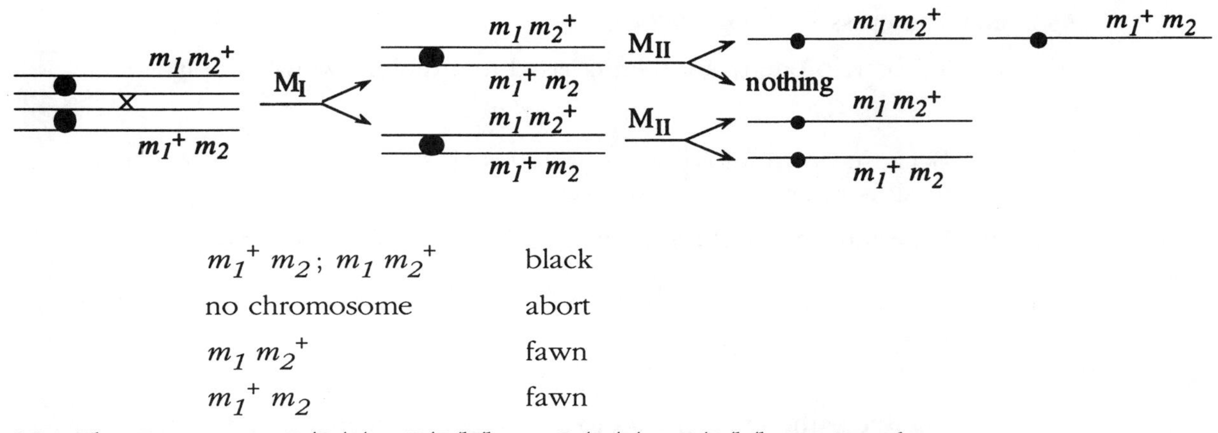

$m_1^+ m_2$; $m_1 m_2^+$	black
no chromosome	abort
$m_1 m_2^+$	fawn
$m_1^+ m_2$	fawn

32. The F_1 cross is *R/R/r/r* ; *B/B/b/b* × *R/R/r/r* ; *B/B/b/b*. Review the answer to question 10 of this chapter to see that the gametes for the *R* gene would be $^1/_6$ *R/R*, $^4/_6$ *R/r*, and $^1/_6$ *r/r* and likewise for the *B* gene would be $^1/_6$ *B/B*, $^4/_6$ *B/b*, and $^1/_6$ *b/b*. Assuming the plants are purple if at least one dominant allele of each gene is present and white if no dominant alleles are present, the following phenotypes are possible:

R/–/–/– ; *B/–/–/–*	purple
r/r/r/r ; *B/–/–/–*	blue
R/–/–/– ; *b/b/b/b*	red
r/r/r/r ; *b/b/b/b*	white

To calculate the frequency of progeny that would be white, consider each gene separately. The combination *r/r/r/r* can be achieved in only one way

$$p(r/r/r/r) = p(r/r) \times p(r/r) = (^1/_6 \times {}^1/_6) = {}^1/_{36}$$

Likewise,

$$p(b/b/b/b) = p(b/b) \times p(b/b) = (^1/_6 \times {}^1/_6) = {}^1/_{36}$$

Therefore

$$p(r/r/r/r; b/b/b/b) = p(r/r/r/r) \times p(b/b/b/b) = (^1/_6 \times {}^1/_6) \times (^1/_6 \times {}^1/_6) = {}^1/_{1296}$$

Because only one combination of gametes for each gene does not have at least one dominant allele.

$$p(\text{of at least one dominant allele}) = 1 - p(\text{no dominant allele}) = 1 - {}^1/_{36}$$
$$= {}^{35}/_{36}$$

Therefore

$$p(\text{blue}) = p(r/r/r/r\,;\ B/–/–/–) = {}^1/_{36} \times {}^{35}/_{36} = {}^{35}/_{1296}$$
$$p(\text{red}) = p(R/–/–/–\,;\ b/b/b/b) = {}^{35}/_{36} \times {}^1/_{36} = {}^{35}/_{1296}$$
$$p(\text{purple}) = p(R/–/–/–\,;\ B/–/–/–) = {}^{35}/_{36} \times {}^{35}/_{36} = {}^{1225}/_{1296}$$

33. a. The cross is *Fr/Fr* × *fr/fr/fr*.

Trisomic progeny are then crossed to a diploid wild-type plant.

Fr/fr/fr × *fr/fr*

Because only diploid progeny of this cross are evaluated, the ratio of fast- to slow-ripening plants will be 1:2.

b. If *Fr* is not located on the trisomic chromosome, the crosses are

Fr/Fr × *fr/fr*

and

Fr/fr × *fr/fr*

Therefore the ratio of fast- to slow-ripening plants will be 1:1.

c. Comparing the ratios of fast-ripening to slow-ripening diploid progeny, *Fr* maps to chromosome 7 because that trisomic cross gives the expected 1:2 ratio.

19 Mechanisms of Recombination

1. Gene conversion may result in a deviation from a 4:4 ratio, with the order unimportant. The following asci show gene conversion: 3, 6.

   Ascus 4 is also produced by gene conversion. To recognize it as such, recall that after meiosis, mitosis should give rise to identical and ordered pairs in *Neurospora*.

2. In the first case l' is being converted to l'^{+}. In the second case, l'' is being converted to l''^{+}. The difference in frequency is due to polarity.

3. A fixed break point is the point at which a DNA strand breaks and begins unwinding as the first step in recombination. The highest level of gene conversion is seen at this point.

   Gene conversion has occurred in cistrons 1, 2, and 3, and the conversions all are from mutant to wild type. Therefore, most likely one piece of heteroduplex DNA extended across the three cistrons:

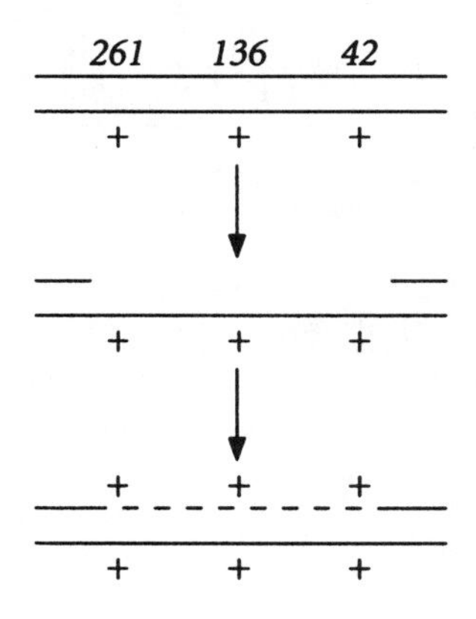

However, there is no evidence that initiation of this heteroduplex took place at a promoter region.

4. The actual mechanism of sister-chromatid exchange induction by mutagens is unknown but would be expected to vary with mutagen effects. The end point must be a break in a DNA strand, very likely as part of postreplication repair.

5. First notice that gene conversion is occurring. In the first cross, a_1 converts (1:3). In the second cross, a_3 converts. In the third cross, a_3 converts. Polarity is possibly involved. The results can be explained by the following map, where hybrid DNA enters only from the left.

$$a_3 \qquad\qquad a_1 \qquad\qquad a_2$$

6. Rewrite the cross and results so that it is clear what they are.

P	$A\,(m_1\, m_2\, m_3^+)\,B \times a\,(m_1^+\, m_2^+\, m_3)\,b$	
F_1	$A\,(m_1\, m_2\, m_3^+)\,B$	parental
	$A\,(m_1\, m_2^+\, m_3^+)\,b$	recombinant
	$a\,(m_1\, m_2\, m_3^+)\,B$	recombinant
	$a\,(m_1^+\, m_2^+\, m_3)\,b$	parental

Next, note the frequency of each allele:

$m_1{:}m_1^+ = 3{:}1$

$m_2{:}m_2^+ = 1{:}1$

$m_3{:}m_3^+ = 1{:}3$

Two gene conversion events have occurred, involving m_1 and m_3.

To understand this at the molecular level, consider the following diagram:

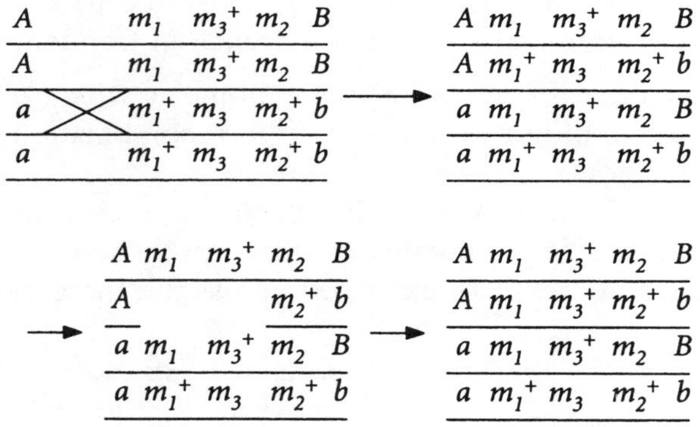

$A \quad m_1 \; m_3^+ \; m_2 \; B$
$A \quad m_1 \; m_3^+ \; m_2 \; B$
$a \quad m_1^+ \; m_3 \; m_2^+ \; b$
$a \quad m_1^+ \; m_3 \; m_2^+ \; b$

→

$A \; m_1 \; m_3^+ \; m_2 \; B$
$A \; m_1^+ \; m_3 \; m_2^+ \; b$
$a \; m_1 \; m_3^+ \; m_2 \; B$
$a \; m_1^+ \; m_3 \; m_2^+ \; b$

→

$A \; m_1 \; m_3^+ \; m_2 \; B$
$A \qquad\quad m_2^+ \; b$
$a \; m_1 \; m_3^+ \; m_2 \; B$
$a \; m_1^+ \; m_3 \; m_2^+ \; b$

→

$A \; m_1 \; m_3^+ \; m_2 \; B$
$A \; m_1 \; m_3^+ \; m_2^+ \; b$
$a \; m_1 \; m_3^+ \; m_2 \; B$
$a \; m_1^+ \; m_3 \; m_2^+ \; b$

A single excision-repair event changed $m_1^+\, m_3$ to $m_1\, m_3^+$.

7. The ratios for a_1 and a_2 are both 3:1. There is no evidence of polarity, which would indicate that gene conversion as part of recombination is occurring. The best explanation is that two separate excision-repair events occurred and, in both cases, the repair retained the mutant rather than the wild type.

8. **a. and b.** A heteroduplex that contains an unequal number of bases in the two strands has a larger distortion than does a simple mismatch. Therefore, the former would be more likely to be repaired. For such a case, both heteroduplex molecules are repaired (leading to 6:2 and 2:6) more often than one (leading to 5:3 or 3:5) or none (leading to 3:1:1:3). The preference in direction (i.e., adding a base rather than subtracting) is analogous to thymine dimer repair. In thymine dimer repair, the unpaired, bulged nucleotides are treated as correct and the strand with the thymine dimer is excised.

A mismatch more often than not escapes repair, leading to a 3:1:1:3 ascus.

Transition mutations would not cause as large a distortion of the helix, and each strand of the heteroduplex should have an equal chance of repair. This would lead to 4:4 (two repairs each in the opposite direction), 5:3 (1 repair), 3:1:1:3 (no repairs or two repairs in opposite directions), and, less frequently, 6:2 (two repairs in the same direction).

c. Because excision repair excises the strand opposite the larger buckle (i.e., opposite the frameshift mutation), the cis transition mutation will also be retained. The nearby genes are converted because of the length of the excision repair.

9. The easiest way to handle these data is to construct the following table, which shows the repair rates for 0, 1, and 2 hybrid DNA molecules:

	0.5 +/+	0.3 +/g_1	0.2 g_1/g_1
0.5 +/+	0.25 (6:2)	0.15 (5:3)	0.1 (4:4)
0.3 +/g_1	0.15 (5:3)	0.09 (3:1:1:3)	0.06 (3:5)
0.2 g_1/g_1	0.1 (4:4)	0.06 (3:5)	0.04 (2:6)

The aberrant asci are all those that are not 4:4. The 4:4 asci occur 20 percent of the time. Correcting for them,

a. 6:2 = 25%/0.8 = 31.25%

b. 2:6 = 4%/0.8 = 5%

c. 3:1:1:3 = 9%/0.8 = 11.25%

d. 5:3 = 30%/0.8 = 37.5%

e. 3:5 = 12%/0.8 = 15%

10. The map is

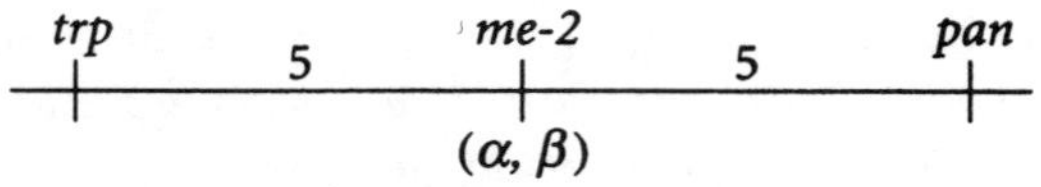

trp —— 5 —— *me-2* —— 5 —— *pan*

(α, β)

Rewrite the crosses and results to be sure that you understand them.

Cross 1: $trp\ (a\ \beta^{+})\ pan^{+} \times trp^{+}\ (a^{+}\ \beta)\ pan$

Cross 2: $trp\ (a^{+}\ \beta)\ pan^{+} \times trp^{+}\ (a\ \beta^{+})\ pan$

Note that all progeny must be $\alpha^{+}\ \beta^{+}$, which requires a crossover between them, and that the order of α and β is unknown.

Consider the first cross. If the sequence is *trp* α β *pan*, then one crossover should lead to a high frequency of ++++. The conventional double crossovers +++– and –+++ should be equally frequent and of lower frequency than ++++. The pattern –++– would result from a triple crossover and would be least frequent. This is summarized in the tabulation below, along with the results if the opposite gene order is true.

		Number of crossovers required	
Pattern	Frequency	*trp* α β *pan*	*trp* β α *pan*
++++	56	1 CO	3 CO
–+++	26	DCO	DCO
+++–	59	DCO	DCO
–++–	16	3 CO	1 CO

For the second cross, the patterns and their interpretation are

		Number of crossovers required	
Pattern	Frequency	*trp* α β *pan*	*trp* β α *pan*
++++	15	3CO	1 CO
–+++	84	DCO	DCO
+++–	23	DCO	DCO
–++–	87	1 CO	3 CO

Both crosses indicate that the sequence is *trp* α β *pan*, but these results are, on the surface, confusing. We see that a double-crossover event does not lead to reciprocal results and, in fact, one double-crossover product occurs as frequently as the single-crossover product. The difficulty is not in the cross but in thinking of the results in terms of a conventional Mendelian cross rather than in terms of gene conversion. Double crossovers are not occurring in the Mendelian sense. In both crosses, "crossing-over" between β and the *pan* allele is occurring at a much higher frequency than expected, which means that β is being converted at a higher level than is α. By convention, that means the polarity runs from *pan* toward *trp*. The asymmetry due to polarity is also seen in the *trp* +:+ *pan* ratios in each cross.

11. Rewrite the original cross:

$$\text{P} \quad A\,x\,y^+ \times a\,x^+\,y$$

The progeny of parental genotypes will be like either of the two parents. The backcrosses are as follows, with the prime indicating progeny generation.

Cross 1: $a'\,x^+\,y \times A\,x\,y^+$ $\rightarrow 10^{-5}$ prototrophs

Cross 2a: $A'\,x\,y^+ \times a\,x^+\,y$ $\rightarrow 10^{-5}$ prototrophs

Cross 2b: $A'\,x\,y^+ \times a\,x^+\,y$ $\rightarrow 10^{-2}$ prototrophs

Recombination is allowing for the higher rate of appearance of prototrophs. Cross 2 is obviously a backcross for some gene affecting the rate of recombination. Whatever that gene is, the allele in the *A* parent blocks recombination (cross 1 and cross 2a), and the allele in the *a* parent allows recombination (cross 2b). It is unlinked to the *his* gene because cross 2 yields results in a 1:1 ratio. The allele that blocks recombination (in *A*) is dominant, whereas the allele that allows recombination is recessive. This is demonstrated by the original cross, in which prototrophs occurred at the lower rate, and by cross 1.

To test this interpretation, one-fourth of the crosses between the $A\,x\,y^+$ and $a\,x^+\,y$ progeny should yield a high rate of recombination and therefore have a high frequency of prototrophs.

20 Transposable Genetic Elements

1. Mutations in *gal* can be generated and from these strains $\lambda dgal$ phage isolated. Through hybridization of denatured $\lambda dgal$ DNA containing the mutation with wild-type $\lambda dgal$ DNA, some of the molecules will be heteroduplexes between one mutant and one wild-type strand. If the mutation was caused by an insertion, the heteroduplexes will show a "looped out" section of single-stranded DNA, confirming that one DNA strand contains a sequence of DNA not present in the other (see Figure 20-9 in the companion text).

   Text Figure 20-8 illustrates a method to compare the densities of gal^+-carrying l phage with gal^--carrying phage. In this experiment, the gal^- phage are denser, indicating that they contain a larger DNA molecule.

   If the *gal* genes are cloned, direct comparison of the restriction maps or even the DNA sequence of mutants compared with wild type will give specific information about whether any are the results of insertions.

2. In replicative transposition, transposable elements move to a new location by replicating into the target DNA, leaving behind a copy of the transposable element at the original site. If, on the other hand, the transposable element excises from its original position and inserts into a new position, this is called conservative transposition.

   To test either mechanism, experiments must be designed so that both the "old" and "new" positions of the transposon can be assayed. If the transposon remains in the old site at the same time that a new copy is detected elsewhere, a replicative mechanism must be in use. If the transposon no longer exists in the old site when a copy is detected elsewhere, a conservative mechanism must be in use.

Figure 20-23 in the companion text describes how replicative transposition can be observed between two plasmids. The same general protocol could be used to detect conservative transposition, but of course the results would be different.

Kleckner and coworkers actually demonstrated conservative transposition by following the movement of a transposon that contained a small heteroduplex within the *lacZ* gene. The DNA of two derivatives of Tn10 carrying different *lacZ* alleles (one being wild-type and the other being mutant) were denatured and allowed to reanneal. In some cases, the DNA molecules that reformed were actually heteroduplexes; one strand contained the $lacZ^+$ allele, and the other strand contained the $lacZ^-$ allele. Transpositions of theses heteroduplexes were then followed. Based on the mechanism of movement, two outcomes are possible. If replicative transposition occurred, the semiconservative nature of DNA replication would generate two genetically different transposons: instead of the heteroduplex $lacZ^+/lacZ^-$ DNA, one would now be all $lacZ^+$ and the other all $lacZ^-$. If transposition was conservative, the *lacZ* gene would still be heteroduplex $lacZ^+/lacZ^-$ after transposition and the first cell division would resolve the heteroduplex. This is what was observed. (The "sectored" colonies are the result of the original cell's still having the $lacZ^+/lacZ^-$ heteroduplex after transposition. After the first division one cell was now $lacZ^+$ and the other was $lacZ^-$.) The experiment is outlined below:

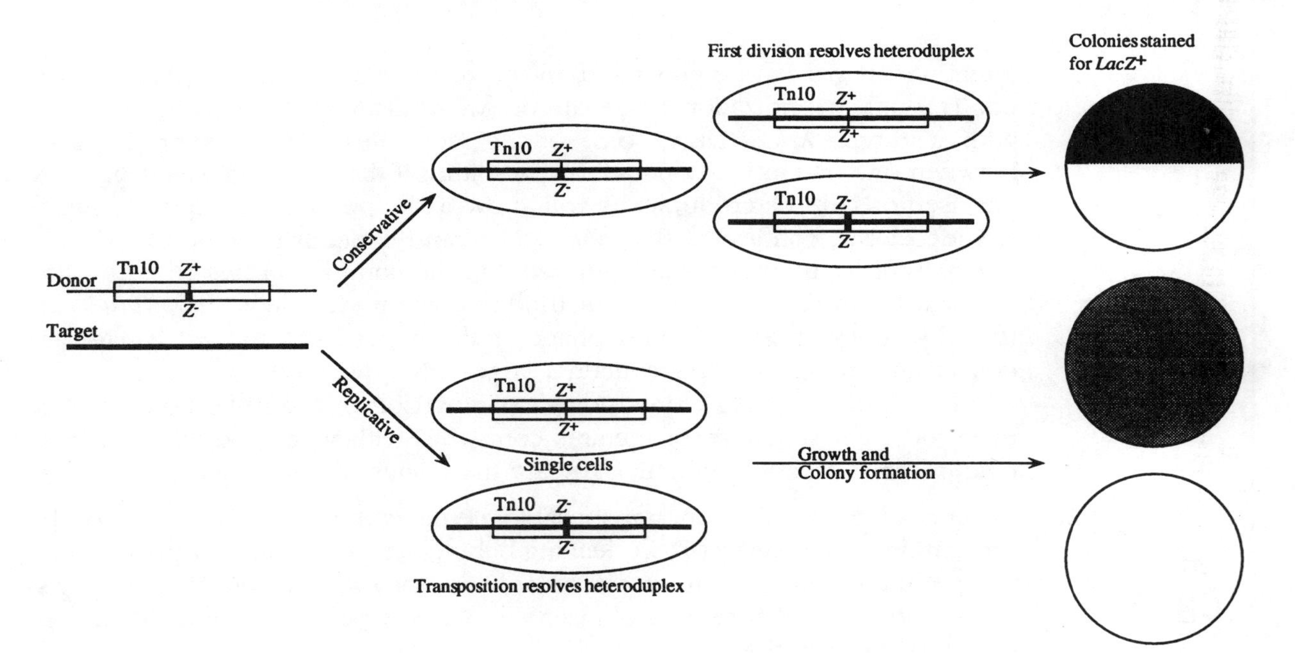

3. R plasmids are the main carriers of drug resistance. These plasmids are self-replicating and contain any number of genes for drug resistance as well as the genes necessary for transfer by conjugation (called the RTF region). It is R plasmid's ability to transfer rapidly to other cells, even those of related species, that allows drug resistance to spread so rapidly. R plasmids acquire drug-resistance genes through transposition. Drug-resistance genes are found flanked by IR

(inverted repeat) sequences and as a unit are known as *transposons*. Many transposons have been identified, and as a set they encode a wide range of drug resistances (see Table 20-3 in the text). Because transposons can "jump" between DNA molecules (e.g., from one plasmid to another or from a plasmid to the bacterial chromosome and vice versa), plasmids can continue to gain new drug-resistance genes as they mix and spread through different strains of cells. It is a classic example of evolution through natural selection. Those cells harboring R plasmids with multiple drug resistances survive to reproduce in the new environment of antibiotic use.

4. Boeke, Fink, and their coworkers demonstrated that transposition of the Ty element in yeast involved an RNA intermediate. They constructed a plasmid using a Ty element that had a promoter that could be activated by galactose, and an intron inserted into its coding region. First, the frequency of transposition was greatly increased by the addition of galactose, indicating that an increase in transcription (and production of RNA) was correlated to rates of transposition. More importantly, after transposition they found that the newly transposed Ty DNA lacked the intron sequence. Because intron splicing occurs only during RNA processing, there must have been an RNA intermediate in the transposition event.

5. P elements are transposable elements found in *Drosophila*. Under certain conditions they are highly mobile and can be used to generate new mutations by random insertion and gene knockout. As such, they are a valuable tool to tag and then clone any number of genes. (See Problem 15 from Chapter 13 for a discussion on cloning by tagging.) P elements can also be manipulated and used to insert almost any DNA (or gene) into the *Drosophila* genome. P element–mediated gene transfer works by inserting the DNA of interest between the inverted repeats necessary for P element transposition and injecting this recombinant DNA along with helper intact P element DNA (to supply the transposase) into very early embryos and screening for (random) insertion among the injected fly's offspring.

6. The a_1 allele is an unstable mutant allele. It contains a "defective" transposable element that requires a trans factor produced by the unlinked *Dt* locus to move (and revert a_1 to A_1, restoring its function). In other words, the a_1 allele is nonautonomous and requires *Dt* to revert. Because Rhoades chose pollen from fully pigmented anthers, the tissue had already undergone a reversion event and was likely A_1/a_1 ; *Dt/Dt*. The pollen from these anthers would be A_1 ; *Dt* and a_1 ; *Dt*. Because each kernel is the result of a separate fertilization, some will be A_1/a_1 ; *Dt/dt* (and fully pigmented) and others will be a_1/a_1 ; *Dt/dt*. The latter can still undergo reversion events during somatic growth, resulting in "dotted" kernels. Each dot would represent the descendants of a cell that had undergone a similar excision–reversion event.

7. The sn^+ patches in an *sn* background and the occurrence of sn^+ progeny from an $sn \times sn$ mating indicate that sn^+ alleles are appearing at relatively high frequencies and that the *sn* alleles are unstable. High reversion rates suggest that the *sn* allele is the result of an insertion of a transposable element that is capable of (frequent) excision.

8. **a. and b.** The soil bacterium *Agrobacterium tumefaciens* contains a large plasmid called the Ti (tumor-inducing) plasmid. When this bacterium infects a plant, a region of the Ti plasmid called the T-DNA is transferred and inserted randomly into the plant's genome. The T-DNA directs the synthesis of plant hormones that cause uncontrolled growth (a tumor) and also directs the plant's synthesis of compounds called *opines*. (These compounds cannot be metabolized by the plant but are used by the bacterium.)

When a piece of "normal" plant tissue is cultured with appropriate nutrients and growth hormones, cells are stimulated to divide in a disorganized manner, forming a mass of undifferentiated cells called a *callus*. These cells will differentiate only into shoots (or roots) if the levels of growth hormones are carefully adjusted. The T-DNA causes undifferentiated growth because it directs the unbalanced synthesis of these same hormones. The fact that some of the infected cultures produced shoots suggests that these cells "lost" the ability to overproduce these hormones. This would be consistent with the loss of the T-DNA (similar to the loss of other transposable elements that is observed in many species). Thus, the A graft would grow normally, and seeds produced by the graft would have no trace of the T-DNA. The fact that cells from the A graft grow like tumor cells when placed on synthetic medium suggests that the medium supplies the high levels of hormones necessary for undifferentiated growth even in the absence of T-DNA.

9. In the *Ac-Ds* system, *Ac* can produce an unstable allele that is autonomous. *Ds* can revert only in the presence of *Ac* and is nonautonomous. In the following, Ac^+ indicates the absence of the *Ac* regulator gene.

Cross 1:

P C/c^{Ds} ; Ac/Ac^+ × c/c ; Ac^+/Ac^+

F$_1$ $^1/_4$ C/c ; Ac/Ac^+ (solid pigment)

$^1/_4$ C/c ; Ac^+/Ac^+ (solid pigment)

$^1/_4$ c^{Ds}/c ; Ac/Ac^+ (unstable colorless or spotted)

$^1/_4$ c^{Ds}/c ; Ac^+/Ac^+ (colorless)

Overall: 2 solid:1 spotted:1 colorless

Cross 2:

P C/c^{Ac} × c/c

F$_1$ $^1/_2$ C/c (solid pigment)

$^1/_2$ C/c^{Ac} (unstable colorless or spotted)

Overall: 1 solid:1 spotted

Cross 3:

P C/c^{Ds} ; Ac/Ac^{+} × C/c^{Ac} ; Ac^{+}/Ac^{+}

F_1 $^1/_8$ C/C ; Ac/Ac^{+} (solid pigment)

$^1/_8$ C/c^{Ac} ; Ac/Ac^{+} (solid pigment)

$^1/_8$ C/C ; Ac^{+}/Ac^{+} (solid pigment)

$^1/_8$ C/c^{Ac} ; Ac^{+}/Ac^{+} (solid pigment)

$^1/_8$ C/c^{Ds} ; Ac^{+}/Ac^{+} (solid pigment)

$^1/_8$ C/c^{Ds} ; Ac^{+}/Ac (solid pigment)

$^1/_8$ c^{Ds}/c^{Ac} ; Ac^{+}/Ac^{+} (unstable colorless or spotted)

$^1/_8$ c^{Ds}/c^{Ac} ; Ac^{+}/Ac (unstable colorless or spotted)

Overall: 3 solid:1 spotted

21 EXTRANUCLEAR GENES

1. Many organelle-encoded polypeptides combine with nucleus-encoded polypeptides to produce active proteins, and these active proteins function in the organelle.

2. Reciprocal crosses showing uniparental or maternal inheritance indicate cytoplasmic inheritance. Cytoplasmic inheritance also can be demonstrated by doing a series of backcrosses, using hybrid females in each case, so that the nuclear genes of one strain are functioning in cytoplasm from the second strain. A heterokaryon test (described in Figure 21-12 of the text) can also demonstrate cytoplasmic inheritance.

3. Maternal inheritance of chloroplasts results in the green-white color variegation observed in *Mirabilis*.

   Cross 1: variegated female × green male → variegated, green, or white progeny

   Cross 2: green female × variegated male → green progeny

   In both crosses, the pollen (male contribution) contains no chloroplasts and thus does not contribute to the inheritance of this phenotype. Eggs from a variegated female plant can be of three types: contain only "green" chloroplasts, contain only "white" chloroplasts, or contain both (variegated). The offspring will have the phenotype associated with the egg's chloroplasts.

4. The crosses are

   Cross 1: stop-start female × wild-type male → all stop-start progeny

   Cross 2: wild-type female × stop-start male → all wild-type progeny

   mtDNA is inherited only from the "female" in *Neurospora.*

5. Both yeast parents contribute mitochondria to the cytoplasm of the resulting diploid cell. Prior to meiosis, cytoplasmic segregation of the mtDNAs occurs as the diploid cells divide mitotically. Subsequent meiosis will then show uniparental inheritance for mitochondria. Therefore, 4:0 and 0:4 asci will be seen.

6. **a.** Petite colonies in yeast are the result of deletions of part of or all the mtDNA. That some petites have also lost the ant^R gene is indicative that the gene is mitochondrial.

   b. Some of the petites have a deleted ant^R gene. Other petites, in which the ant^R gene was retained, had a deletion in a different region.

7. The genetic determinants of R and S are showing maternal inheritance and are therefore cytoplasmic. It is possible that the gene that confers resistance maps to either the mtDNA or the cpDNA.

8. In *Chlamydomonas* the cpDNA is uniparentally inherited from the mat^+ parent, whereas the mtDNA is uniparentally inherited from the mat^- parent. In these crosses, the cpDNA from the mat^- *Chlamydomonas* is lost.

   Cross 1: all Morph 1; 2 kb and 3 kb bands

   Cross 2: all Morph 2; 3 kb and 5 kb bands

   Possible restriction sites for these DNA morphs are

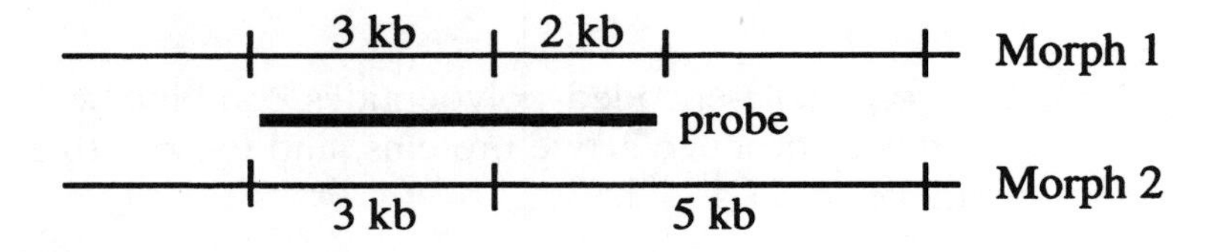

A sketch of the autoradiogram would look like

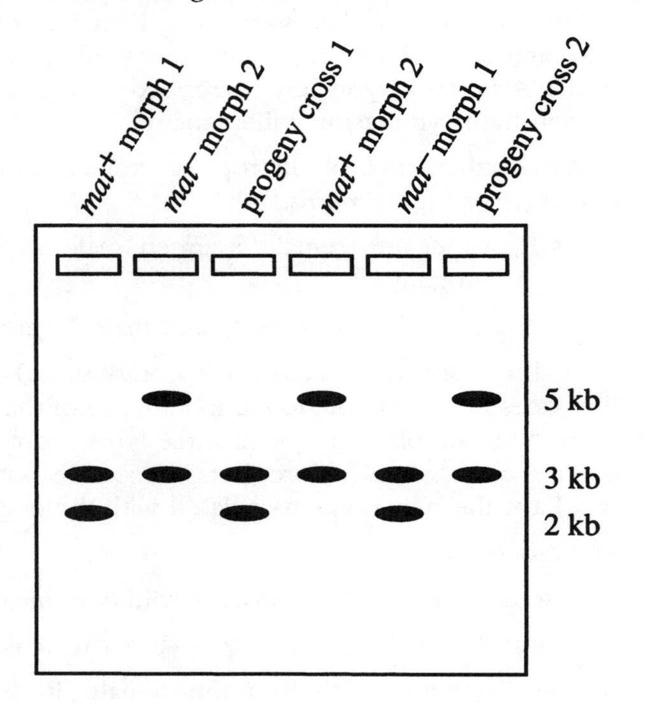

9. Both yeast parents contribute mitochondria to the cytoplasm of the resulting diploid cell. Recombination of mtDNA can take place in these heteroplasmons, but cytoplasmic segregation results in tetrads that contain either one of the parental mtDNA genotypes or one of the reciprocal recombinant types.

$oli^R\ cap^R$	or	$oli^S\ cap^S$	or	$oli^R\ cap^S$	or	$oli^S\ cap^R$
$oli^R\ cap^R$		$oli^S\ cap^S$		$oli^R\ cap^S$		$oli^S\ cap^R$
$oli^R\ cap^R$		$oli^S\ cap^S$		$oli^R\ cap^S$		$oli^S\ cap^R$
$oli^R\ cap^R$		$oli^S\ cap^S$		$oli^R\ cap^S$		$oli^S\ cap^R$

10. Both crosses show maternal inheritance of a chloroplast gene. Variegation must result from a mixture of normal and prazinizan chloroplasts. The variegated phenotype may be due to a rare (or minor) male contribution of cpDNA to the zygote.

11. If the mutation is in the chloroplast, reciprocal crosses will give different results, whereas if it is a dominant nuclear mutation, reciprocal crosses will give the same results.

12. This pattern is observed when a maternal recessive nuclear gene determines phenotype (called maternal effect). The crosses are

P *d/d* dwarf female × *D/D* normal male

F_1 *D/d* dwarf (all dwarf because mother is *d/d*)

F_2 $^3/_4$ *D/–*:$^1/_4$ *d/d* normal (all normal because mother is *D/d*)

F_3 $^3/_4$ normal (mother is *D/–*):$^1/_4$ dwarf (mother is *d/d*)

13. After the initial hybridization, a series of backcrosses using pollen from B will result in the desired combination of cytoplasm A and nucleus B. With each cross the female contributes all the cytoplasm and half the nuclear contents, whereas the male contributes half the nuclear contents.

14. **a.** Maternal inheritance is suggested.

b. The net result was that a line carrying the nuclear genes of *E. hirsutum* contained the cytoplasm of *E. luteum.* This demonstrated that the very tall progeny were not the result of a nuclear gene.

15. Let male sterility be symbolized by MS.

a. A line that was homozygous for *Rf* and contained the male-sterility factor would result in fertile males. When this line was crossed for two generations with females from a line not carrying the restorer gene, the male-sterility trait would reappear.

b. The F_1 would carry the male-sterility factor and would be heterozygous for the *Rf* gene. Therefore, it would be fertile.

c. The cross is *Rf/rf* MS × *rf/rf* (no cytoplasmic transmission). The progeny would be $^1/_2$ *Rf/rf* MS (fertile) and $^1/_2$ *rf/rf* MS (sterile).

d. (i)

P	*Rf-1/rf-1* ; *Rf-2/rf-2* × *rf-1/rf-1* ; *rf-2/rf-2* MS (female)	
F_1	$^1/_4$ *Rf-1/rf-1* ; *Rf-2/rf-2* MS	fertile
	$^1/_4$ *Rf-1/rf-1* ; *rf-2/rf-2* MS	fertile
	$^1/_4$ *rf-1/rf-1* ; *Rf-2/rf-2* MS	fertile
	$^1/_4$ *rf-1/rf-1* ; *rf-2/rf-2* MS	male sterile

(ii)

P	*Rf-1/Rf-1* ; *rf-2/rf-2* × *rf-1/rf-1* ; *rf-2/rf-2* MS (female)	
F_1	100% *Rf-1/rf-1* ; *rf-2/rf-2* MS	fertile

(iii)

P	*Rf-1/rf-1* ; *rf-2/rf-2* × *rf-1/rf-1* ; *rf-2/rf-2* MS (female)	
F_1	$^1/_2$ *Rf-1/rf-1* ; *rf-2/rf-2* MS	fertile
	$^1/_2$ *rf-1/rf-1* ; *rf-2/rf-2* MS	male sterile

(iv)

P	*Rf-1/rf-1* ; *Rf-2/Rf-2* × *rf-1/rf-1* ; *rf-2/rf-2* MS (female)	
F_1	$^1/_2$ *Rf-1/rf-1* ; *Rf-2/rf-2* MS	fertile
	$^1/_2$ *rf-1/rf-1* ; *Rf-2/rf-2* MS	fertile

16. The suggestion is that the mutants are heteroplasmons and contain both wild-type (*sm-s*) and mutant (*sm-r*) organelle DNA. Subsequent mitotic divisions can lead to cytoplasmic segregation such that daughter cells contain only one organelle DNA type.

17. Because nuclei do not combine in the heterokaryon, isolation of $PABA^-$ progeny that are red indicates that the red phenotype is inherited cytoplasmically.

18. **a., b., and c.** Notice that the results are not reciprocal, which indicates an extranuclear gene. Notice also that in the first cross, there is a 1:1 ratio of phenotypes, indicating a nuclear gene. This may be a case where the same function has a different location in different species. The two genes are incompatible in the hybrid. Let *cy* = cytoplasmic factor in *N. sitophila*, A^c = the nuclear allele in *N. crassa*, and *A* = the nuclear allele in *N. sitophila*. Aconidial is then A^c *cy*. The crosses are

$A\ cy \times A^c \rightarrow {}^1/_2\ A\ cy$ (normal):$^1/_2\ A^c\ cy$ (aconidial)

$A^c \times A$ (no *cy* contribution from male) $\rightarrow$ all normal (*A* and A^c)

Neither parent was aconidial, because the sporeless phenotype requires the interaction of a nuclear allele A^c from one species with a cytoplasmic factor (*cy*) from the other species.

19. Let poky be symbolized by *c*. Let the nuclear suppressor of poky be symbolized by *n*. To do these problems, you cannot simply do the crosses in sequence. For instance, the parental genotypes in cross (a) must be written taking cross (c) into consideration.

Cross	Progeny	Phenotype
a. $(c^+)\ n^+ \times (c)\ n^+$	all $(c^+)\ n^+$	all nonpoky
b. $(c^+)\ n \times (c)\ n^+$	$^1/_2\ (c^+)\ n$:$^1/_2\ (c^+)\ n^+$	all nonpoky
c. $(c)\ n^+ \times (c^+)\ n^+$	all $(c)\ n^+$	all poky
d. $(c)\ n^+ \times (c^+)\ n$	$^1/_2\ (c)\ n^+$ (= D)	all poky
	$^1/_2\ (c)\ n$ (= E)	all nonpoky
e. $(c)\ n \times (c^+)\ n$	all $(c)\ n$	all nonpoky
f. $(c)\ n \times (c^+)\ n^+$	$^1/_2\ (c)\ n$	all nonpoky
	$^1/_2\ (c)\ n^+$	all poky

20. ***Unpacking the Problem***

1. mtDNA is mitochondrial DNA.
2. 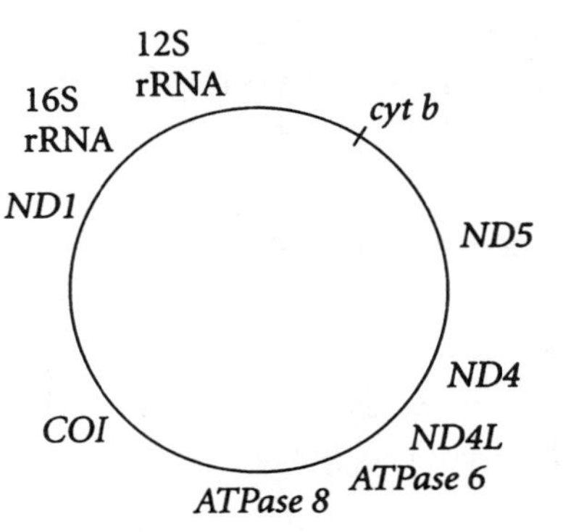

3. A cytoplasmic petite is a small, mutant yeast colony that uses fermentation to grow rather than oxidative phosphorylation because of defective mitochondrial DNA. It is "cytoplasmic" because the defect is in the cytoplasmically located mitochondria rather than in a nuclear gene, and it is "petite" because it grows slowly and, in comparison with wild type, produces smaller colonies per unit time.
4. At the mitochondrial level the mtDNA either is completely missing or contains a large deletion.
5. The wild-type DNA of mitochondria is circular. Petite DNA is represented as an arc to emphasize that, although circular DNA is found within the mitochondria of petites, most of the information has been lost owing to deletion and that the DNA present consists of tandem repeats of a small segment of the wild-type sequence.
6. Compare restriction fragments of a petite with wild-type DNA.

7. An *mit*$^-$ mutant is equivalent to a point mutation. Like petites, the *mit*$^-$ mutant grows by fermentation, producing small colonies. Unlike petites, they can form true revertants.

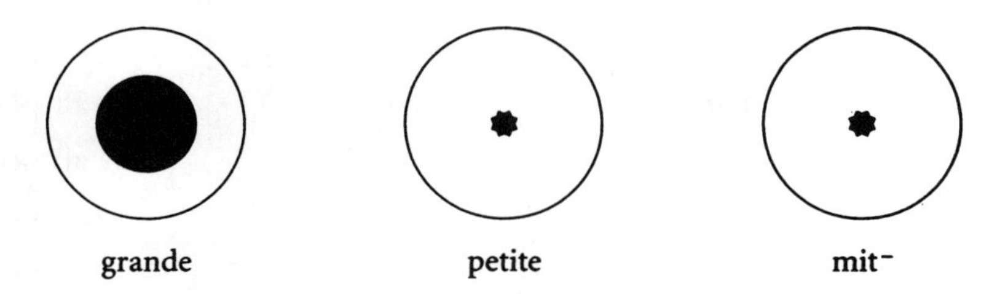

8. *mit*$^-$ strains are not drug-resistant. Drug resistance is frequently due to alterations within ribosomal components, but these changes must maintain function. *mit*$^-$ strains do not have functional mitochondria, so ribosome sensitivity or resistance is not relevant.
9. A heteroplasmon is also called a *cytohet* (meaning "cytoplasmic heterozygote"). The terms denote the fact that the cytoplasm contains mitochondria of two genotypes.
10. In this problem, heteroplasmons were made by fusing (mating) *mit*$^-$ and petite strains. Separate auxotrophic markers for each strain would be useful to force the formation and selection of prototrophic heteroplasmons.
11. Fused indicates that the two cells have joined membranes to form one cell. In yeast, this is easily accomplished through mating two haploid cells to form a diploid (heteroplasmon) cell.
12. The yeast cells divide (bud), and colonies eventually form on growth medium.
13.

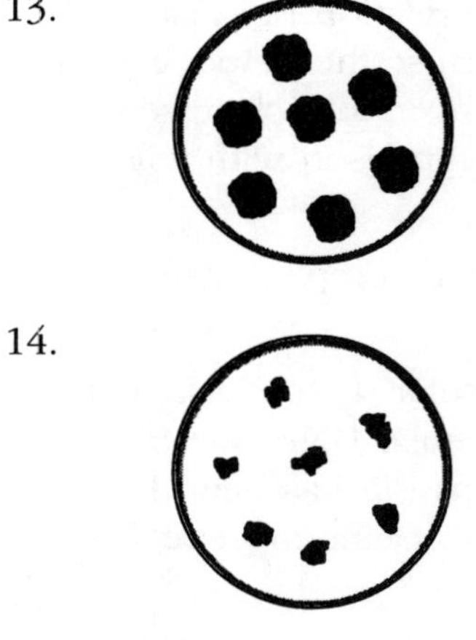

14.

15. The "+" indicates grande growth, and the "–" indicates petite growth. Recombination produces wild-type mtDNA, and this allows grande growth.
16. No; 1 and 10 are the same.
17. No; petites E and F show the same behavior with the *mit*$^-$ mutants.

18. If two petites can produce the same results with the same mit^- mutant, then the two have retained or lost overlapping regions. If two mit^- mutants show the same behavior in these crosses, they have mutations in the same region of the mitochondrial chromosome.

Solution to the Problem

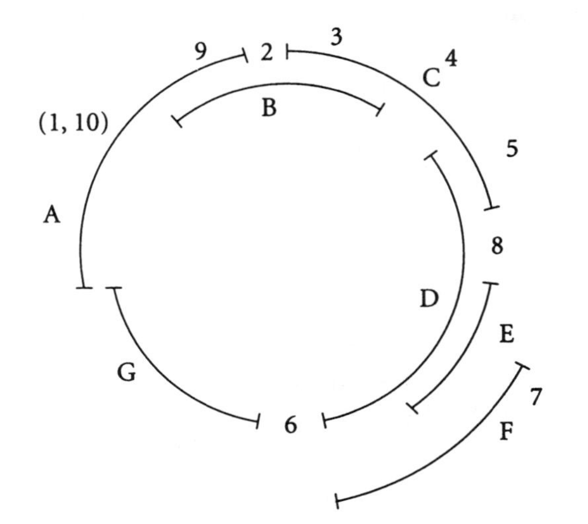

21. **a. and b.** Each meiosis shows uniparental inheritance of the antibiotic resistance, suggesting cytoplasmic inheritance for this trait. Mating type shows 2:2 segregation typical of a nuclear gene.

c. Because ant^R is probably mitochondrial and because petites have been shown to result from deletions in the mitochondrial genome, ant^R may be lost in some petites.

22. **a.** The first tetrad shows a 2:2 pattern, indicating a nuclear gene, whereas the second tetrad shows a 0:4 pattern, indicating maternal inheritance and a cytoplasmic factor (mitochondrial).

b. The nuclear gene should always show a 1:1 segregation pattern. Segregation of the mitochondrial gene would also produce an ascus that is all $cyt2^+$.

c. Both produce proteins involved with the electron-transport chain. The products of the two genes might be required at different steps of this chain, or the two proteins might combine to form a functional complex.

23. Both yeast parents contribute mitochondria to the cytoplasm of the resulting diploid cell. Recombination of mtDNA can take place in these heteroplasmons, but cytoplasmic segregation results in tetrads that contain either one of the parental mtDNA genotypes or one of the reciprocal recombinant types. In this case, some tetrads will show only strain-1 type, some will show only strain-2 type, and some will be recombinant. Because these restriction sites are polymorphic, recombination will result in novel-sized fragments. For each, tetrads will be all of a single pattern.

24. a. No; during diploid budding all the progeny receive one type of mtDNA owing to cytoplasmic segregation. Here, all progeny cells receive both types of 2-μm DNA.

b. One *Eco*RI site in a circular molecule will produce one linear piece of DNA that migrates to the 2-μm position. With two *Eco*RI sites, two pieces will be produced, and their sum will be 2-μm. Therefore, ascospores from the diploid buds will show three bands on a Southern blot.

2 μm	—	Band 1
< 2 μm	—	Band 2
< 2 μm	—	Band 3

Bands 2 and 3 total the size of band 1.

25. Because there are two fragments generated with both restriction enzymes, each must cut the circular plasmid twice. For both, the mt rRNA is on the larger of the two resulting fragments. Because A results in one larger and one smaller fragment compared with B, the A sites are closer to each other than the B sites.

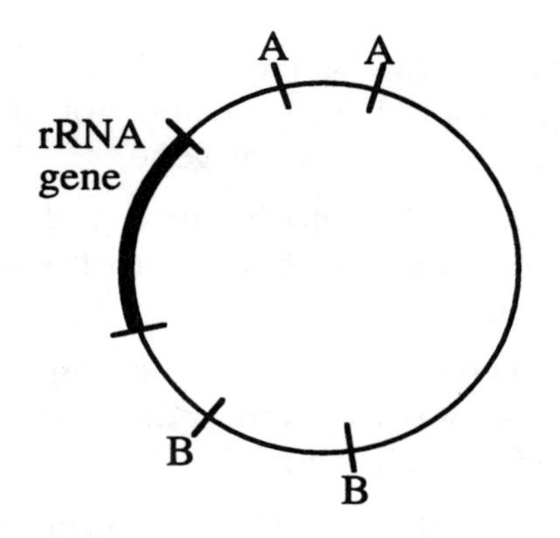

26. First, prepare a restriction map of the mtDNA using various restriction enzymes. Using the assumption of evolutionary conservation, on a Southern blot hybridize known probes from yeast or another organism in which the genes have already been identified.

27. a. Progeny plants inherited only normal cpDNA (lane 1), only mutant cpDNA (lane 2), or both (lane 3). In order to get homozygous cpDNA, seen in lanes 1 and 2, segregation of chloroplasts had to occur.

b. Based on other systems (e.g., yeast) both *Gryllus* and *Drosophila* would be expected to show segregation of mitochondria. This is not observed in these rare females. Therefore, it must be hypothesized that their mitochondria all contain two genomes, one of which carries the mutation and one of which is normal. Segregation of mitochondria may not occur in these species.

28. The two repeats, center to center, are separated by 83 kb. If the two direct repeats lined up as below and then experienced a recombination event, the two smaller circles would be 83 kb and 135 kb, each containing a single copy of the direct repeat:

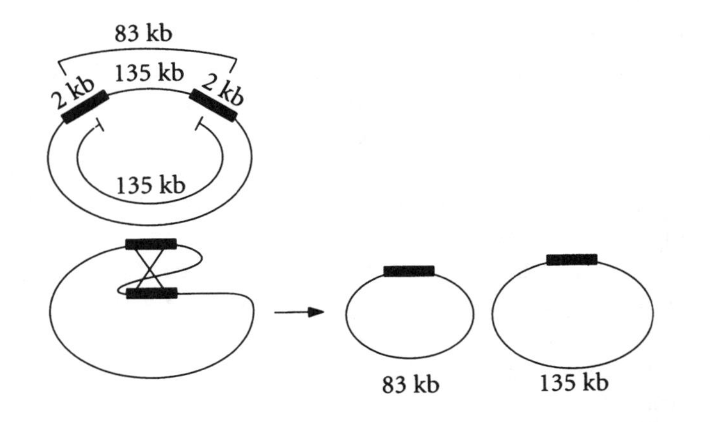

29. a. Both the gametophyte and the sporophyte are closer in shape to the mother than the father. Note that a size increase occurs in each type of cross.

b. Gametophyte and sporophyte morphology are affected by extranuclear factors. Leaf size may be a function of the interplay between nuclear genome contributions.

c. If extranuclear factors are affecting morphology whereas nuclear factors are affecting leaf size, then repeated backcrosses could be conducted, using the hybrid as the female. This would result in the cytoplasmic information's remaining constant whereas the nuclear information becomes increasingly like that of the backcross parent. Leaf morphology should therefore remain constant, whereas leaf size would decrease toward the size of the backcross parent.

30. a. The complete absence of male offspring is the unusual aspect of this pedigree. In addition, all progeny that mate carry the trait for lack of male offspring. If the male lethality factor were nuclear, the male parent would be expected to alter this pattern. Therefore, cytoplasmic inheritance is suggested.

b. If all females resulted from chance alone, then the probability of this result is $(^1/_2)^n$, where n = the number of female births. In this case n is 72. Chance is an unlikely explanation for the observations.

The observations can be explained by cytoplasmic factors by assuming that the proposed mutation in mitochondria is lethal only in males.

Mendelian inheritance cannot explain the observations, because all the fathers would have had to carry the male-lethal mutation in order to observe such a pattern. This would be highly unlikely.

31. a. The pedigree clearly shows maternal inheritance.

b. Most likely, the mutant DNA is mitochondrial.

32. Recall that each cell has many mitochondria, each with numerous genomes. Also recall that cytoplasmic segregation is routinely found in mitochondrial mixtures within the same cell.

The best explanation for this pedigree is that the mother in generation I experienced a mutation in a single cell that was a progenitor of her egg cells (primordial germ cell). By chance alone, the two males with the disorder in the second generation were from egg cells that had experienced a great deal of cytoplasmic segregation prior to fertilization, whereas the two females in that generation received a mixture.

The spontaneous abortions that occurred for the first woman in generation II were the result of extensive cytoplasmic segregation in her primordial germ cells: aberrant mitochondria were retained. The spontaneous abortions of the second woman in generation II also came from such cells. The normal children of this woman were the result of extensive segregation in the opposite direction: normal mitochondria were retained. The affected children of this woman were from egg cells that had undergone less cytoplasmic segregation by the time of fertilization so that they developed to term but still suffered from the disease.

22 Regulation of Cell Number: Normal and Cancer Cells

1. **a.** Cyclins bind to, activate, and direct CDKs to phosphorylate specific cellular targets and by doing so control the cell cycle. Normal cell division requires the sequential production and then removal of different cyclins. The activity of the cyclin-CDK heterodimer is also regulated through p21. Overproduction of one of the cyclins could disrupt the orderly process of cell division, but it would be limited by the amount of CDK present as well as the state of the p21 "brake."

   b. A nonsense mutation would lead to a decrease of the normal protein product. If that protein were part of the receptor for a growth factor, which stimulates cell proliferation, then cell division could not be triggered. This would likely be recessive and would slow cell proliferation, not accelerate it.

   c. Overproduction of FasL will signal adjacent cells through their Fas cell-surface receptors, which, in turn, leads to Apaf activation. This, in turn, causes proteolysis and activation of the initiator caspase, ultimately leading to apoptosis of the cell. Although this would be dominant, it would lead to excess cell death, not proliferation.

   d. Cytoplasmic tyrosine-specific protein kinase phosphorylates proteins in response to signals received by the cell. These phosphorylations lead to activation of the transcription factors for the next step in the cell cycle. If the active site is disrupted, then phosphorylations will not occur and transcription factors for the next step will not be activated. This would likely be recessive and would slow cell proliferation, not accelerate it.

e. If the enhancer causes large amounts of the apoptosis inhibitor to be expressed in the liver, the normal pathway of cell death will be blocked. These liver cells (the enhancer is liver-specific) will have an unusually long lifetime in which to accumulate various mutations that could lead to cancer. This chromosomal rearrangement would be dominant.

2. Once apoptosis is initiated, a self-destruct switch has been thrown: endonucleases and proteases are released, DNA is fragmented, and organelles are disrupted. This obviously is a "terminal" state from which the cell will not have a need or chance to reuse the machinery of destruction. On the other hand, the various proteins needed for the regulation and execution of the cell cycle will be needed again if the cell continues to divide. By recycling many of these, the cell obviously conserves its resources (proteins are energetically expensive to make) and recycling also allows for more rapid divisions because the cell does not have to spend time remaking all the pieces.

3. **a.** This would be dominant. The misexpression of FasL from one allele would be dominant to the normal expression of the wild-type FasL allele. In this case, each liver cell would signal its neighboring cells to undergo apoptosis.

b. No. It would lead to excess cell death, not proliferation.

4. (1) Certain cancers are inherited as highly penetrant simple Mendelian traits.

(2) Most carcinogenic agents are also mutagenic.

(3) Various oncogenes have been isolated from tumor viruses.

(4) A number of genes that lead to the susceptibility to particular types of cancer have been mapped, isolated, and studied.

(5) Dominant oncogenes have been isolated from tumor cells.

(6) Certain cancers are highly correlated to specific chromosomal rearrangements.

5. Normal Ras is a G-protein that activates a protein kinase, which in turn phosphorylates a transcription factor. If it were simply deleted, no cancer could develop, because cell division would not occur. If it were simply duplicated, an excess of the G-protein could not cause cancer, because it must be activated before it can activate the protein kinase, and presumably the enzyme that activates normal Ras is closely regulated and would not activate too many copies. However, if it were to have a point mutation, it might now bind GTP, even in the absence of normal control signals, and be in a state of permanent activation. As a positive regulator of cell growth, this mutant Ras would continually promote cell proliferation.

In contrast, normal *c-myc* is a transcription factor. If the gene were to be duplicated, too much transcription factor could lead to malignancy.

6. Inhibition of apoptosis can contribute to tumor formation by allowing cells to have an unusually long lifetime in which to accumulate various mutations that lead to cancer. Also, the normal role of apoptosis in removing abnormal cells and, through p53, killing cells that have "damaged" DNA would also be inhibited.

7. **a.** Mutations in a tumor suppressor gene are recessive and due to loss of function. That function can be restored by the introduction of a wild-type allele.

b. Mutations in an oncogene are dominant and due to gain of function (overexpression or misexpression). The normal function will not inhibit these mutants, and the introduced gene would be ineffective in restoring the normal phenotype.

8. **a.** Type A diabetes is most likely due to a defect in the pancreas. The pancreas normally makes insulin, and type A diabetes can be treated by supplying insulin. Type B diabetes is most likely due to a target cell defect because type B is unresponsive to exogenous insulin.

b. Type B diabetes appears to be caused by a defect in the target cell. A number of genes are responsible for the receptor and the subsequent cascade of changes that occur in leading to a change in transcription. Any of these genes could have a mutant form.

9. p53 detects and is activated by DNA damage. When activated, p53 activates p21, an inhibitor of the cyclin-CDK complex necessary for the progression of the cell cycle. If the DNA damage is repairable, this system will eventually deactivate p53 and allow cell division. However, if the damage is irreparable, p53 would stay active and would activate the apoptosis pathway, ultimately leading to cell death. It is for this reason that the "loss" of p53 is often associated with cancer.

23 The Genetic Basis of Development

1. Sex determination in *Drosophila* is autonomous at the cellular level. The *Sxl* gene is permanently turned on or remains off early in development in response to the balance of X (NUM):A (DEM) proteins. Because the X:A ratio in these flies is intermediate (0.67, not 1.0 or 0.5), the stochastic interaction of the NUM and DEM gene products at these intermediate levels will result in some cells developing as female and other cells developing as male.

2. Normally, the *tra* gene in the female is active, whereas in the male it is not active. The active *tra* form of the gene product results in a change in the *dsx* product, shifting development toward the female. If the *dsx* product is not altered, development proceeds along the male line. A mutation in the gene that results in chromosomal females developing as phenotypic, but sterile, males must involve an inactive *tra* product. XX "males" homozygous for the *tra* mutation could be transplanted with wild-type male germ cells (pole cells) very early in development, which should result in normal gonad development.

3. In humans, a single copy of the Y chromosome is sufficient to shift development toward normal male phenotype. The extra copy of the X chromosome is simply inactivated. Both mechanisms seem to be all-or-none rather than to be based on concentration levels.

4. Because maleness is based on the presence of androgens produced by the developing testes and femaleness is based on the absence of those androgens, what seems to be crucial here is whether the migrating germ cells organize testes. Although what determines this is unknown, it may be that a minimal

number of XY cells are required to organize a testis. If, in the mosaic, not enough of these cells exist, then development will be female. If a sufficient number exist, development will be male.

5. The concentration of Sxl is crucial for female development and dispensable for male development. The dominant Sxl^M male-lethal mutations may not actually kill all males but simply produce an excessive amount of gene product so that only females (fertile XX and XY) result. The reversions may eliminate all gene product, resulting in XX (sterile) and XY males. The reversions would be recessive because, presumably, a single normal copy of the gene may produce enough gene product to "toggle the switch" in development to female.

6. **a.** There must be a diffusible substance produced by the anchor cell that affects development of the six cells. The cell that receives the most signal or has the strongest response to the substance is 1°, and the 3° represents a lack of response due to a low concentration or absence of the diffusible substance.

b. Remove the anchor cell and the six equivalent cells. Arrange the six cells in a circle around the anchor cell. All six cells will develop the same phenotype, which will depend on the distance from the anchor cell.

7. **a.** The results suggest that ABa and ABp are not determined at this point in their development. Also, future determination and differentiation of these cells are dependent on their position within the developing organism.

b. Because an absence of EMS cells leads to a lack of determination and differentiation of AB cells, the EMS cells must be at least in part responsible for AB-cell development, either through direct contact or by the production of a diffusible substance.

c. Most descendants of the AB cells do not become muscle cells when P2 is present; all descendants of the AB cells become muscle cells when P2 is absent. Therefore, P2 must prevent some AB descendants from becoming muscle cells.

8. Because the receptor is defective, testosterone cannot signal the cell and initiate the cascade of developmental changes that will switch the embryo from the "default" female development to male development. Therefore, the phenotype will be female.

9. A number of experiments could be devised. A comparison of amino acid sequence between mammalian gene products and insect gene products would indicate which genes are most similar to one another. Using cloned cDNA sequences from mammalian genes for hybridization to insect DNA would also indicate which genes are most similar to one another.

10. **a.** The anterior 20% of the embryo is normally devoted to the head and thorax regions. The bicaudal phenotype results in the loss of these regions and in the loss of Al through A3. The gap proteins are responsible for the induction of the pair-rule proteins, which ultimately set the number of segments, and the homeotic proteins, which set the identity of the segments. Obviously, the gap proteins are improperly regulated to produce the bicaudal phenotype.

Normal regulation of the gap proteins is accomplished by differential sensitivity to the differing concentrations of the maternally derived morphogens. Because the anterior portion of the embryo has been removed, high concentrations of the morphogens in these regions have also been removed. This results in the abnormal segment number and identity that are observed.

b. Wild-type NOS protein suppresses the translation and is responsible for the observed gradient of HB-M. In the absence of NOS function, the normal shallow A to P gradient of HB-M is lost. Because the gap genes respond differentially to the BCD:HB-M ratio, abnormal regulation of the gap genes occurs, which leads to reduced segmentation. This results simply in a broader head and thorax, and no mirror-image phenotype is possible.

11. If you diagram these results, you will see that deletion of a gene that functions posteriorly allows the next-most-anterior segments to extend in a posterior direction. Deletion of an anterior gene does not allow extension of the next-most-posterior segment in an anterior direction. The gap genes activate *Ubx* in both thoracic and abdominal segments, whereas the *abd-A* and *Abd-B* genes are activated only in the middle and posterior abdominal segments. The functioning of the *abd-A* and *Abd-B* genes in those segments somehow prevents *Ubx* expression. However, if the *abd-A* and *Abd-B* genes are deleted, *Ubx* can be expressed in these regions.

12. Proper *ftz* expression requires *Kr* in the fourth and fifth segments and *kni* in the fifth and sixth segments.

13. It may be that the wild-type allele in the embryo produces a gene product that can inhibit the gene product of the rescuable maternal-effect lethal mutations, whereas the nonrescuable maternal-effect lethal mutations produce a product that cannot be inhibited.

Alternatively, the nonrescuable maternal-effect lethal mutations may produce a product that is required very early in development, before the developing fly is producing any proteins, whereas the rescuable maternal-effect lethal mutations may act later in development when embryo protein production can compensate for the maternal mutation.

14. a. The determination of anterior-posterior portions of the embryo is governed by a concentration gradient of *bcd*. The concentration is highest in the anterior region and lowest in the posterior region. The furrow develops at a critical concentration of *bcd*. As bcd^+ gene dosage (and, therefore, BCD concentration) decreases, the furrow shifts anteriorly; as the gene dosage increases, the furrow shifts posteriorly.

b.

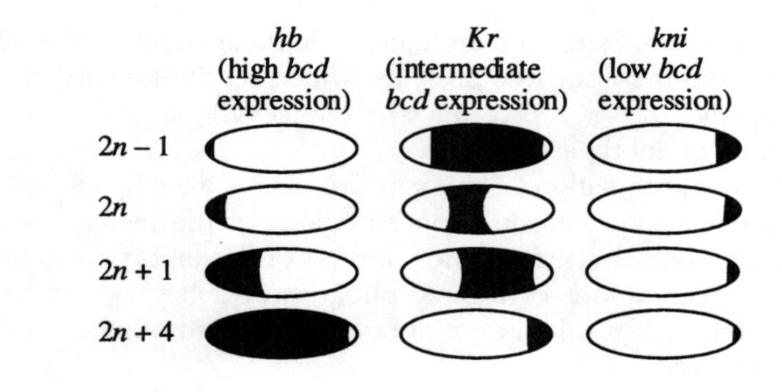

15. The anterior-posterior axis would be reversed.

24 POPULATION GENETICS

1. The frequency of an allele in a population can be altered by natural selection, mutation, migration, nonrandom mating, and genetic drift (sampling errors).

2. There are a total of (2)(384) + (2)(210) + (2)(260) = 1708 alleles in the population. Of those, (2)(384) + 210 = 978 are *A1* and 210 + (2)(260) = 730 are *A2*. The frequency of *A1* is 978/1708 = 0.57, and the frequency of *A2* is 730/1708 = 0.43.

3. The given data are $q^2 = 0.04$ and $p^2 + 2pq = 0.96$. Assuming Hardy-Weinberg equilibrium, if $q^2 = 0.04$, $q = 0.2$ and $p = 0.8$. The frequency of *B/B* is $p^2 = 0.64$, and the frequency of *B/b* is $2pq = 0.32$.

4. This problem assumes that there is no backward mutation. Use the following equation (from Box 24-3 in the text).

$$p_n = p_o e^{-n\mu}$$

That is, $p_{50,000} = (0.8)e^{-(5 \times 10^4)(4 \times 10^{-6})} = (0.8)e^{-0.2} = 0.65$

5. **a.** If the variants represent different alleles of gene *X*, a cross between any two variants should result in a 1:1 progeny ratio (because the organism is haploid). All the variants should map to the same locus. Amino acid sequencing of the variants should reveal differences of one to just a few amino acids.

   b. There could be another gene (gene *Y*), with five variants, that modifies the gene *X* product post-transcriptionally. If so, the easiest way to distinguish between the two explanations would be to find another mutation in *X* and do a dihybrid cross. For example, if there is independent assortment,

P X^1 ; Y^1 × X^2 ; Y^2

F_1 1 X^1 ; Y^1:1 X^1 ; Y^2:1 X^2 ; Y^1:1 X^2 ; Y^2

If the new mutation in X led to no enzyme activity, the ratio would be

2 no activity:1 variant one activity:1 variant two activity

The same mutant in a one-gene situation would yield 1 active:1 inactive.

6. **a.** If the population is in equilibrium, $p^2 + 2pq + q^2 = 1$. Calculate the actual frequencies of p and q in the population and compare their genotypic distribution to the predicted values. For this population:

$p = [406 + {}^1/_2(744)]/1482 = 0.52$

$q = [332 + {}^1/_2(744)]/1482 = 0.48$

The genotypes should be distributed as follows if the population is in equilibrium:

$L^M/L^M = p^2(1482) = 401$ Actual: 406

$L^M/L^N = 2pq(1482) = 740$ Actual: 744

$L^N/L^N = q^2(1482) = 341$ Actual: 332

This compares well with the actual data, so the population is in equilibrium.

b. If mating is random with respect to blood type, then the following frequency of matings should occur:

$L^M/L^M \times L^M/L^M = (p^2)(p^2)(741) = 54$ Actual: 58

$L^M/L^M \times L^M/L^N$ or $L^M/L^N \times L^M/L^M = (2)(p^2)(2pq)(741) = 200$ Actual: 202

$L^M/L^N \times L^M/L^N = (2pq)(2pq)(741) = 185$ Actual: 190

$L^M/L^M \times L^N/L^N$ or $L^N/L^N \times L^M/L^M = (2)(p^2)(q^2)(741) = 92$ Actual: 88

$L^M/L^N \times L^N/L^N$ or $L^N/L^N \times L^M/L^N = 2(2pq)(q^2)(741) = 170$ Actual: 162

$L^N/L^N \times L^N/L^N = (q^2)(q^2)(741) = 39$ Actual: 41

Again, this compares nicely with the actual data, so the mating is random with respect to blood type.

7. **a. and b.** For each, p and q must be calculated and then compared with the predicted genotypic frequencies of $p^2 + 2pq + q^2 = 1$.

Population	p	q	*Equilibrium?*
1	1.0	0.0	Yes
2	0.5	0.5	No
3	0.0	1.0	Yes
4	0.625	0.375	No
5	0.375	0.625	No
6	0.5	0.5	Yes
7	0.5	0.5	No
8	0.2	0.8	Yes
9	0.8	0.2	Yes
10	0.993	0.007	Yes

c. The formulas to use are $q^2 = \mu/_s$ and $s = 1-W$ (from *Genetics in Process* 17-8 in the text).

$$4.9 \times 10^{-5} = 5 \times 10^{-6}/s\ ;\quad s = 0.102, \text{ so } W = 0.898$$

d. For simplicity, assume that the differences in survivorship occur prior to reproduction. Thus, each genotype's fitness can be used to determine the relative percentage each contributes to the next generation.

Genotype	Frequency	Fitness	Contribution	A	a
A/A	0.25	1.0	0.25	0.25	0.0
A/a	0.50	0.8	0.40	0.20	0.20
a/a	0.25	0.6	0.15	0.0	0.15
				0.45	0.35

$$p' = 0.45/(0.45 + 0.35) = 0.56$$

$$q' = 0.35/(0.45 + 0.35) = 0.44$$

Alternatively, the formulas to use are

$$p' = p\frac{pW_{AA} + qW_{Aa}}{\overline{W}}$$

$$\overline{W} = p^2W_{AA} + 2pqW_{Aa} + q^2W_{aa}$$

$$p' = (0.5)[(0.5)(1.0) + (0.5)(0.8)]/[(0.25)(1.0) + (0.5)(0.8) + (0.25)(0.6)]$$

$$= (0.5)(0.9)/(0.8) = 0.56$$

8. **a.** Assuming the population is in Hardy-Weinberg equilibrium and that the allelic frequency is the same in both sexes, we can directly calculate the frequency of the colorblindness allele as $q = 0.1$. (Because this trait is sex-linked, q is equal to the frequency of affected males.) Colorblind females must be homozygous for this X-linked recessive trait, so their frequency in the population is equal to $q^2 = 0.01$.

b. There would be 10 colorblind men for every colorblind woman (q/q^2).

c. For this condition to be true, the mothers must be heterozygous for the trait and the fathers must be colorblind ($X^C/X^c \times X^c/Y$). The frequency of heterozygous women in the population will be $2pq$, and the frequency of colorblind men will be q. Therefore, the frequency of such random marriages will be $(2pq)(q) = 0.018$.

d. All children will be phenotypically normal only if the mother is homozygous for the noncolorblind allele ($p^2 = 0.81$). The father's genotype does not matter and therefore can be ignored.

e. There are several ways of approaching this problem. One way to visualize the data, however, is to construct the following

Mother \ Father	0.4 X^C	0.6 X^c	Y
0.8 X^C	0.32 X^C/X^C	0.48 X^C/X^c	0.8 X^C/Y
0.2 X^c	0.08 X^C/X^c	0.12 X^c/X^c	0.2 X^c/Y

As can be seen, the frequency of colorblind females will be 0.12 and colorblind males 0.2.

f. From analysis of the results in (e), the frequency of the colorblind allele will be 0.2 in males (the same as in the females of the previous generation) and $^1/_2(0.08 + 0.48) + 0.12 = 0.4$ in females.

9. The frequency of a phenotype in a population is a function of the frequency of alleles that lead to that phenotype in the population. To determine dominance and recessiveness, do standard Mendelian crosses.

10. Assume that proper function results from the right gene products in the proper ratio to all other gene products. A mutation will change the gene product, eliminate the gene product, or change the ratio of it to all other gene products. All three outcomes upset a previously balanced system. While a new and "better" balance may be achieved, this is less likely than being its deleterious.

11. Wild-type alleles are usually dominant because most mutations result in lowered or eliminated function. To be dominant, the heterozygote has approximately the same phenotype as the dominant homozygote. This will typically be true when the wild-type allele produces a product and the mutant allele does not.

Chromosomal rearrangements are often dominant mutations because they can cause gross changes in gene regulation or even cause fusions of several gene products. Novel activities, overproduction of gene products, etc., are typical of dominant mutations.

12. Prior to migration, $q^A = 0.1$ and $q^B = 0.3$ in the two populations. Because the two populations are equal in number, immediately after migration, $q^{A+B} = {}^1/_2(q^A + q^B) = {}^1/_2(0.1 + 0.3) = 0.2$. At the new equilibrium, the frequency of affected males is $q = 0.2$, and the frequency of affected females is $q^2 = (0.2)^2 = 0.04$. (Colorblindness is an X-linked trait.)

13. For a population in equilibrium, the probability of individuals' being homozygous for a recessive allele is q^2. Thus for small values of q, few individuals in a randomly mating population will express the trait. However, if two individuals share a close common ancestor, there is an increased chance of homozygosity by descent, because only one "progenitor" need be heterozygous.

For the following, it is assumed that the allele in question is rare. Thus the chance of both "progenitors" being heterozygous will be ignored.

a. For a parent-sib mating, the pedigree can be represented as follows

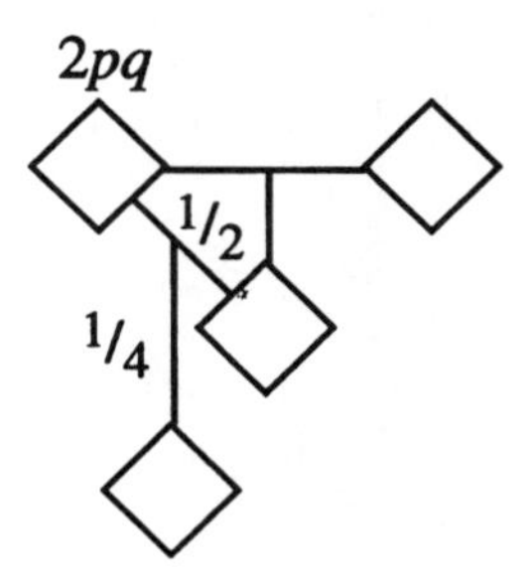

In this example, it is only the chance of the incestuous parent's being heterozygous that matters. Thus, the chance of the descendant's being homozygous is

$$2pq(^1/_2)(^1/_4) = {}^{pq}/_4$$

If q is very small, then p is nearly 1.0 and the chance of an affected child can be represented as approximately $^q/_4$. (Again, this should be compared to the expected random-mating frequency of q^2.)

b. For a mating of first cousins, the pedigree can be represented as follows:

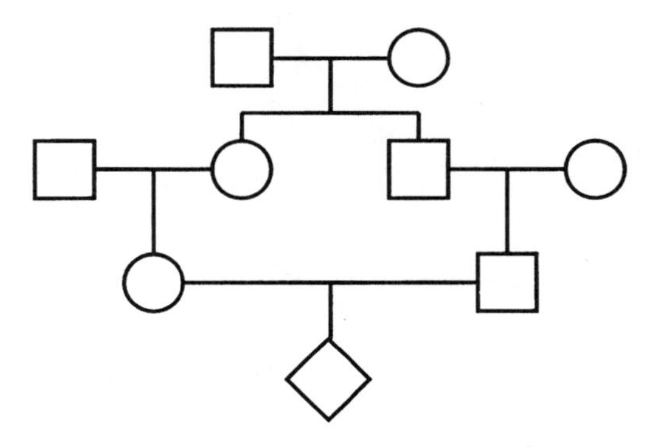

The probability of inheriting the recessive allele if *either* grandparent is heterozygous can be represented as follows:

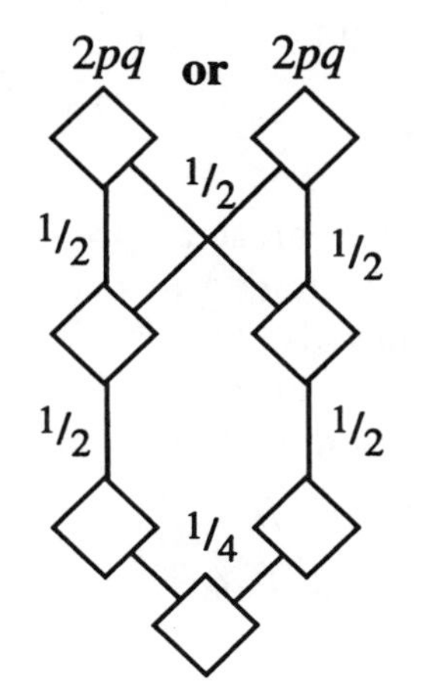

Thus the chance of this child's being affected is

$$2pq\,(^1/_2)(^1/_2)(^1/_2)(^1/_2)(^1/_4) + 2pq\,(^1/_2)(^1/_2)(^1/_2)(^1/_2)(^1/_4) = {}^{pq}/_{16}$$

Again, if q is rare, p is nearly 1.0, so the chance of homozygosity by descent is approximately $^q/_{16}$.

c. An aunt-nephew (or uncle-niece) mating can be represented as

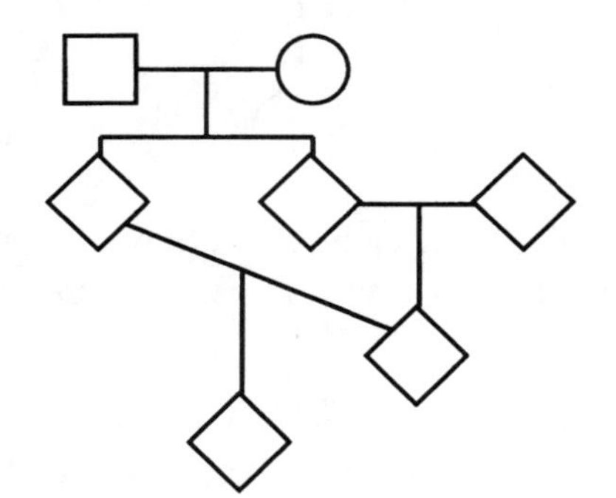

Following the possible inheritance of the recessive allele from either grandparent,

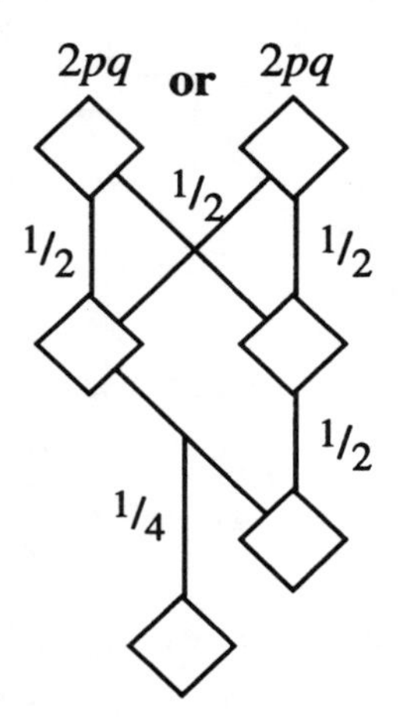

The chance of this child's being homozygous is

$2pq(^1/_2)(^1/_2)(^1/_2)(^1/_4) + 2pq(^1/_2)(^1/_2)(^1/_2)(^1/_4) = {}^{pq}/_8$, or for rare alleles approximately $^q/_8$.

14. Albinos appear to have had greater opportunity to mate. They may have been considered lucky and encouraged to breed at very high levels in comparison with nonalbinos. They may also have been encouraged to mate with each other. Alternatively, in the tribes with a very low frequency, albinos may have been considered very unlucky and destroyed at birth or prevented from marriage.

15. The allele frequencies are

$$f(A) = 0.2 + {}^1/_2(0.60) = 50\%$$

$$f(a) = {}^1/_2(0.60) + 0.2 = 50\%$$

Positive assortative mating: the alleles will randomly unite within the same phenotype. For $A/-$, the mating population is 0.2 A/A + 0.6 A/a. The allelic frequencies within this subpopulation are

$$f(A) = [0.2 + {}^1/_2(0.6)]/0.8 = 0.625$$
$$f(a) = {}^1/_2(0.6)/0.8 = 0.375$$

The phenotypic frequencies that result are

$A/-$: $p^2 + 2pq = (0.625)^2 + 2(0.625)(0.375) = 0.39 + 0.47 = 0.86$

a/a: $q^2 = (0.375)^2 = 0.14$

However, assuming that all contribute equally to the next generation and this subpopulation represents 0.8 of the total population, these figures must be adjusted to reflect this weighting:

$A/-$: $(0.86)(0.8) = 0.69$
a/a: $(0.14)(0.8) = 0.11$

The a/a contribution from the other subpopulation will remain unchanged because there is only one genotype, a/a. Their weighted contribution to the total phenotypic frequency is 0.20. Therefore, after one generation, the phenotypic frequencies will be $A/- = 0.69$ and $a/a = 0.20 + 0.11 = 0.31$, and the genotypic frequencies will be $f(A) = 0.5$ and $f(a) = 0.5$. Over time, these allelic frequencies will stay the same, but the frequency of heterozygotes will continue to decrease until there are two separate populations, A/A and a/a, which will not interbreed.

Negative assortative mating: mating is between unlike phenotypes. The two types of progeny will be A/a and a/a. A/A will not exist. A/a will result from all $A/A \times a/a$ matings and half the $A/a \times a/a$ matings. These matings will occur with the following relative frequencies

$A/A \times a/a = (0.2)(0.2) = 0.04$
$A/a \times a/a = (0.6)(0.2) = 0.12$

Because these are the only matings that will occur, they must be put on a 100 percent basis by dividing by the total frequency of matings that occur:

$A/A \times a/a$: $0.04/0.16 = 0.25$, all of which will be A/a
$A/a \times a/a$: $0.12/0.16 = 0.75$, half A/a and half a/a

The phenotypic frequencies in this generation will be

A/a: $0.25 + 0.75/2 = 0.625$

a/a: $0.75/2 = 0.375$

In the next generation, because all matings are now between heterozygotes and homozygous recessives, the final allelic frequencies of $f(A) = 0.25$ and $f(a) = 0.75$ will be obtained and the population will be 50% A/a and 50% a/a.

16. Many genes affect bristle number in *Drosophila*. The artificial selection resulted in lines with mostly high-bristle-number alleles. Some mutations may have occurred during the 20 generations of selective breeding, but most of the response was due to alleles present in the original population. Assortment and recombination generated lines with more high-bristle-number alleles.

Fixation of some alleles causing high bristle number would prevent complete reversal. Some high-bristle-number alleles would have no negative effects on fitness, so there would be no force pushing bristle number back down because of those loci.

The low fertility in the high-bristle-number line could have been due to pleiotropy or linkage. Some alleles that caused high bristle number may also have caused low fertility (pleiotropy). Chromosomes with high-bristle-number alleles may also carry alleles at different loci that caused low fertility (linkage). After artificial selection was relaxed, the low-fertility alleles would have been selected against through natural selection. A few generations of relaxed selection would have allowed low-fertility-linked alleles to recombine away, producing high-bristle-number chromosomes that did not contain low-fertility alleles. When selection was reapplied, the low-fertility alleles had been reduced in frequency or separated from the high-bristle loci, so this time there was much less of a fertility problem.

17. a. The needed equations are

$$p' = p\frac{pW_{AA} + qW_{Aa}}{\overline{W}}$$

$$\overline{W} = p^2W_{AA} + 2pqW_{Aa} + q^2W_{aa}$$

$p' = 0.5\ [(0.5)(0.9) + 0.5(1.0)]/[(0.25)(0.9) + (0.5)(1.0) + (0.25)(0.7)] = 0.53$

b. The needed equation is

$$\hat{p} = \frac{W_{a/a} - W_{A/a}}{(W_{a/a} - W_{A/a}) + (W_{A/A} - W_{A/a})}$$

$$= \frac{0.7 - 1.0}{(0.7 - 1.0) + (0.9 - 1.0)}$$

$$= 0.75$$

18. The needed equation is

$$q^2 = \mu/s$$

or $\quad s = \mu/q^2 = 10^{-5}/10^{-3} = 0.01$

19. Affected individuals $= B/b = 2pq = 4 \times 10^{-6}$. Because q is almost equal to 1.0, $2p = 4 \times 10^{-6}$. Therefore, $p = 2 \times 10^{-6}$.

$$\mu = hsp = (1.0)(0.7)(2 \times 10^{-6}) = 1.4 \times 10^{-6}$$

where h = degree of dominance of the deleterious allele.

20. The probability of not getting a recessive lethal genotype for one gene is $1 - {}^{1}/_{8} = {}^{7}/_{8}$. If there are n lethal genes, the probability of not being homozygous for any of them is $({}^{7}/_{8})^{n} = {}^{13}/_{31}$. Solving for n, an average of 6.5 recessive lethals are predicted.

If the actual percentage of "normal" children is less, owing to missed in utero fatalities, the average number of recessive lethals would be higher.

21. **a.** The formula needed is

$$\hat{q} = \sqrt{\mu/s}$$

$$= 4.47 \times 10^{-3}$$

so,

$$\text{Genetic cost} = sq^2 = 0.5(4.47 \times 10^{-3})^2 = 10^{-5}$$

b. Using the same formulas as part a,

$$\hat{q} = 6.32 \times 10^{-3}$$

$$\text{Genetic cost} = sq^2 = 0.5(6.32 \times 10^{-3})^2 = 2 \times 10^{-5}$$

c.

$$\hat{q} = 5.77 \times 10^{-3}$$

$$\text{Genetic cost} = sq^2 = 0.3(5.77 \times 10^{-3})^2 = 10^{-5}$$

25 QUANTITATIVE GENETICS

1. There are many traits that vary more or less continuously over a wide range. For example, height, weight, shape, color, reproductive rate, metabolic activity, etc., vary quantitatively rather than qualitatively. Continuous variation can often be represented by a bell-shaped curve, where the "average" phenotype is more common than the extremes. Discontinuous variation describes the easily classifiable, discrete phenotypes of simple Mendelian genetics: seed shape, auxotrophic mutants, sickle-cell anemia, etc. These traits show a simple relationship between genotype and phenotype.

2. **a.** Broad heritability measures that portion of the total variance that is due to genetic variance. The equation to use is

H^2 = the genetic variance/phenotypic variance
where genetic variance = phenotypic variance – environmental variance

$$H^2 = \frac{s_p^2 - s_e^2}{s_p^2}$$

Narrow heritability measures that portion of the total variance that is due to the additive genetic variation. The equation to use is

$$h^2 = \frac{\text{additive genetic variance}}{\text{additive genetic variance + dominance variance + environmental variance}}$$

$$h^2 = \frac{s_a^2}{s_a^2 - s_d^2 + s_e^2}$$

Shank length:

$$H^2 = (310.2 - 248.1)/(310.2) = 0.200$$

$$h^2 = 46.5/(46.5 + 15.6 + 248.1) = 0.150$$

Neck length:

$$H^2 = (730.4 - 292.2/(730.4) = 0.600$$

$$h^2 = 73.0/(73.0 + 365.2 + 292.2) = 0.010$$

Fat content:

$$H^2 = (106.0 - 53.0)/(106.0) = 0.500$$

$$h^2 = 42.4/(42.4 + 10.6 + 53.0) = 0.400$$

b. The larger the value of h^2, the greater the difference between selected parents and the population as a whole and the more that characteristic will respond to selection. Therefore, fat content would respond best to selection.

c. The formula needed is

$$\text{Selection response} = h^2 \times \text{selection differential}$$

Therefore, selection response = (0.400)(10.5% – 6.5%) = 1.6% decrease in fat content, or 8.9% fat content.

3. **a.** The probability of any gene's being homozygous is $^1/_2$ (e.g., for *A*: *A/A* or *a/a*), and the probability of being heterozygous (or not homozygous) is also $^1/_2$. Thus, the probability for any one gene's being homozygous whereas the other two are heterozygous is $(^1/_2)^3$. Because there are three ways for this to happen (homozygosity at *A* or at *B* or at *C*), the total probability is

$p(\text{homozygous at 1 locus}) = 3(^1/_2)^3 = {}^3/_8$

The same logic can be applied to any two genes' being homozygous

$$p(\text{homozygous at 2 loci}) = 3(^1/_2)^3 = {}^3/_8$$

There are two ways for all three genes to be homozygous, so

$$p(\text{homozygous at 3 loci}) = 2(^1/_2)^3 = {}^2/_8$$

b. $p(\text{0 capital letters}) = p(\text{all homozygous recessive}) = (^1/_4)^3 = {}^1/_{64}$

$p(\text{1 capital letter}) = p(\text{1 heterozygote and 2 homozygous recessive}) = 3(^1/_2)(^1/_4)(^1/_4) = {}^3/_{32}$

$p(\text{2 capital letters}) = p(\text{1 homozygous dominant and 2 homozygous recessive})$

or

$p(\text{2 heterozygotes and 1 homozygous recessive})$

$= 3(^1/_4)^3 + 3(^1/_4)(^1/_2)^2 = {}^{15}/_{64}$

$p(\text{3 capital letters}) = p(\text{all heterozygous})$

or

$p(\text{1 homozygous dominant, 1 heterozygous, and 1 homozygous recessive})$

$= (^1/_2)^3 + 6(^1/_4)(^1/_2)(^1/_4) = {}^{10}/_{32}$

p(4 capital letters) = p(2 homozygous dominant and 1 homozygous recessive)

or

p(1 homozygous dominant and 2 heterozygous)

$= 3(^1/_4)^3 + 3(^1/_4)(^1/_2)^2 = {}^{15}/_{64}$

p(5 capital letters) = p(2 homozygous dominant and 1 heterozygous)

$= 3(^1/_4)^2(^1/_2) = {}^3/_{32}$

p(6 capital letters) = p(all homozygous dominant) $= (^1/_4)^3 = {}^1/_{64}$

4. For three genes there are a total of 27 genotypes that will occur in predictable proportions. For example, there are three genotypes that have two genes that are heterozygous and one gene that is homozygous recessive (*A/a* ; *B/b* ; *c/c*, *A/a* ; *b/b* ; *C/c*, *a/a* ; *B/b* ; *C/c*). The frequency of this combination is $3(^1/_2)(^1/_2)(^1/_4) = {}^3/_{16}$, and the phenotypic score is 3 + 3 + 1 = 7. For all the genotypes possible, the total distribution of phenotypic scores is as follows:

Score	Proportion
3	$^1/_{64}$
5	$^3/_{32}$
6	$^3/_{64}$
7	$^3/_{16}$
8	$^3/_{16}$
9	$^{11}/_{64}$
10	$^3/_{16}$
11	$^3/_{32}$
12	$^1/_{64}$

And the plot of these data will be

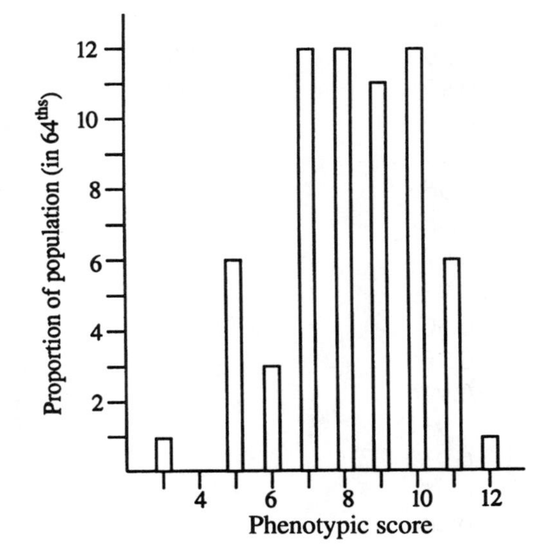

5. The population described would be distributed as follows:

3 bristles	$^{19}/_{64}$
2 bristles	$^{44}/_{64}$
1 bristle	$^{1}/_{64}$

The 3-bristle class would contain 7 different genotypes, the 2-bristle class would contain 19 different genotypes, and the 1-bristle class would contain only 1 genotype. It would be very difficult to determine the underlying genetic situation by doing controlled crosses and determining progeny frequencies.

6. **a.** Solving the formula for values of x over the stated range for each genotype gives the following data:

x	1	2	3
0.03	0.90		
0.04	0.91		
0.05	0.93		
0.06	0.93		
0.07	0.94		
0.08	0.95		
0.09	0.96		
0.10	0.97		0.90
0.11	0.97		0.92
0.12	0.98	0.90	0.93
0.13	0.98	0.92	0.94
0.14	0.99	0.94	0.95
0.15	0.99	0.95	0.96
0.16	0.99	0.96	0.97
0.17	1.00	0.98	0.98
0.18	1.00	0.98	0.98
0.19	1.00	0.99	0.99
0.20	1.00	1.00	0.99
0.21	1.00	1.00	1.00
0.22	1.00	1.00	1.00
0.23	1.00	1.00	1.00
0.24	0.99	1.00	1.00
0.25	0.99		1.00
0.26	0.99		1.00
0.27	0.98		1.00
0.28	0.98		0.99
0.29	0.97		0.99
0.30	0.97		0.98
0.31	0.96		0.98
0.32	0.95		0.97
0.33	0.94		0.96
0.34	0.93		0.95
0.35	0.93		0.94
0.36	0.91		0.93
0.37	0.90		0.92
0.38			0.90

Plotting these data give the following curves:

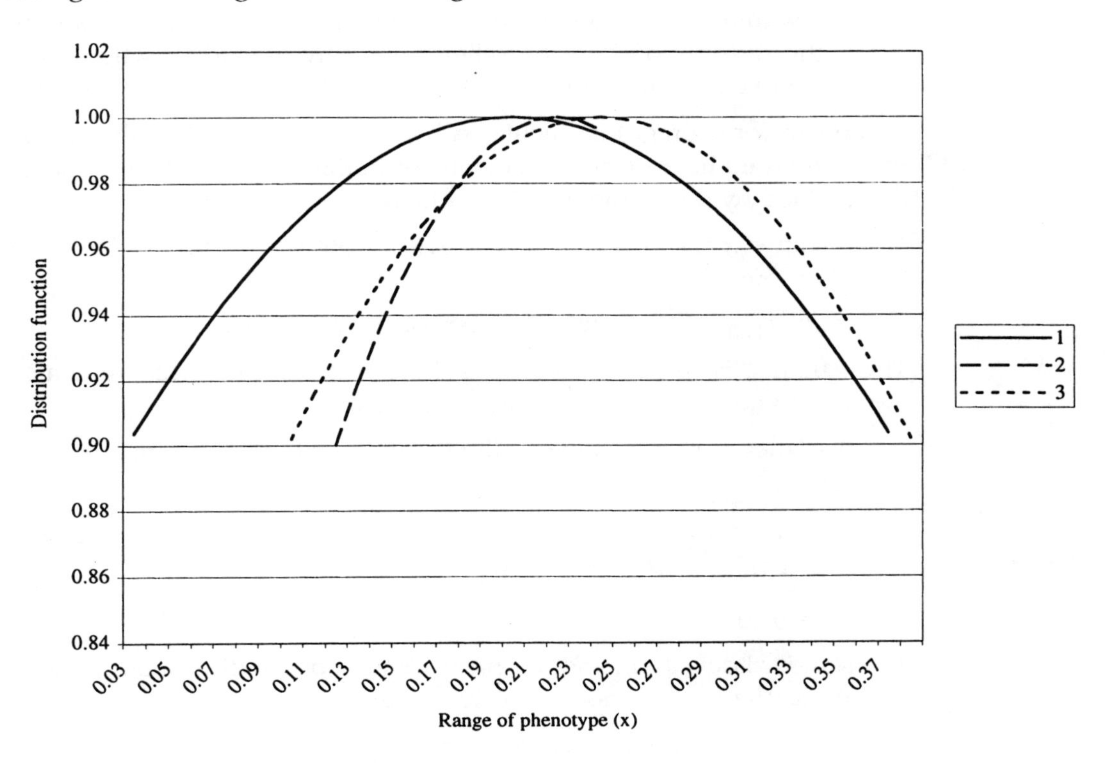

b. Because the three genotypes are equally frequent, the average distribution across the entire range of phenotypes will be

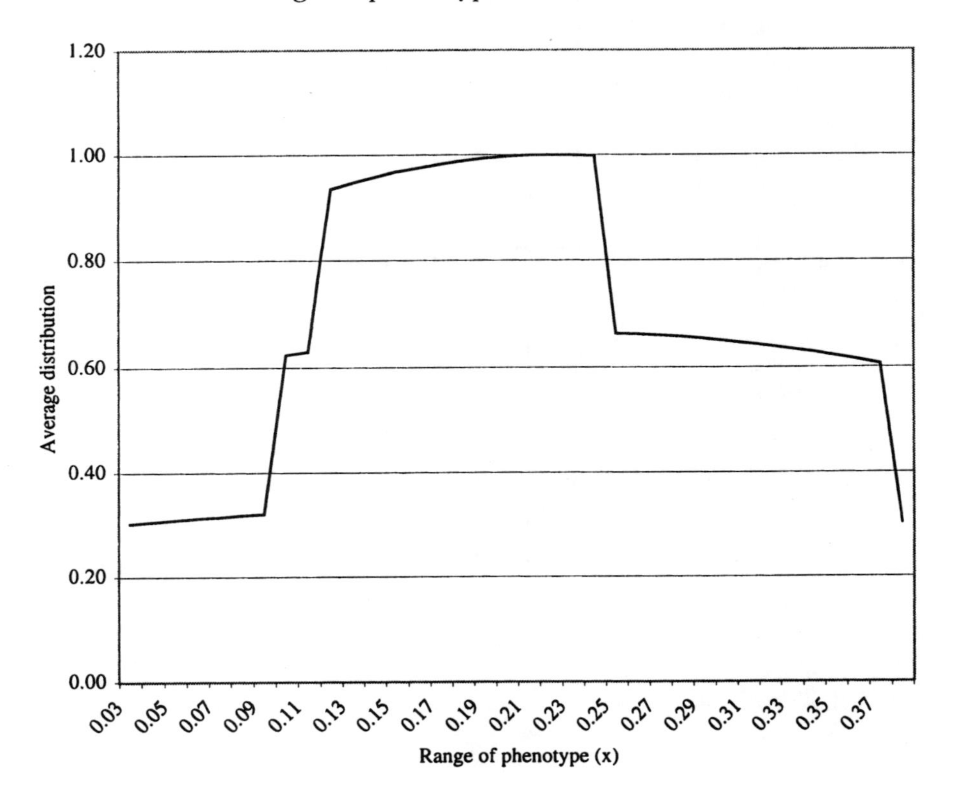

There are regions of the overall phenotypic distribution where the variation within a given genotype does not overlap, and this gives sharp steps to the distribution. On the other hand, any individual whose phenotype lies between values of 0.12 to 0.24 could have any of the three genotypes.

7. The mean (or average) is calculated by dividing the sum of all measurements by the total number of measurements, or in this case, the total number of bristles divided by the number of individuals.

Mean = $\bar{x}$ = [1 + 4(2) + 7(3) + 31(4) + 56(5) + 17(6) + 4(7)]/(1 + 4 + 7 + 31 + 56 + 17 + 4)

= $^{564}/_{120}$ = 4.7 average number of bristles/individual

The variance is useful for studying the distribution of measurements around the mean and is defined in this example as

variance = s^2 = average of the (actual bristle count – mean)^2

$$s^2 = {}^1/_n\Sigma(x_i - \bar{x})^2$$

$$= {}^1/_{120}\Sigma[(1 - 4.7)^2 + (2 - 4.7)^2 + (3 - 4.7)^2 + (4 - 4.7)^2 + (5 - 4.7)^2 + (6 - 4.7)^2 + (7 - 4.7)^2]$$

$$= 0.26$$

The standard deviation, another measurement of the distribution, is simply calculated as the square root of the variance:

standard deviation = $s = \sqrt{0.26} = 0.51$

8. **a.**

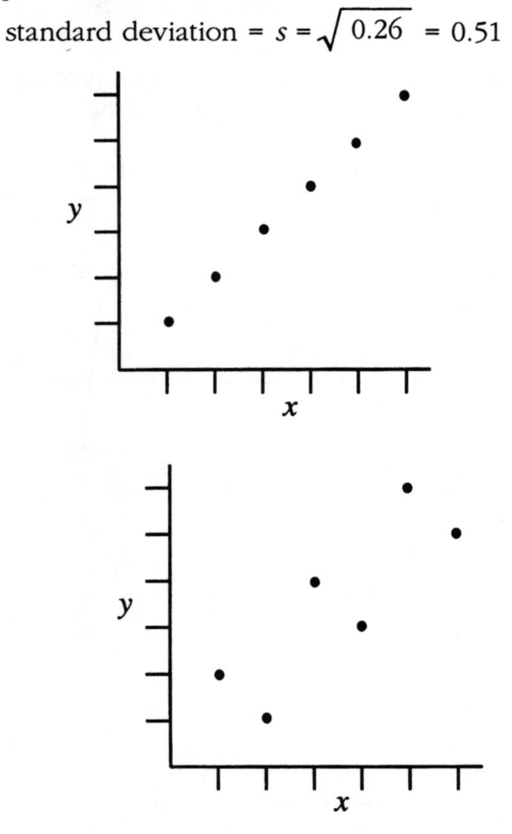

b.

c.

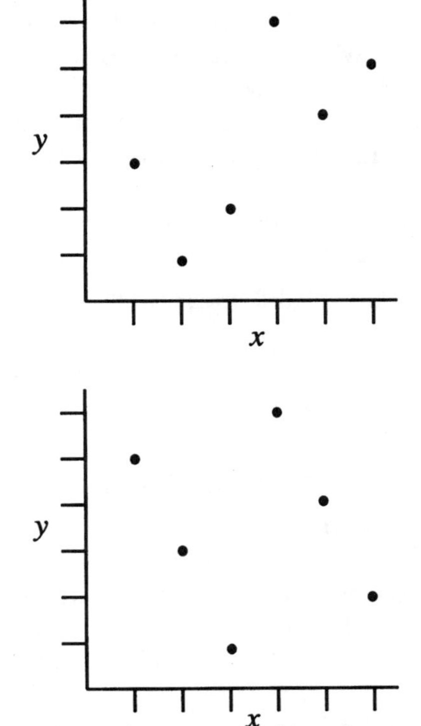

d.

Use the following formulas to calculate the correlation coefficient (r_{xy}) between x and y:

$$r_{xy} = \frac{\text{cov } xy}{s_x s_y} \qquad \text{and} \qquad \text{cov } xy = {}^1/_N \sum x_i y_i - \overline{xy}$$

a. cov xy = $^1/_6$ [(1)(1) + (2)(2) + (3)(3) + (4)(4) + (5)(5) + (6)(6)] – ($^{21}/_6$)($^{21}/_6$)
= 2.92

$$\text{standard deviation } x = s_x = \sqrt{{}^1/_N \sum (x_i - \bar{x})^2} = 1.71$$

$$\text{standard deviation } y = s_y = \sqrt{{}^1/_N \sum (y_i - \bar{y})^2} = 1.71$$

Therefore, r_{xy} = 2.92/(1.71)(1.71) = 1.0. The other correlation coefficients are calculated in a like manner.

b. 0.83

c. 0.66

d. –0.20

9. **a.** H^2 has meaning only with respect to the population that was studied in the environment in which it was studied. Even if a trait shows high heritability, it does not imply the trait is unaffected by its environment. The only acceptable analysis is to study directly the norms of reaction of the various genotypes in the population over the range of projected environments. Because it is so difficult to fully replicate a human genotype so that it might be tested in different environments, there is no known norm of reaction for any human quantitative trait.

b. Neither H^2 nor h^2 is a reliable measure that can be used to generalize from a particular sample to a "universe" of the human population. They certainly should not be used in social decision making (as implied by the terms *eugenics* and *dysgenics*).

c. Again, H^2 and h^2 are not reliable measures, and they should not be used in any decision making with regard to social problems.

10. The following are unknown: (1) norms of reaction for the genotypes affecting IQ, (2) the environmental distribution in which the individuals developed, and (3) the genotypic distributions in the populations. Even if the above were known, because heritability is specific to a specific population and its environment, the difference between two different populations cannot be given a value of heritability.

11. First, define alcoholism in behavioral terms. Next, realize that all observations must be limited to the behavior you used in the definition and all conclusions from your observations are applicable only to that behavior. In order to do your data gathering, you must work with a population in which familiality is distinguished from heritability. In practical terms, this means using individuals who are genetically close but who are found in all environments possible.

12. Before beginning, it is necessary to understand the data. The first entry, h/h h/h, refers to the II and III chromosomes, respectively. Thus, there are four h (high bristle number) sets of alleles in two or more genes on separate chromosomes. The next entry is h/l h/h. Chromosome II is now heterozygous and chromosome III is still homozygous, etc.

The effect of substituting one l chromosome II for an h chromosome II, and therefore going from homozygous h/h to heterozygous h/l, can be seen in the differences along the rows in the first two columns. The average change is (2.9 + 3.1 + 2.7)/3 = 2.9. When chromosome II goes from heterozygous h/l to homozygous l/l, the average change is (3.2 + 5.2 + 6.8)/3 = 5.1.

The effect of substituting one l chromosome III for an h chromosome III can be seen in the differences between rows: 25.1 – 23.0 = 2.1; 22.2 – 19.9 = 2.3; and 19.0 – 14.7 = 4.3 (average change 2.9). And going from h/l to l/l for chromosome III gives an average change of (11.2 + 10.8 + 12.4)/3 = 11.5 bristles.

Here is a summary of these results:

	Chromosome II	Chromosome III
h/h to h/l	2.9	2.9
h/l to l/l	5.1	11.5

Each set of alleles for both chromosomes is expressed in the phenotype, but that expression varies with the chromosome. Chromosome III appears to have a stronger effect on the phenotype than does chromosome II. (Compare h/h h/h with both l/l h/h and h/h l/l. The difference in the first case is 6.1, and in the second case, 13.3.) Finally, there is partial dominance of h over l for both chromosomes. The change from h/h to h/l is less than the change from h/l to l/l.

13. a.

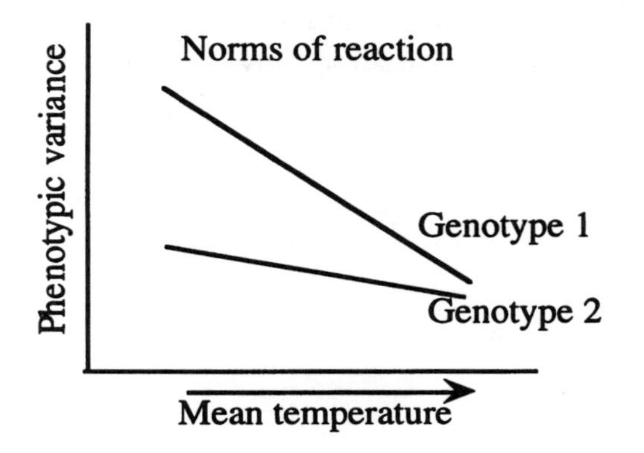

b. Broad heritability is defined as

$$H^2 = \frac{s_g^2}{s_p^2}$$

Assuming that the genetic variance stays the same, phenotypic variance must decrease if H^2 increases. Therefore, the same plot as (a) will satisfy the conditions.

c. To satisfy the conditions that genetic variance increases as H^2 decreases, phenotypic variance must also increase. Therefore the plot will be

d.

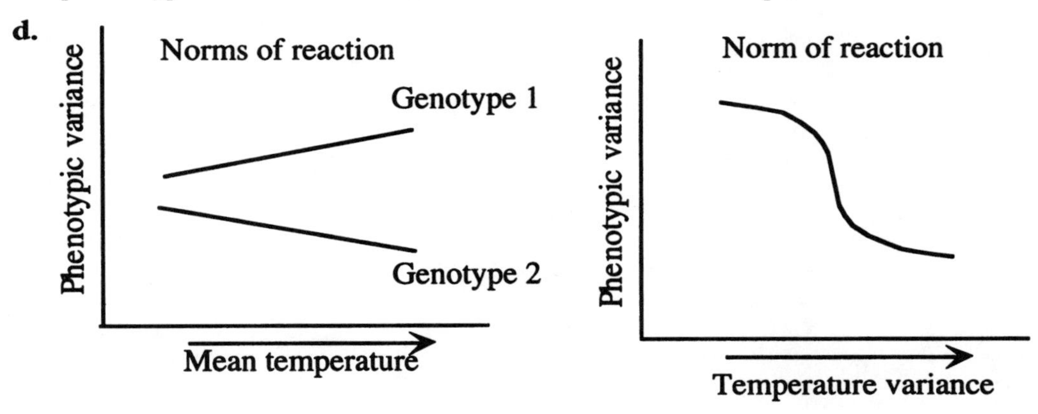

14. a. The regression line shows the relationship between the two variables. It attempts to predict one (the son's height) from the other (the father's height.) If the relationship is perfectly correlated, the slope of the regression line should approximate 1. If you assume that individuals at the extreme of any spectrum are homozygous for the genes responsible for these phenotypes, then their offspring are more likely to be heterozygous than are the original individuals. That is, they will be less extreme. Also, there is no attempt to include the maternal contribution to this phenotype.

b. For Galton's data, regression is an estimate of heritability (h^2), *assuming* that there were few environmental differences between all fathers and all sons both individually and as a group. However, there is no evidence given to determine if the traits are familial but not heritable. This data would indicate genetic variation only if the relatives do not share common environments more than nonrelatives do.

26 Genetics and Evolution

1. A transformational scheme of evolution predicts that all members of a species will change over time. By sharing similar environments and life experiences, each member is altered in its lifetime and its progeny inherit these alterations. For example, if all giraffes stretch their necks during their lifetimes to reach the ever higher foliage, all offspring will inherit this acquired "stretched" neck and begin their lives where their parents have left off. Over time, longer-and longer-necked giraffes would evolve.

   A variational scheme of evolution predicts that not all members of a species contribute equally to future generations. Heritable differences among members of the same species result in some being more "fit" and able to produce more offspring than others. Over time, one type of individual is replaced by another. For example, using an economic metaphor, there is currently an explosion of internet and "dot.com" companies. A variation scheme of evolution would suggest that some of these companies will grow, evolve, and prosper while others, based on intrinsic (heritable) differences (management, capitalization, etc.) will become extinct.

2. The three principles are (1) organisms within a species vary from one another, (2) the variation is heritable, and (3) different types leave different numbers of offspring in future generations.

3. The variation required by Darwin's proposed mechanism of evolution cannot exist or be maintained if "blending" or nonrandom segregation occurs. In either case, populations will become homogeneous and variation will be rapidly lost. The "particulate" nature of genes described by Mendel allows for

segregation of traits generation after generation. In this way, even currently detrimental (and recessive) alleles can be retested as the environment and circumstances change.

4. A population will not differentiate from other populations by local inbreeding if

$$\mu \geq 1/N$$

so

$$N \geq 1/\mu$$

$$N \geq 10^5$$

5. The rate of loss of heterozygosity in a closed population is $1/(2N)$ per generation.

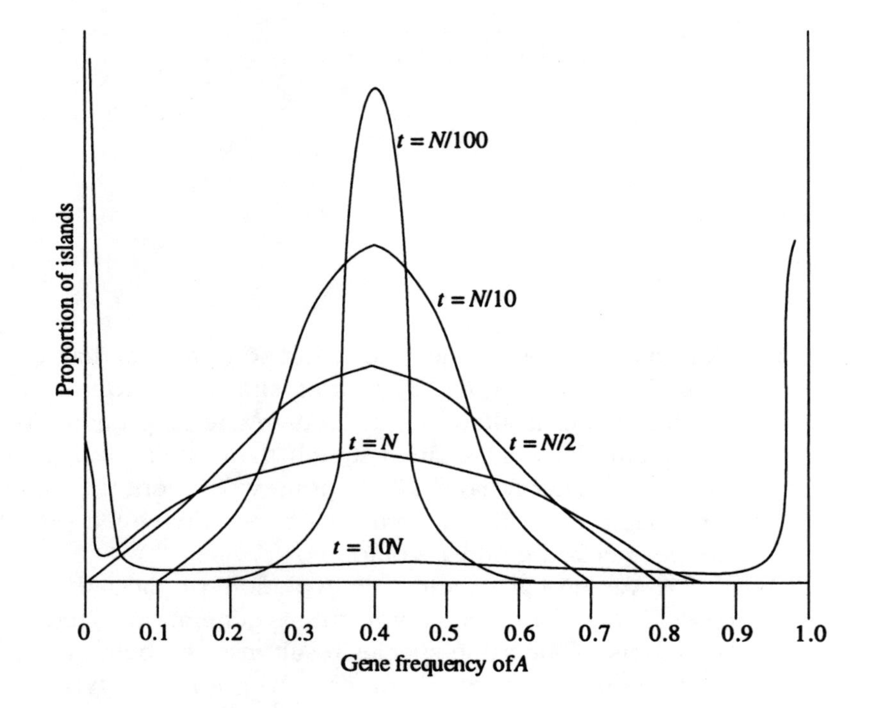

6. A population will not differentiate from other populations by local inbreeding if

the number of migrant individuals ≥ 1 per generation

For (a), migration is not sufficient to prevent local inbreeding, so the results are roughly the same as seen in Problem 5. In the case of (b), there is one migrant per generation, so the populations will not differentiate and allelic frequencies will remain the same in all populations.

7. The mean fitness is calculated by summing the frequency of each progeny class times its fitness. For example, the frequency of $A/A \cdot B/B$ is $p(A)^2 \times p(B)^2$, or (0.64)(0.81) = 0.52. This frequency is multiplied by its fitness to give (0.52 × 0.95) = 0.49. Summing for all classes, the mean fitness of population 1[$p(A)$ = 0.8, $p(B)$ = 0.9] is 0.92. Because selection acts to increase the mean fitness,

the frequency of both A and B should increase in the next generation (the $A/A \cdot BB$ class has a fitness of 0.95).

The mean fitness of population 2[$p(A) = 0.2$, $p(B) = 0.2$] is 0.73. Again, the frequency of both A and B should increase.

There is a single adaptive peak at $A/A \cdot B/B$. By inspection, the fitness is lowest at $a/a \cdot b/b$ and highest at $A/A \cdot B/B$. The allelic frequencies at the peak are 1.0 for both A and B.

8. The mean fitness for population 1 $p(A) = 0.5$, $p(B) = 0.5$] is 0.825. The mean fitness for population 2[$p(A) = 0.1$, $p(B) = 0.1$] is 0.856. There are four adaptive peaks: at $p(A) = 0.0$ or 1.0 and $p(B) = 0.0$ or 1.0. (The mean fitness will be 0.90 at any of these points.) With population 1, the direction of change for both $p(A)$ and $p(B)$ will be random. Both higher or lower frequencies of either allele can result in increased mean fitness (although there are some combinations that would lower fitness). Because population 2 is already near an adaptive peak of $p(A) = 0.0$ and $p(B) = 0.0$, both $p(A)$ and $p(B)$ should decrease to increase the mean fitness.

9. Figure 26-12 of the text shows the frequency distribution of haploid chromosome numbers among dicots. Above a chromosome number of 12, even numbers are much more common than odd numbers. This is evidence of frequent polyploidization during plant evolution. For example, if a species of plant with an odd haploid chromosome number undergoes a "doubling" event, the chromosome number becomes even.

10. The α and ß gene families show remarkable amino acid sequence similarities (see Table 26-4 of the text). Within each gene family, sequence similarities are greater and, in some cases, member genes have identical intron–exon structure.

11. All human populations have high i, intermediate I^A, and low I^B frequencies. The variations that do exist between the different geographical populations is most likely due to genetic drift. There is no evidence that selection plays any role regarding these alleles.

12. A *geographical race* is a population that is genetically distinguishable from other local populations but is capable of exchanging genes with those other local populations.

A *species* is a group of organisms that exchanges genes within the group but cannot do so with other groups.

Populations that are geographically separated will diverge from each other as a consequence of a combination of unique mutations, selection, and genetic drift. For populations to diverge enough to become reproductively isolated, spatial separation sufficient to prevent any effective migration is usually necessary.

13. To test the species distinctness, it is necessary to be able to manipulate and culture *D. pseudoobscura* and *D. persimilis* in captivity. If they cannot be cultured in the laboratory, their species distinctness cannot be established. The mating behavior compatibility of the different *Drosophila* can be tested by placing a mixture of males of both forms with females of one of the forms to see whether there are any female mating preferences. The same experiment can then be repeated with mixed females and one sort of male and with mixtures of males and females of both forms. From such experiments, patterns of mating preference can be observed. Even if there is some small amount of mating of different forms, this may occur only because of the unnatural conditions in which the test is being carried out. On the other hand, no mating of any kind may occur, even between the same forms, because the necessary cues for mating are missing, in which case nothing can be concluded.

If matings between different forms do occur, the survivorship of the interpopulation hybrids can be compared with that of the intrapopulation matings. If hybrids survive, their fertility can be tested by attempting to backcross them to the two different parental strains. As in the case of the mating tests, under the unnatural conditions of the laboratory, some survivorship or fertility of species hybrids is possible even though the isolation in nature is complete. Any clear reduction in observed survivorship or fertility of the hybrids is strong presumptive evidence that they belong to different species.

14. 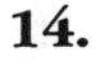

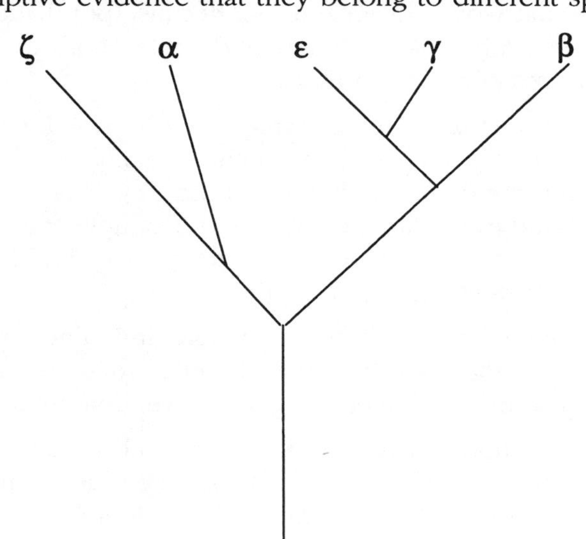

15. For polymorphic sites within a species, let nonsynonymous = a and synonymous = b. For polymorphic sites between the species, let nonsynonymous = c and synonymous = d. If divergence is due to neutral evolution, then

$$a/b = c/d$$

If divergence is due to selection, then

$$a/b < c/d$$

However, in this example, $a/b = 20/50 > c/d = 2/18$, which fits neither expectation.

Because the ratio of nonsynonymous to synonymous polymorphisms (a/b) is relatively high, the gene being studied may encode a protein tolerant of substitution (like fibrinopeptides, see Figure 26-18 of the text). The relatively fewer species differences may suggest that speciation was a recent event, so few polymorphisms have been fixed in one species that are not variants in the other.